Electrino Physics

ELECTRON   PION

Figure 1. An electron is composed of two semions orbiting about each other.

Figure 2. A net zero spin pion is composed of two orbiting pairs of orbiting quartons.

by Gordon L. Ziegler

About the covers: The drawings illustrate the electrino structure of key particles in the particle physics model by Gordon L. Ziegler. Quartons, semions, and unitons are different flavors of electrinos in the model. The viewgraphs were hand drawn by John Blacklaw, Illustrator and Orneda F. Ziegler. Other drawings in the book were done by Richard Cowley and Bruce Picket.

Electrino
PHYSICS

from
the beginning studies of light
to
particle structure and a unified field theory

from
balancing decay schemes
to
reversing the order to disorder arrow in
the second law of thermodynamics

from
particle structure
to
the masses of charged leptons

by Gordon L. Ziegler

ISBN: Softcover   978-1-4931-4864-6
               eBook   978-1-4931-4865-3

Last revised November 3, 2011.

This book draws largely from
Formulating the Universe, Volumes I-III,
by Gordon L. Ziegler,
owned by Benevolent Enterprises.
Material used with permission.

Author:
Gordon L. Ziegler
P.O. Box 1162
Olympia, WA 98507-1162 USA
e-mail: ben_ent100@msn.com

This book was printed in the United States of America.

Rev. date: 11/23/2013

To order additional copies of this book, contact:
Xlibris LLC
1-888-795-4274
www.Xlibris.com
Orders@Xlibris.com
143747

PREFACE

This volume is based on aether relativity and the postulate that a smooth symmetric charge distribution cannot have detectable spin—or consequently charges come in ±e, ±e/2, ±e/4, and ±e/8—the Electrino Hypothesis—and not in ±2e/3 and ±e/3 as in the Quark Hypothesis. In Appendix B, the structures of all known particles are induced totally without quarks and gluons. The Electrino Hypothesis is sufficient to compose all known particles.

The physics world is searching for a unified field theory and unified particle theory. This volume contains the foundation of both. Gravity and the strong force are united to the electro-magnetic force at the Planck mass, which in imaginary units is the mass of a whole elementary particle in this model. It takes 61 elementary particles in the quark-lepton model to construct all known particles. By contrast, the particle fusion aspect of this model means that all the copies of all the particles in the Universe could be ionized and fused from a single particle. This volume begins the derivation of these things.

Chapter 1 recounts the particle-wave controversy of the centuries as a prototype synthesis of the aether-relativity controversy in Chapter 2. A thought experiment in this chapter falsifies both the principle of relativity in the absolute and the principle of equivalence. The aether-relativity controversy is resolved by deriving from first principles Special Quasi-Relativity in an Aether in Chapter 3, and General Quasi-Relativity in an Aether in Chapter 4. General Quasi-Relativity is obtained by inserting a field of escape velocities in and out, about a gravitational body, in Special Quasi-Relativity, obtaining the Schwarschild Line Element in the space about a gravitational body. A model of gravity and inertia is developed in Chapter 5. An aether model of particle physics is derived in Chapter 6, with special attention to whole elementary particles, including electrons and photons. Elementary particle fusion is briefly introduced in Chapter 6, along with the quantization of spin and a string-like character for elementary particles. A unified field theory is presented in Chapter 7, with a further unification of physics from a single definition in Chapter 8. This model has all forces united to the parent force gravity. The relationship is shown between charge and gravity.

This model could be tested by e^-e^- collisions or e^+e^+ collisions at 1.878 GeV or more in the center of mass frame. Benefits to society from the model could be gravity-free and inertia-less travel, new reactors releasing energy from matter (without radioactive wastes)(see Chapter 15), the testing of a new Grand Unification Theory (GUT), and the reversal of the order to disorder arrow in the second law of thermodynamics (see Chapter 16).

In Chapters 10 and 11 and Appendix A, a new type of pictorial equation is presented which accounts for the elementary particles in their various states. As such, the new system, called chonomics, is very powerful.

Chapter 12 explains how to create new anti-matter through the fusion of electrons or how to create new matter through the fusion of positrons.

Chapter 13 tells how to calculate relativity with real masses—elementary masses in orbital systems.

Chapter 14 derives a new mechanism for the interstellar red shift—the dual photon. The universe may be found to be older than calculated under the Big Bang theory.

Chapter 15 presents two very different calculations for the power to be obtained from the fusion of the electrons in 1.0 Amp beams at 2.0 GeV in the Center of Mass Frame. According to the calculation, we would expect, from our experience with electron-positron annihilation, the resultant power would be scarcely detectable. According to the more natural calculation, the resultant power would be a staggering net 2.0 billion Watts (two million kilowatts). Since the electrino fusion model of elementary particles is a new model, not a person in the world should now be able to determine a priori the correct calculation or correct efficiency of electron fusion. That should be determined by experiment. Let us hope the much higher power is correct. Then electrino fusion power reactors would be possible, as well anti-matter rockets. This chapter also presents a new model of supernova. Chapter 16 tells how to reverse the order to disorder arrow in the second law of thermodynamics. These feats could be accomplished with a properly designed accelerator facility.

The next work of this book is to induce the structure of "elementary" particles in Appendix B. This is a large undertaking requiring extreme precision. The final work of this book is to calculate from first principles the masses of charged leptons (in Chapter 21), laying the foundation for future work in predicting the masses of all the particles and calculating the many fundamental constants. It would be exciting to be able to calculate the mass of any given particle, given its chonomic structure. Should this be accomplished, the model would be the most advanced model of particle physics. The predictions could be used as tests of the model.

CONTENTS

Chapter 1

PARTICLE OR WAVE?

This volume derives a unified field theory and the basis for a unified particle theory. Formulating a unified theory is a feat not possible without taking advantage of the insight gained through centuries of physical experiment and model building. This chapter and the next shall glean preliminary insights from the evolution of physics theories.

From 1650 to 1925 intense speculation prevailed regarding the nature of light and the mechanism of light transmission. Scientific thought oscillated between two opposing theoretical models—the corpuscular theory and the wave theory—before a harmonious, though paradoxical, duality was seen to explain light and matter.[1]

A. Wave

In several countries of Europe in the 17th and 18th centuries scientists studied diffraction, or the spreading of light into shadows. It was first observed in Italy in the 17th century. In England a worker observed the interference colors of thin films, like an oil film on a wet road surface, or in the iridescent colors of a butterfly's wing. He felt that light consisted of vibrations propagated at high speed. A scientist in Holland, Christian Huygens, improved the wave theory.

B. Particle

In England Sir Isaac Newton didn't attach much importance to the small amount of spreading of light and believed that rectilinear propagation could not be reconciled with the wave theory.

1

Polarization phenomena (which can be accounted for by transverse wave motion in a single plane) discovered in the 17th century by a Danish physicist, Erasmus Bartholin, and by Huygens were not consistent with the theory of longitudinal waves (waves vibrating in the direction of propagation, like compression waves in a coiled spring), which was the only wave theory then considered. Newton therefore supported the corpuscular theory, although he did not reject the wave theory completely.[2]

Isaac Newton (1642-1727), the discoverer of the law of gravity, calculus, and Newtonian mechanics, championed the particle model of light. He believed light is like a barrage of minute speeding bullets,[3] and that the laws applicable to material objects could account also for light behavior. Moving objects tend to go in straight lines because of inertia; light travels in straight lines. An obstacle in the path of a spray of bullets blanks out the spray behind it, just as the illuminated object casts a shadow. The angle of incidence equals the angle of reflection for a ball striking a wall and also for light reflected from a smooth surface (specular reflection). Newton was able to explain diffuse reflection of light by means of microscopic roughness of the reflecting surface. Newton's particle model made the inverse square law of light intensity intuitively obvious. It also brilliantly anticipated light pressure, and accounted for absorption heating. However, the particle model did not do so well with refraction.

Refraction is the bending of light at the surface boundary of diverse media, such as air and water, or a vacuum and glass. This change of light ray direction is always such that

$$\frac{\sin \theta_i}{\sin \theta_r} = n_{12}, \tag{1-1}$$

where $\sin \theta_i$ is the sine of the angle of incidence, $\sin \theta_r$ is the sine of the angle of refraction, and n_{12} is the relative index of refraction of the two transparent media. If the light goes from a vacuum to a refractive medium, the law is

$$\frac{\sin \theta_i}{\sin \theta_r} = n_m, \tag{1-2}$$

where n_m is the index of refraction of the medium.

Newton could account for this observed law of light (Snell's law) only by assuming that the particles of light receive a little push at the surface of the medium and travel faster in the refractive medium than in air or a vacuum.[4] In spite of the opposing wave model, this view was dominant for over a century. But later careful measurements showed that light goes slower in a refractive medium such as water or glass than in air or a vacuum, not faster.[5] According to Newton's particle theory,

$$n_m = \frac{v_m}{c}, \tag{1-3}$$

where v_m is the speed of light in the medium and c is the speed of light in a vacuum. However, the precise opposite was found to be true. In actuality,

$$n_m = \frac{c}{v_m}. \tag{1-4}$$

Because of this discrepancy, the particle model of light

appeared to fail.

C. Wave

"Thomas Young, a physician who began to work on sound and light in the closing years of the eighteenth century,"[6] began the swing back from the corpuscular theory to the wave theory. "In 1800 Young published his first thoughts on a wave theory of light as a small part of a paper discussing some experiments on sound. In the section of this paper titled 'Of the Analogy between Light and Sound', Young proposed a wave theory of light rather similar to Huygen's theory."[7]

Young continued his work in the next few years and in a volume published in 1807 he discussed his famous two-slit interference experiment which is so often cited as proving the existence of waves of light. Young also attempted provisional explanations of inflection or diffraction, and did some excellent work on the colors of thin plates and Newton's rings. Later, in 1809, he defended Huygens' theory of double refraction, along with several modifications of his own, against Laplace's corpuscular theory of double refraction.[8]

Young's experiments were brilliantly conceived and executed, but the arguments which he gave in support of his *theory* did not much surpass what Huygens had accomplished. Apparently, through lack of training, Young was not able to bring the sophisticated theoretical developments of recent science to work in his favor. The Newtonian school was still exceedingly strong in the beginning of the nineteenth century, and Young's theory did not attract many followers. Herschel and Laplace continued to develop optics in the corpuscular manner after

Young's fundamental papers on the wave theory and its experimental foundations.

There were some reasons for this other than simple scientific inertia. In the years 1808-10 E. L. Malus performed a number of experiments on the intensity of light that was reflected from a transparent body's surface. Malus analyzed the light using a double refracting crystal of Iceland spar and noted that light reflected from the surface of transparent media possessed the same property of having "Sides" which Newton had noticed in connection with doubly refracted rays. Malus gave the name "polarization" to this property, and attributed it, on the basis of a particle theory of light, to light corpuscles having their sides all turned in the same direction, much as a magnet turns a series of needles all to the same side. Subsequent to Malus' publication, the French physicist Biot developed a more complex corpuscular theory of polarization.[9]

David Brewster conducted several experiments in England in the years 1814-19. Some of these experiments were similar to those of Malus. The results Brewster obtained were more significant to the future of optics. One such result was a formula connecting the angle of complete polarization of the reflected ray with the refractive index of the media--a relation which Malus had sought but could not determine. Brewster also discovered biaxial crystals in which there were two axes along which double refraction did not occur, rather than one axis as in Iceland spar. Immediately, as a result of Brewster's discovery, Huygens' analysis of double refraction and the wave theory of light in general were called into serious question. Huygens' construction no longer sufficed to account for the refraction in the more complex biaxial crystals.

It was clear that polarization was a problem
which was exceedingly difficult to explain on the
basis of the wave theory, as long as light was
conceived on the analogy with sound.   Sound
waves, as *longitudinal* waves, could not account for
the   "sidedness"   displayed   in   polarization
phenomena.   Young, purportedly reflecting on
Brewster's experiments and on the results of an
experiment carried out in France by Arago and
Fresnel,[10] was the first to suggest a possible
explanation of polarization on the basis of the wave
theory. In a letter to Arago dated 17 January, 1817,
Young proposed that if light waves were conceived
of as *transverse* waves, they could admit of
polarization.   Not long after, in another letter to
Arago, Young compared light waves to the motions
of a cord which has one of its ends agitated in a
plane.

Arago showed this letter to Fresnel who at once
seized upon the hypothesis of transverse waves as
one with which he could explain polarization.
Subsequently, Fresnel made this hypothesis the
basis of his most influential dynamical theories of
double refraction and reflection and refraction.[11]

In the second quarter of the nineteenth century, the
wave model of light became firmly established.   The
advantages of the corpuscular theory over the wave theory
were removed when the transverse wave concept of light
was adopted. Polarization of light could then be explained
by the wave theory.   Sharply defined shadows could be
accounted for in the wave theory if the wavelength of light
was very short.   The wave theory predicted the correct
speed relation for light in a refractive medium (Eq. 1-4),
thus surpassing Newton's particle model of light.[12]   Indeed,
"the wave theory was generally accepted and seemed

capable of explaining all known optical phenomena, though, with hindsight, it can now be seen that there were some important difficulties."[13]   The scientific tug-of-war between the particle and wave models of light, however, was intense at times. During one such time, Poisson, a believer in the particle model, thought he could disprove the wave model of light by its own internal inconsistency. He showed mathematical wave theory predicted there should be a bright spot in the center of a shadow formed by obstructing a well-collimated monochromatic beam of light. Poisson's natural intuition, based on everyday experience with shadows, told him that such a thing wouldn't be. He sent an assistant into the laboratory to prove the nonexistence of the phenomenon. Amazingly enough, the bright spot was there--in the center of the shadow. This supported the wave theory. Ironically, the phenomenon was dubbed, "Poisson's bright spot."

The fascination of the scientific world with electricity and magnetism led initially to a refinement and advancement of the wave theory of light. Maxwell showed that it is mathematically possible for electromagnetic waves to propagate through free space. "Maxwell found the speed of propagation of these waves to be equal to the ratio of the electromagnetic to the electrostatic units of charge. This ratio was so nearly equal to the measured speed of light that he concluded that light must itself be an electromagnetic disturbance of the type described by his equations."[14] The wave model became electro-magnetic.

D. Particle

With the advent of electronics, the photoelectric effect was discovered. This phenomenon brought a great surprise: light ejected electrons from metals not like a resonance from a continuous electro-magnetic wave, but like a beam of tiny particles striking the metal at random.[15]

In the wave model, reducing the intensity of the light beam should make the photoelectric effect suddenly stop as the photo-power per surface area value drops below the threshold power density required to eject the electrons from the metal. But instead it was found that reducing the frequency of light had that effect, while reducing the intensity of light (where the frequency of light was high enough to eject electrons) reduced only the number of electrons ejected—this phenomenon holding true down to extremely low levels of light. A wave model could not explain this discovery, but a particle model could. How could light act in so many ways like a wave, and yet in this new instance prove to act like particles? Science seemed doomed to an eternal particle-wave dilemma.

E. Duality

Einstein succeeded in explaining this photoelectric effect by making a remarkable assumption, namely, that the energy in a light beam travels through space in concentrated bundles, called *photons*. The energy E of a single photon is given by

$$E = h\nu. \qquad\qquad (1\text{-}5)$$

. . . Einstein's hypothesis suggests that light traveling through space behaves not like a wave at all but like a particle. Millikan, whose experiments verified Einstein's ideas in every detail, spoke of Einstein's "bold, not to say reckless, hypothesis."[16]

Einstein's photons were seen to be "quanta" of light energy (or discrete bundles of energy). Thus the photon concept fit in nicely with the already developing quantum theory of the atom, in which energy states in the atom were separated by discrete differences of energy. While the photons, or light quanta, acted like particles in the photoelectric effect, the photons had frequencies, and thus wavelike characteristics. Particle-wave duality was emerging from the light-nature mystery.

In 1924 Louis de Broglie of France reasoned that (a) nature is strikingly symmetrical in many ways; (b) our observable universe is composed entirely of light and matter; (c) if light has a dual, wave-particle nature, perhaps matter has also. Since matter was then regarded as being composed of particles, de Broglie's reasoning suggested that one should search for a wave-like behavior for matter.

De Broglie's suggestion might not have received serious attention had he not predicted what the expected wavelength of the so-called matter waves would be. We recall that about 1680 Huygens put forward a wave theory of light that did not receive general acceptance, in part because Huygens was not able to state what the wavelength of the light was. When Thomas Young rectified this defect in 1800, the wave theory of light started on its way to acceptance.

De Broglie assumed that the wavelength of the predicted matter waves was given by the same relationship that held for light, namely, ...

$$\lambda = \frac{h}{p}, \tag{1-6}$$

which connects the wavelength of a light wave with the momentum of the associated photons. The dual nature of light shows up strikingly in this equation and also in [Eq. 1-5] $(E = h\nu)$. Each equation contains within its structure both a wave concept (ν and λ) and a particle concept (E and p). *De Broglie predicted that the wavelength of matter waves would also be given by [Eq. 1-6], where p would now be the momentum of the particle of matter. . . .*

In 1926 Elsasser pointed out that the wave nature of matter might be tested in the same way that the wave nature of X-rays was first tested, namely, by allowing a beam of electrons of the appropriate energy to fall on a crystalline solid. The atoms of the crystal serve as a three-dimensional array of diffracting centers for the electron "wave"; we should look for strong diffracted peaks in certain characteristic directions, just as for X-ray diffraction.

This idea was tested by C. J. Davisson and L. H. Germer in this country and by G. P. Thomson in Scotland.[17]

De Broglie's conjecture was strikingly confirmed by experimental results. In just several years' time his predicted phenomenon was extensively utilized in various kinds of research. From various quarters in the scientific world, reports came telling how de Broglie's theory explained previously unaccounted for data, or telling of a new mystery that the de Broglie theory solved. The breakthrough was incredible. In the photoelectric effect, waves were seen to act like particles; and now sub-atomic sized particles were seen to act like waves. Because of this, electron diffractometers could be developed as well as electron microscopes. Many basic mysteries could now be understood. Scientists had cause to rejoice, for the profound duality of nature had at last demonstrated itself.

What was needed was not a particle model for light or matter, nor a wave model, but a new model which could accurately harmonize the particle-like and wave-like characteristics of light and matter. The basics of the model quickly took shape. The new model of nature, called quantum mechanics, has had extensive successes. It was discovered that both the particle-like and the wave-like characteristics of electrons are essential for the existence of the atom. That is, if only the particle properties of the electrons were active, the electrons in the universe would quickly spiral into the nuclei of the atoms (due to the loss of energy through electromagnetic radiation). Instead of orderly electron clouds orbiting atomic nuclei, there would be only a nuclear plasma such as there is in the core of the sun. There could be no chemical bonds or chemicals. Thus there could be no plant or animal life. Indeed, the quantum nature of matter is essential for our existence.

There were many other such successes of the quantum mechanical model of the universe. The quantum nature of matter beautifully accounted for the atomic spectra, and has recently been utilized to explain even such mysterious phenomena as superconductivity, superfluidity, paramagnetism, and diamagnetism. In quantum mechanics there is a kind of simplicity in complexity, and order and mathematical beauty. The realization and acceptance of the profound duality of nature has unlocked untold mysteries and catapulted man into the atomic, the electronic, and the space age. A radiation physics textbook remarks about this particle-wave dualism in atomic and nuclear physics:

> It may seem strange that, having found the wave theory of electromagnetic radiation inadequate to explain certain physical phenomena, part of the wave model should be incorporated into the quantum model of electromagnetic radiation. This dualism, however, seems to be inherent in the "explanations" of atomic and nuclear physics. Mass and energy, particle and wave in the case of electromagnetic energy, and . . . wave and particle in the case of sub-atomic particulates, *all seem to be part of a dualism in nature*; either aspect of this dualism

can be demonstrated in the laboratory by appropriate experiments. (Italics supplied.)[18]

A worthwhile observation that can be made from the resolution of this important mystery of light is that, in the quest of the ultimate truth, it is quite natural for scientific thought to oscillate between two seemingly diametrically opposed viewpoints as scientific data and knowledge increase. This oscillation itself should alert us to the possibility of a resolution of the mystery through a synthesis of the opposite ideas. This observation is important, for in the next chapter we shall observe another oscillation of scientific thought regarding another important mystery of light. The author proposes a synthesis of the two seemingly opposing ideas. The author's model of a unified universe is founded on the theorized profound paradoxical duality in this other conundrum of light. This is the mystery: Is there an absolute reference frame with an aether, or is there relativity without an aether?

---

[1]"Light," *Encyclopaedia Britannica*, 15th edition (Chicago: Encyclopaedia Britannica, Inc., 1974), Macropaedia, Volume 10, pp. 929, 930.

[2]*Ibid.*, p. 929.

[3]Physical Science Study Committee, *College Physics* (Printed in U.S.A.: Raytheon Education Company, 1968), p. 74.

[4]*Ibid.*, pp. 76-79.

[5]*Ibid.*, pp. 82-84.

[6]Kenneth F. Schaffner, *Nineteenth-Century Aether Theories* (Oxford: Pergamon Press, 1972), p. 11.

[7]*Ibid.*

[8]*Ibid.*, pp. 82-84.

[9]*Ibid.*

[10]This is the experiment in which two pencils of light polarized in planes at right angles to one another cannot be made to interfere under any condition of path-length difference. The results were not published until 1819 though the experiment had been done several years earlier.

[11]*Ibid.*

[12]*College Physics, op. cit.*, pp. 105-107.

[13]*College Physics, op. cit.*, p. 929.

[14]Robert B. Leighton, *Principles of Modern Physics* (New York: McGraw-Hill Book Company, Inc., 1959), p. 3.

[15]*College Physics, op. cit.*, pp. 592-594

[16]David Halliday, Robert Resnick, *Physics* For Students of Science and Engineering, Part II, Second Edition (New York: John Wiley & Sons, Inc., 1962), pp. 1090, 1091.

[17]*Ibid.*, pp. 1108, 1109 (Italics in original.).

[18]Herman Cember, *Introduction to Health Physics* (Oxford: Pergamon Press, 1969), p. 34.

Problem Set 1

1. Why are two eyes better than one? Why are two ears

better than one? Why are two hands better than one? Why are two feet better than one?

2. Why is particle-wave duality advantageous for matter and light?

3. Kepler tried unsuccessfully to model all celestial motion as circles within circles. Circles have single centers or foci. How many foci did Kepler determine general planetary orbits require in each orbit? What is the geometric shape of such orbits?

4. What duality explains hyperfine splitting of atomic spectra?

5. Are mysteries always dualities? Give two examples of mysteries that compose three or more aspects.

6. With the knowledge you have gained in this chapter and in this problem set, draft a "mystery resolution principle."

Chapter 2

AETHER OR RELATIVITY?

Intimately involved in the particle-wave riddle in the nature of light mystery is the question of whether or not there is a luminiferous aether (or ether), or light wave medium. If light is a particle, there would not need to be a light wave medium. The particles of light, once emitted, would travel in free space unimpeded until they struck some obstacle or object and were absorbed or reflected. On the other hand, if light is a wave, it would only stand to reason that light is a wave of some medium, just as waves on the sea are water waves, and sound is waves of atoms, whether gaseous, liquid, or solid. Is there, or is there not, a medium for light waves?

The centuries-long wave-particle controversy (which we studied in the last chapter) was, at the same time a debate over whether there is or is not an aether, or light wave medium. For the most part, whenever there was a belief that light is a wave, then there was a belief in a light wave medium, or aether. And, for the most part, whenever there was a belief that light is a particle, then there was a getting away from the concept of an aether of space.

A. Wave-Aether Concept

Both the particle and wave concepts go back to ancient times, yet the wave-aether concept appears to have been more highly developed in ancient times than the particle theory. The word "aether" itself is an ancient word--a Latinization of the Greek root αιθηρ, "which means upper air or sky and relates to the refined fire of the empyrean."[1] "From Empedocles and Aristotle through Descartes and Newton to Maxwell and Lodge, the age-old concept of a

15

plenum or an aether has given one solution to the hoary philosophical problem of action-at-a-distance, across seemingly empty space."[2]

"In its long history, the aether has had a number of different tasks assigned to it, such ascriptions occasionally leading scientists to perhaps multiply aethers beyond necessity."[3] But the purest goal of aether theory has been to find a single aether which can account for all known forces at a distance.

> It [the aether] must be a medium which can be effective for transmitting all the types of physical action known to us; it would be worse than no solution to have one medium to transmit gravitation, another to transmit electric effects, another to transmit light and so on. Thus the attempt to find out a constitution for the aether will involve a synthesis of intimate correlation of the various types of physical agencies, which appear so different to us mainly because we perceive them through different senses.[4]

> > It should also be noted . . . that such a "unified field theory" would have been a mechanical theory, thus accomplishing a complete unification of physics that has often been an inspiration and goal to physicists from Oersted and Faraday to Einstein and Wheeler.[5]

There have been many and varied aether theories by Descartes,[6] Huygens,[7] Newton,[8, 9] Young,[10] Fresnel,[11] Stokes, Cauchy, Green, MacCullagh, Fitzgerald, Larmor,[12] and Maxwell.[13, 14, 15]

B. Corpuscular Theory

At first Newton had aether theories for the transmission of heat through a vacuum[16] and "as a background for absolute mechanical motion in the Universe."[17] Later Newton had less and less place for an aether in his system of physics, until in 1717 "he equated the Aether with the Vacuum and rested content with a corpuscular theory of light."[18] Newton's growing authority in physics "brought hard days on the proponents of the wave theory of light during the eighteenth century, and accordingly on the development and acceptance of aether theories in which the aether functioned as the light medium."[19]

Due to the influence of Newton and others, the aether theory was almost entirely rejected in the eighteenth century. But it came back again in the nineteenth century. "It is a common practice so far as optics is concerned to call the seventeenth century an era of plenitude in nature, the eighteenth century an era of the corpuscularian theory of light, and the nineteenth century an era dominated by the undulatory or wave theory of light."[20] We recognize this progression of thought as a scientific opinion oscillation from, away from, and back to an aether theory.

C. Wave-Aether Theory

Thomas Young brought back the wave-aether theory of light.[21] Young was the one who resolved the wave model problem with light polarization by suggesting that light may be a transverse wave instead of a longitudinal wave like sound. Fresnel seized upon the transverse wave concept and worked out essentially a "kinematical" theory of wave motion. Fresnel believed that the aether, while being molecular with forces acting between the particles, acted like an elastic solid. This theory, called the "elastic solid theory," assumes that the aethereal medium is so constituted as to permit transverse waves to be propagated through it, but not longitudinal waves.[22]

There was one more major adjustment to optical aether theories before aether theories came on hard times again. Breakthroughs in electricity and magnetism led to the concept of an electrical aether in space which could support electrostatic and electromagnetic wavefronts propagating at the speed of light. This theoretical discovery was made by James Clerk Maxwell (1831-1879), Scotch physicist, through the insight gained by his remarkable synthesis of knowledge on electric and magnetic effects into a concise set of elegant formulas, which have since been named in his honor. His equations showed that electromagnetic waves ought to exist in empty space, and that they ought to travel with the speed of light. With this added insight, scientists concluded that light was indeed an electromagnetic wave of the appropriate frequency, and that therefore the luminiferous aether was somehow electromagnetic in character. Some final contributions to the electrical aether theory put forth the view that electrostatic effects in space are due to polarization of the aether, while magnetic effects are due to the velocity of the aether.

Since Maxwell's equations showed that electromagnetic waves ought to exist in empty space (that is, space devoid of atoms and molecules), some scientists consider that Maxwell began the trend toward the belief that an aether is not necessary. But Maxwell did not so regard the matter. He believed firmly in an electrical aether. Maxwell's equations contained terms for "displacement currents" and "electric displacements" in a vacuum as well as in solids.[23, 24] These he interpreted in terms of his theory of "aether strains."

D. Aether Experiment

The more concrete concepts of the nature of the luminiferous aether that resulted from Maxwell's work, together with the realization that electromagnetic waves ought to exist in free space, set the stage for an experimental attempt to detect the aether and measure its velocity relative to the earth. Such an experiment was conducted by Michelson (1881) and by Michelson and Morley (1887).

Michelson and Morley constructed an elaborate interferometer mounted on a large stone floating in Mercury. (See Figs. 2-1 and 2-2.) Light from a monochromatic light source was made to strike a half-silvered mirror a. Half of the light passed through the half-silvered mirror to mirror c. Half of the light was reflected from the half-silvered mirror to mirror b. Light from b passed through the half-silvered mirror a to d, and light from c reflected from half-silvered mirror a to d. (See Fig. 2-2.) The paths were equal in length. If the system was at rest relative to the aether, the light traveling the two paths should interfere constructively.

Based on the Galilean transformation (see Eqs. 2-1) and Newtonian mechanics, the reasoning of Michelson and Morley was that if the system (and the earth) moved with velocity V with respect to the aether along the sac axis, the time of flight for the light beam along aca would be longer than the time of flight for the light beam along aba. (The mathematical details are in Figs. 2-2 through 2-9.) Thus the two beams should be out of phase and should interfere destructively.

The displacement to be expected from the Michelson-Morley apparatus was 0.4 fringe (if the earth were traveling through an aether).

The actual displacement was certainly less than the twentieth part of this, and probably less than the fortieth part. But since the displacement is proportional to the square of the velocity, the relative velocity of the earth and the ether is probably less than one-sixth the earth's orbital velocity, and certainly less than one-fourth [See Fig. 2-10].[25]

The experimental results of A. A. Michelson and E. W. Morley were contrary to what we would expect, based on the Galilean transformation. The experiments have since been oft repeated (with variations) with different wavelengths of light, with starlight, with extremely monochromatic light from a modern laser, at high altitudes, under the earth's surface, on different continents, and at different seasons, over a period of some 80 years. We can say that without length contraction and time dilation the ether drift is zero to a precision which is best expressed by saying that the speeds of light upstream and downstream are equal within the variation of less than 10^3 cm/sec, or of 1 part in 1000 of the earth's orbital velocity about the sun.[26]

The null result of the Michelson-Morley experiment was nothing short of a big scientific scandal that sent scientists scrambling for a suitable explanation. The Michelson-Morley experiment was a continental divide—a major watershed—in the concepts not only of light, but also of space and time and physical dynamics, or mechanics. Most scientists abandoned the aether theory of light altogether. But interestingly enough, we see that after this major shock, aether theories are coming back again.

Many attempts were made to explain the null result of the Michelson-Morley experiment without altogether giving up the idea of an aether: It was suggested that bodies moving through space might drag aether along locally. Another idea was that the velocity of light adds vectorially to that of the source. But both these ideas were fraught with difficulties.[27]

The most serious proposal advanced was that bodies which move through the aether suffer a change of shape just sufficient to make the speed of light

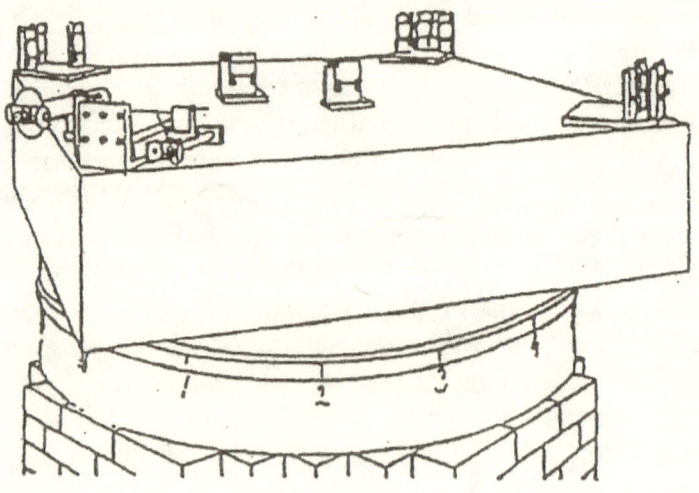

Fig. 2-1.* The Michelson-Morley interferometer was mounted on a large stone, floating on Mercury. It was designed so that it could be rotated on the base and aligned with different angle markings.

*Figures 2-1 to 2-10 are adapted from Charles Kittel, *et. al.*, *Mechanics*: Berkeley Physics Course–Volume 1 (New York: McGraw-Hill, 1965), pp. 332, 333, 335.

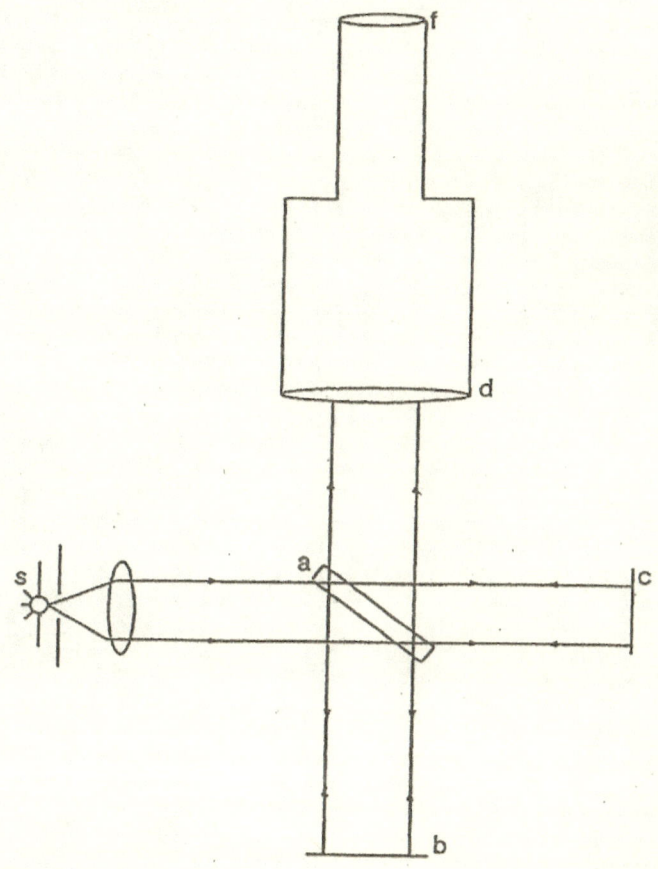

Fig. 2-2. The Michelson-Morley experimental interferometer consists of light source s, half-silvered mirror a, mirrors b and c, and telescope detector d; f represents focus of telescope. If interferometer is at rest in ether, an interference pattern between beams aba and aca . .

.

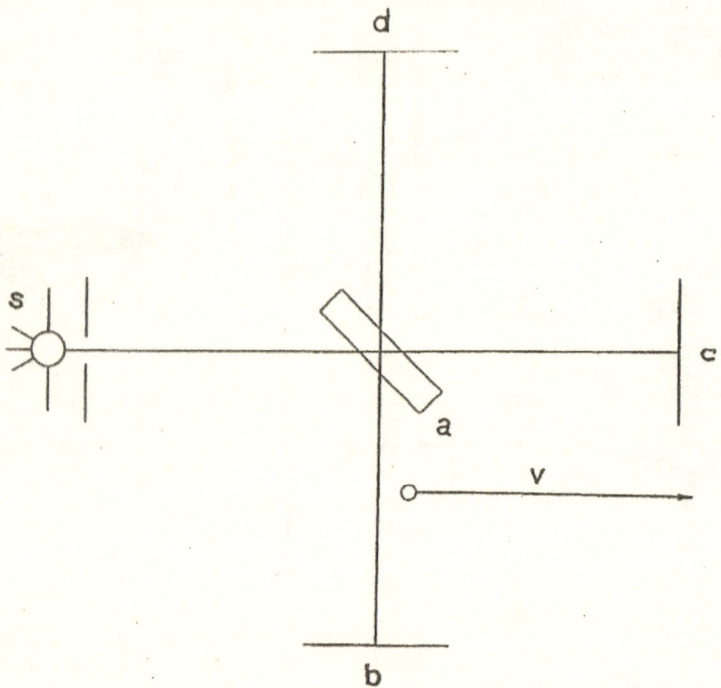

Fig. 2-3. . . . is observed at d. If apparatus (and earth) have velocity V with respect to hypothetical aether, we would expect interference pattern to change at d, since the times to traverse aba, aca would now change by different amounts. To see this . . .

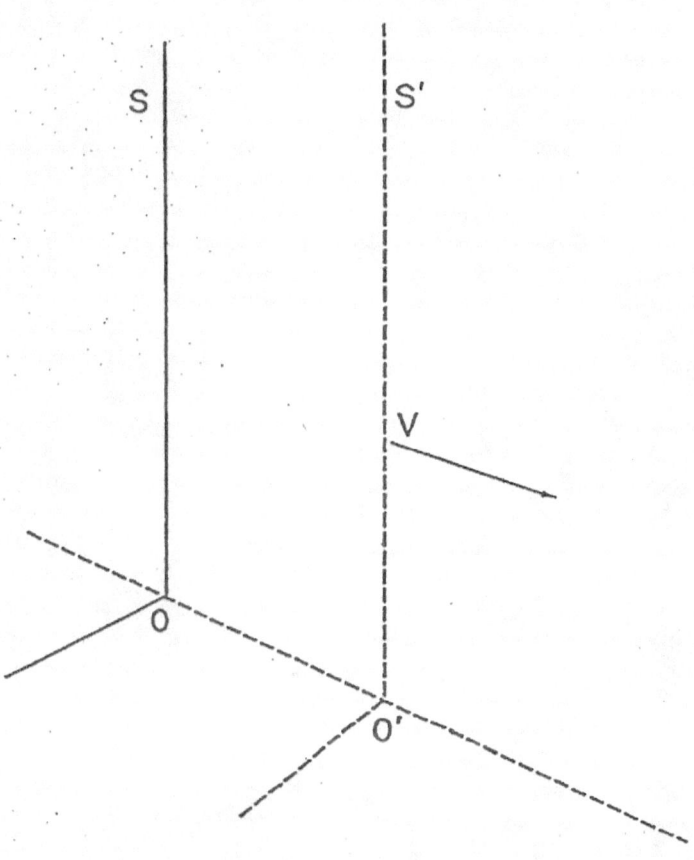

Fig. 2-4. . . . consider a Galilean frame S' moving with earth and interferometer. S is a Galilean frame at rest in aether.

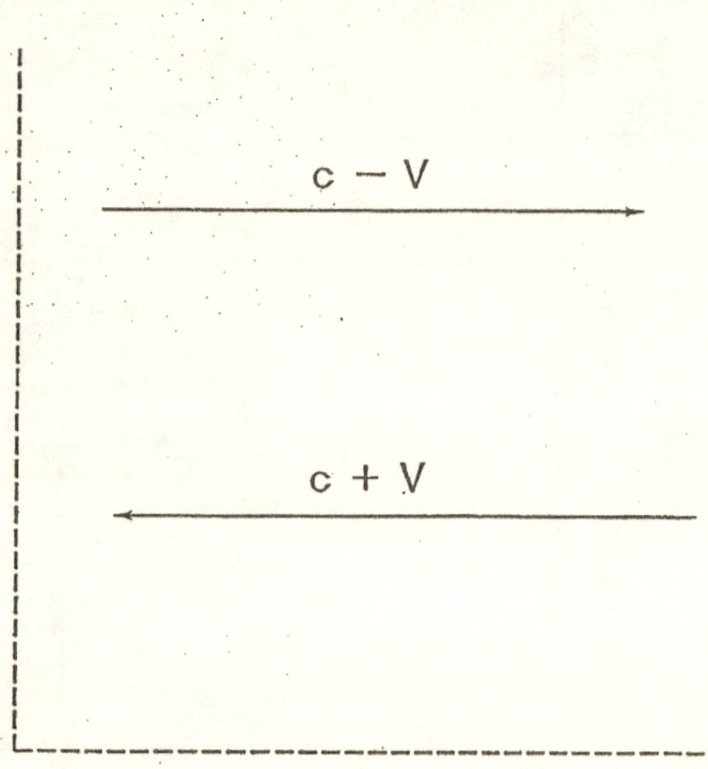

Fig. 2-5. According to the Galilean transformation, light moving to right has speed c - V in S'; light moving to left has speed c + V in S'.

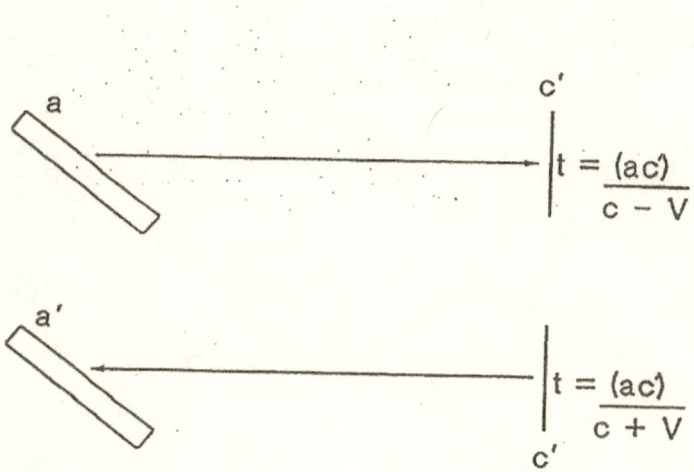

Fig. 2-6. Thus the time to go from a to c' and back to a' is

$$\Delta t(ac'a') = \frac{(ac')}{c-V} + \frac{(ac')}{c+V'},$$

where (ac') denotes the distance between a and c'.

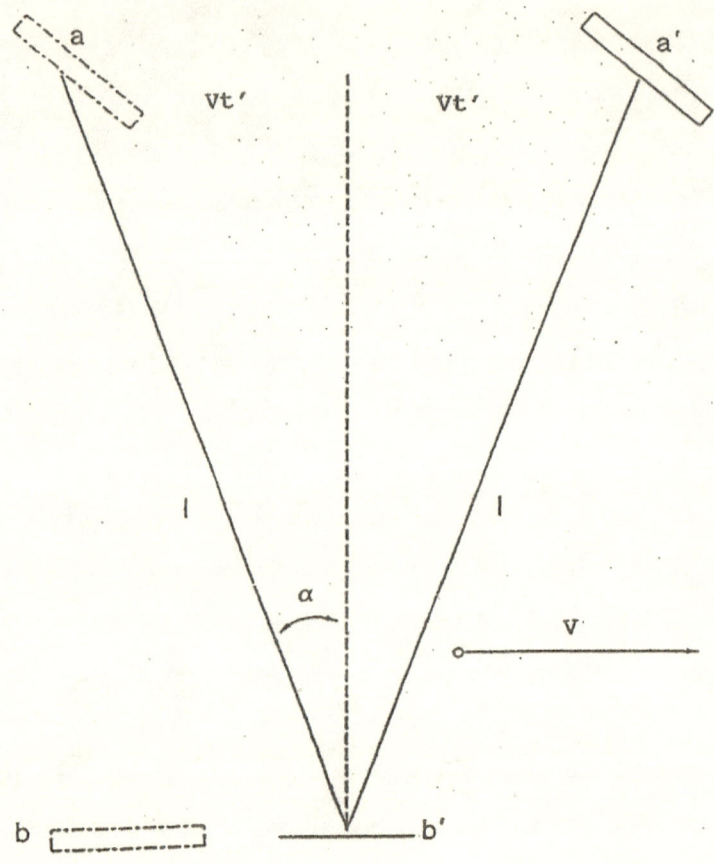

Fig. 2-7. What is the time $\Delta t(ab'a) = 2t'$ to go from a to b and back to a? In the Galilean frame S at rest in aether, the interferometer has velocity V to right; light has speed c.

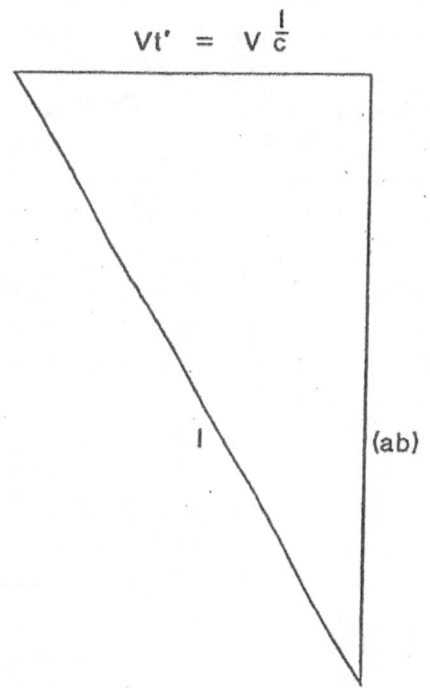

$$Vt' = V\frac{l}{c}$$

l

(ab)

Fig. 2-8.

$$l^2 = \frac{V^2}{c^2}l^2 + (ab)^2$$

$$\therefore l = \frac{(ab)}{(1 - V^2/c^2)^{1/2}}$$

$$\Delta t(aba') = 2t' = 2(ab)\sqrt{c^2 - V^2}.$$

Thus even if $(ab) = (ac)$, the Galilean transformation leads us to expect a . . .

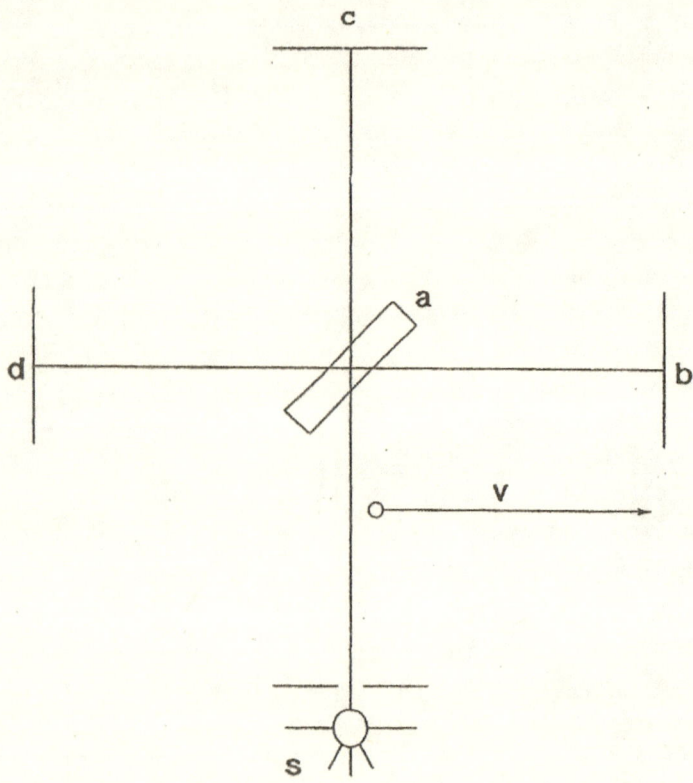

Fig. 2-9. . . . shift in the interference pattern if the interferometer changes its velocity with respect to the aether. None was observed.

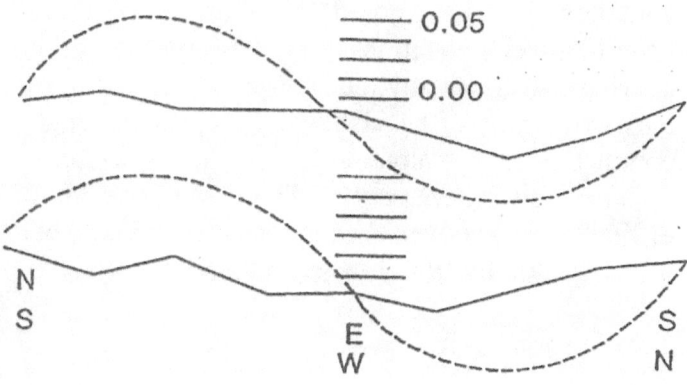

Fig. 2-10. Michelson and Morley wrote: "The results of the observations are expressed graphically [in the figure]. The upper is the curve for the observations at noon, and the lower that for the evening observations. The dotted curves represent *one-eighth* of the theoretical displacements. It seems fair to conclude from the figure that if there is any displacement due to the relative motion of the earth and the luminiferous ether, this cannot be much greater than 0.01 of the distance between the fringes." (Michelson and Morley, *Am. J. Sci.* **34**, 333 (1887).) The vertical axis is the displacement of the fringes; the horizontal axis refers to the orientation of the interferometer relative to an east-west line.

appear to be the same in various directions. The change that is needed is a contraction in the direction of motion of the body, such that a body of length L_o at rest is shrunk to a length $L_o(1 - V^2/c^2)^{1/2}$ when in motion at speed V parallel to its length. This effect, called the Lorentz-FitzGerald contraction, was supposed to follow from follow from Maxwell's equations; no one was successful in proving that it actually did, however. Furthermore, later experiments[28] have shown that a simple length contraction is not alone sufficient; a *time-dilation* effect is also necessary. This would be most difficult for Maxwell's equations to provide, but follows naturally from Einstein's theory.[29]

E. Two Theories of Relativity

In 1904, one year before Einstein published his Special Theory of Relativity, Hendrik Antoon Lorentz originated a theory of relativity.

We must make a distinction between Einstein SR and Lorentzian Relativity (LR). Both Lorentz in 1904 and Einstein in 1905 chose to adopt the principle of relativity discussed by Poincare in 1899, which apparently originated some years earlier in the 19th century. Lorentz also popularized the famous transformations that bear his name, later used by Einstein. However, Lorentz's relativity theory assumed an aether, a preferred frame, and a universal time. Einstein did away with the need for these. But it is important to realize that none of the 11 independent experiments said to confirm the validity of SR experimentally distinguish it from LR—at least not in Einstein's favor.

Experiment	Description	Year
Bradley	Discovery of aberration of light	1728
Fresnel	Light suffers drag from the local medium	1817
Airy	Aberration independent of the local medium	1871
Michelson-Morley	Speed of light independent of Earth's orbital motion	1881
De Sitter	Speed of light independent of speed of source	1913
Sagnac	Speed of light depends on rotational speed	1913
Kennedy-Thorndike	Measured time also affected by motion	1932
Ives-Stilwell	Ions radiate at frequencies affected by their motion	1941
Frisch-Smith	Radioactive decay of mesons is slowed by motion	1963
Hafele-Keating	Atomic clock changes depend on Earth's rotation	1982
GPS	Clocks in all frames continuously synchronized	1997

Table 2-1. Independent experiments bearing on Special Relativity

Several of the experiments bearing on various aspects of SR (see Table 2-1) gave results consistent with both SR and LR. But Sagnac in 1913, Michelson following the Michelson-Gale confirmation of the Sagnac effect for the rotating Earth in 1925 (not an independent experiment, so not listed in Table 2-1), and Ives in 1941, all claimed at the time they published that their results were experimental contradictions of Einstein SR because they implied a preferred frame. In hindsight, it can be argued that most of the experiments contain some aspect that makes their interpretation simpler in a preferred frame, consistent with LR. In modern discussions of LR, the preferred frame is not universal, but rather coincides with the local gravity field. Yet, none of these experiments is impossible for SR to explain....

These experiments confirm the original aether-formulated relativity principle to high precision. However, the issue of the need for a preferred frame in nature is, charitably, not yet settled. Certainly,

However, the issue of the need for a preferred frame in nature is, charitably, not yet settled. Certainly, experts do not yet agree on its resolution. But of those who have compared both LR and SR to the experiments, most seem convinced that LR more easily explains the behaviour of nature.[30]

F. Einstein's Special Relativity

Einstein's solution to the Michelson-Morley null result riddle stemmed from the principle of relativity.

It has long been accepted there is no meaning to absolute translational motion through space. Now this idea is supported by our everyday experience with the laws of mechanics as they apply inside a uniformly moving automobile, train, or airplane; for in such a "moving" reference frame these dynamical laws appear the same as in a laboratory "at rest" on the earth. One can describe the situation analytically by saying the laws of motion are *covariant*—i.e., they retain the same form—with respect to a transformation of coordinates to a uniformly moving system. Now this property of the laws of mechanics leads naturally to the supposition that the universe may be so constituted that it is impossible by any kind of experiment whatever to detect absolute motion through space. This hypothesis is called the *principle of relativity*.

Now the principle of relativity was an accepted theory of physics for over two centuries. But when J. C. Maxwell, in 1865, formulated his dynamical theory of the electromagnetic field, it appeared that absolute motion through space might be detectable by *optical* means.

Now it seems quite clear that, if space is not really empty, but filled with a rigid medium, there might be some meaning to absolute motion after all. And it even appears possible our speed through this medium might be measured by comparing the speed of light in different directions. Such an optical experiment was carried out by Michelson (1881) and by Michelson and Morley (1887).[31]

The null result of the Michelson-Morley experiment led Einstein to assert again with force the principle of relativity—the idea that it is impossible by any kind of experiment whatever to detect absolute motion through space. Previous explanations of the principle of relativity had led to the so-called Galilean transformation:

$$x' = x - Vt \quad y' = y \quad z' = z \quad t' = t. \tag{2-1}$$

This transformation was the basis for classical mechanics. The principle of relativity with the added assumption that the speed of light is constant to all observers led Einstein and others to the Lorentz transformation:

$$x' = \frac{x - Vt}{(1 - V^2/c^2)^{1/2}} \quad y' = y \quad z' = z \quad t' = \frac{t - (V/c^2)x}{(1 - V^2/c^2)^{1/2}}. \tag{2-2}$$

This transformation is the basis of Einstein's Special Theory of Relativity in which there is relative mass increase, time dilations, and length contraction for objects traveling an appreciable velocity compared to the speed of light as seen by a "stationary" observer.

The Special Theory of Relativity was developed by Einstein in 1905 as a result of his consideration of objects moving relative to one another with constant velocity. "In 1915 Einstein developed the theory of general relativity in which he considered objects accelerated with respect to one another."[32] The principle that simplified this problem was that of equivalence—"that forces produced by gravity are in every way equivalent to forces produced by acceleration, so that it is theoretically impossible to distinguish between gravitational and accelerational forces by experiment."[33] In Einstein's General Theory of Relativity, Newton's simple hypothesis that every object attracts every other object in direct proportion to its mass is replaced by the relativistic hypothesis that the continuum is curved in the neighborhood of massive objects. Einstein's revised law of gravity states simply that the world lines of every object is a geodesic in the continuum. A geodesic line is the shortest distance between two points, but in curved space it is not generally a straight line. In the same way, geodesic lines on the surface of the earth are great circles, which are not straight lines on any ordinary map.[34]

While more accurate, relativistic mechanics is much more complicated mathematically than classical mechanics.

Now the famous apocryphal statement that only ten people in the world understood Einstein's theory referred to the complex tensor algebra and Riemannian geometry of general relativity; by comparison, special relativity can be understood by any college student who has studied elementary calculus.[35]

Any operation of and experiment with modern elementary particle accelerators is a test of Relativity. Several classic tests of the General Theory of Relativity have been performed with good results. These will be presented in detail in Chapter 4.

G. Special Relativity's Weaknesses

Except for popularity in scientific publications, Einstein's aetherless theory of relativity hasn't done quite as well as Lorentz's aether model of relativity. Without any

preferred reference frame, Einstein's theory hasn't accounted for the experimental data as simply as Lorentz's theory
with a preferred reference frame. Scientists can argue that
Einstein's theory can account for the data. But other scientists can also argue that it cannot account for all the data
parsimoniously. That principle is an accepted principle of
physics. It is a postulate in this book in Chapter 6. This
principle should have weight in this contest.

Another area of difference between SR and LR is the
twin or clock paradox. In the twin paradox, one of two
twins rockets away from earth a distance into space, turns
around, then rockets back. To the traveling twin, the stay at
home twin appears to make a round trip in the opposite
direction. Which one will age more than the other, or will
they age alike? Will each twin claim that the other aged
less because of time dilation? Or will there be a difference
in the aging? In SR there is a twin or clock paradox.

How does the resolution of the "twin paradox" compare in
LR and SR?

In LR, the answer is simple: Now the Earth frame at
the outset, and the dominant local gravity field in general,
constitutes a preferred frame. So then the high-speed
traveler always comes back younger, and there is no true
reciprocity of perspective for his or other frames.

Now in SR, the answer is not so simple; yet an
explanation exists. We find the reciprocity of frames
required by SR when Einstein assumed that all inertial
frames were equivalent introduces a second affect on
"time" in nature that is not reflected in clock rates alone.
We might call this effect "time slippage" so we can discuss
it. Now time slippage represents the difference in time for
any remote event as judged by observers (even
momentarily coincident ones) in different inertial frames.[36]

The author alluded to above goes on to give several numerical examples of time slippage for the traveling twin when he turns around, in order that complete reciprocity of frames may be maintained in Einstein's Special Theory of Relativity. Lorentzian Relativity doesn't need any time slippage. It has a preferred frame. Perhaps scientists should make one more unmanned lunar orbit mission with atomic clocks and return of the module to earth for comparison of the atomic clocks with ground clocks, to see if there is any time slippage, or to see if the clocks are as in LR with a preferred reference frame.

The problem is caused in Einstein's SR because he did not believe there was any way of determining an absolute space in special relativity. He postulated that one uniformly moving reference frame was as good as another. The theory of relativity was thought to be consistent no matter what the rest velocity was assumed to be. Therefore twin A should experience length contraction and time dilation as computed by twin B if twin B assumes he is at rest during his flight.

Some physicists try to harmonize the clock paradox through general relativity, as though the accelerations of twin B affect the positions of his clock. "Cyclotron experiments have shown that, even at accelerations of 10^{19} g (g = acceleration of gravity at the Earth's surface), clock rates are unaffected. Only speed affects clock rates, but not acceleration per se."[37]

The clock paradox can be put in better perspective by studying triplets instead of twins. Let triplet X stay at home on planet earth. Let triplet Y rocket out in space similar to twin B, and let triplet Z rocket out in space in the opposite direction as triplet Y. After a period of coasting, let triplets Y and Z simultaneously decelerate and rocket back to earth for a close-encounter fly by the earth and each other. Let X, Y, and Z compare their clocks. Then Y and Z should continue to coast for awhile, then simultaneously decelerate and rocket back to earth for another close-encounter and fly by. Let them again compare their clocks. Finally let Y and Z continue coasting for awhile, simultaneously decelerating and accelerating back to planet earth, where they land and compare their clocks with triplet X.

Who should have slower clocks? X, Y, or Z? Z should expect Y to have time dilation relative to him according to Einstein's theory. Also Y should expect Z to have time dilation relative to him.   One cannot appeal to the accelerations and general relativity to harmonize this contradiction, for the accelerations are symmetrical. Triplet X would expect Y and Z to be time dilated equally.

None of this is a problem in LR, which has a preferred reference frame.

H. Clock Paradox Experiment

Confusion over the clock paradox was so long-lasting that men decided only experiment could settle the question.

The first experiment to resolve this question was by Hafele and Keating in their contracircumglobar journey in jet aircraft with portable cesium beam atomic clocks.[38, 39] Other experiments were later made in orbiting vehicles.

Hafele and Keating put portable cesium beam atomic clocks (very precise clocks that can measure nanosecond differences of time intervals lasting days) in commercial jet aircraft on around-the-world flights. One set of clocks was put aboard a westward around-the-world flight. Another set of clocks was put aboard an eastward around-the-world flight. These were compared with the MEAN(USNO) clock of the U.S. Naval Observatory for the earth-bound clock. Hafele and Keating kept careful flight logs as to altitudes, velocities and directions as a function of time for both flights. From these flight data Hafele and Keating calculated theoretical time differences through special and general relativity. The general relativity time differences were due to different clock altitudes in the earth's gravitational field. The special relativity time differences were due to different plane speeds and directions and the earth's rotation rate and different latitudes in the flights. The theoretical time differences were compared with measured time differences with good results.

Notice data in Tables 2-2 and 2-3. They are predicted and measured time differences between the "stationary" earth clock (which actually travels eastward at about 1000 miles per hour due to the rotation of the earth) and the clock on the jet flying eastward around the earth and the jet flying westward around the earth. Not the earth clock, but the clock in the jet flying westward ran the fastest, showing that it was more nearly at rest than the other clocks.

I. Falsification of Principles of Relativity and Equivalence

These tests have been thought to be proofs of Einstein's special and general theories of relativity, for the *formulas* of special and general relativity were once again proven correct. But somehow a couple of important points have been missed. These experiments show that it *is* possible to measure absolute velocity in space through atomic clock experiments. (The westward flying clocks were more nearly at rest than the earth because of the earth's daily rotation.) Thus *the principle of relativity*—the supposition that the universe may be so constituted that it is impossible by any kind of experiment whatever to detect absolute motion through space—*is false*! The principle is almost true. Almost no phenomena will measure absolute motion in space. Inertia will not show absolute motion. But relativity is not absolute. Space is absolute. Absolute motion can be detected with atomic clock experiments.

Table 2-2
Predicted relativistic time differences (nsec).*

Effect	Direction	
	East	West
Gravitational	144 ± 14	179 ± 18
Kinematic	-184 ± 18	96 ± 10
Net	-40 ± 23	275 ± 21

Predicted relativistic time differences for atomic clocks in Hafele and Keating's contracircumglobar journey in jet aircraft.

*J. C. Hafele and Richard E. Keating, "Around-the-World Atomic Clocks: Predicted Relativistic Time Gains," *Science*, Vol. 177, July 14, 1972, p. 167.

Table 2-3
Observed relativistic time differences..**

Δt (nsec)

	Eastward*	Westward*
Mean ± S.D.	-59 ± 10	273 ± 7
Predicted ± Error Est.	-40 ± 23	275 ± 21

Observed relativistic time differences for atomic clocks in Hafele and Keating's contracircumglobar journey in jet aircraft.

*Negative signs indicate that upon return the time indicated on the flying clocks was less than the time indicated on the MEAN(USNO) clock of the U.S. Naval Observatory.

**J. C. Hafele and Richard E. Keating, "Around-the-World Atomic Clocks: Observed Relativistic Time Gains," *Science*, Vol. 177, July 14, 1972, p. 168.

With the falsification of the principle of relativity, Einstein's aetherless theory of special relativity is undermined.   Interestingly enough, the Hafele-Keating experiment, correctly interpreted, also falsifies the principle of equivalence, the foundation of Einstein's general theory of relativity.   The principle of equivalence is that it is impossible by any means to detect the difference between gravity and a corresponding acceleration.   But with the Hafele-Keating experiment, we see from experimental data, not just theory, that atomic clocks go at constant rates at constant altitudes in a gravitational field, but go at constant rates at constant velocities, not constant accelerations. With sufficient experiments with atomic clocks it *is* possible to differentiate between accelerations and gravitational fields.[40]   [Look this endnote up.] Thus Einstein's whole Theory of General Relativity is based on a faulty foundation. Again the principle of equivalence is nearly true. Accelerometers cannot differentiate between gravity and inertia. But atomic clocks can. Therefore the principle of equivalence is partial, not absolute.

Some have argued that the motions of the jets around the earth and of the earth itself were accelerated motions, not covered by special relativity. But the special relativistic formulas did harmonize well with the measured results, and besides there is no acceleration term in time dilation—only velocity terms and gravity terms. Gravity is thus more like a gradient of velocities in space than like an acceleration.

In LR, symmetry (covariance) sometimes may need to be sacrificed "to retain causality."[41] Thus, if LR is true, the covariance of physical laws (one postulate of SR) may be relative, limited, not absolute.

Now the single most important difference is that, in SR, nothing can propagate faster than c in forward time. In contrast in LR, electromagnetic based forces and clocks would cease to operate at speeds of c or higher. However no problem in principle exists in attaining any speed whatever in forward time using forces such as gravity that retain their efficiency at high speeds.[42] The results of the Hafele-Keating experiment should have been more of a scientific scandal than the Michelson-Morley experiment. But scientists have been slow to discern the true character of those results.

J. Aether Quasi-Relativity

How could Einstein's theories of relativity provide correct formulas for the Hafele-Keating experiment, when his principles of relativity and of equivalence are false? Actually, while Einstein made a big point out of the principle of relativity and ascribed his theory to it, his Special Theory of Relativity was not based on the principle of relativity, but on the assumption that there is a covariance of physical laws and that the speed of light is constant to all observers (separate from the principle of relativity).

The Hafele-Keating experiment shows that there is an absolute space after all. Is there also an aether? Several recent experimental and theoretical results can best be accounted for by an aether system. A hard vacuum seems to have immense positive pressure. Particles like protons are seen to be bubbles in the vacuum. In the bag model of quarks it takes energy to push aside ordinary space—about

55 million ev/cubic fermi.[43]   Another interesting recent concept is the "decay of the vacuum"—the electrical ionization of the vacuum in the presence of a strong localized electric charge—such as the nucleus of a hypothetical atom with $Z = 173$ or more.[44]

The easiest way to account for such phenomena would be to theorize a fluid of near-zero rest mass particles in the vacuum that can be electrically polarized and ionized—a luminiferous aether so to speak. Is there an aether after all?

Recently there has developed renewed interest in an aether, with new evidence of the existence of an aether consisting of a sea of neutrinos and gravitons. Estimates have been made of the density of this aether.[45]

Perhaps there is an absolute space and an aether after all.   But what about relativity?   Einstein's formulas for special and general relativity proved correct in the Hafele-Keating experiment as well as in many other experiments. Certainly we cannot go back to classical mechanics even with an aether. What we need is not an absolute reference frame with an aether merely, or relativity without an aether, but relativity up to a point (quasi-relativity) in an aether. We may see a grand aether-relativity duality in nature similar to and parallel to the particle-wave duality.

It was once popular to believe in an absolute space defined by an aether. But this became unpopular after the null result of the Michelson-Morley experiment. Careful interferometry failed to detect an expected aether drift across the surface of the planet. Einstein accounted for the null result of the Michelson-Morley experiment by assuming the speed of light is constant to all observers and deriving his Special Theory of Relativity.   If there is covariance of physical laws and the constancy of c, there will be a null result to every Michelson-Morley type of experiment with or without an aether simply because the speed of light is constant to all observers. Einstein did not say that an aether was incompatible with relativity. He said, "the introduction of a 'luminiferous aether' will prove to be superfluous inasmuch as the view here to be developed will not require an 'absolute stationary space' provided with special properties, nor assign a velocity vector to a point of the empty space in which electromagnetic processes take place."[46] His calculations of special relativity may not have required an absolute space, but experiment has.   Also, general relativity introduces a preferred coordinate system if special relativity does not. If Einstein had studied equivalence of gravity and inertia before relativity, he might have had a different view of the need for an absolute space.

There's another evidence of an aether and an absolute space. For the last 30 years of his life, Albert Einstein attempted to derive a unified field theory in which gravity is united to the electric force. He failed. Many others have subsequently tried and failed to do this. Scientists have also likened the difficulty of uniting relativity and particle physics to attempts to unite fire and ice. This book shows in both points the difficulty is with Einstein's aetherless theory of relativity. With quasi-relativity in an aether the solutions to the above two problems are possible and can be understood by high school and college physics students.

---

[1]Loyd S. Swenson, Jr., *The Ethereal Aether* (Austin: University of Texas Press, 1972), p. xii.

[2]*Ibid.*, p. xii

[3]Kenneth F. Schaffner, *Nineteenth-Century Aether Theries* (Oxford: Pergamon Press, 1972), p. 7.

[4]Sir Joseph Larmor, *Aether and Matter* (1900), as quoted by Schaffner, *op. cit.*, pp. 3.4.

[5]Schaffner, *op. cit.*, p. 4.

[6]*Ibid.*, p. 7.

[7]*Ibid.*, p. 9.

[8]*Ibid.*, pp. 8, 9.

[9]Swenson, *op. cit.*, p. 6.

[10]Schaffner, *op. cit.*, pp. 11, 12.

[11]*Ibid.*, pp. 14-17.

[12]*Ibid.*, various pages throughout the book.

[13]Edward M. Purcell, *Electricity and Magnetism*, Berkeley Physics Course, Volume 2 (New York: McGraw-Hill Book Company, 1965), p. 263.

[14]William Taussig Scott, *The Physics of Electricity and Magnetism* (New York: John Wiley & Sons, Inc., 1966), p. 120.

[15]Charles Kittel, Walter D. Knight, and Malvin A. Ruderman, *Mechanics*, Berkeley Physics Course, Volume 1 (New York: McGraw-Hill Book Company, 1965), p. 330.

[16]Schaffner, *op. cit.*, p. 8.

[17]Swenson, *op. cit.*, p. 6.

[18]*Ibid.*, p. 10.

[19]Schaffner, *op. cit.*, pp. 9, 10.

[20]Swenson, *op. cit.*, pp. 4, 5.

[21]Schaffner, *op. cit.*, p. 11.

[22]*Ibid.*, pp. 16, 17.

[23]Purcell, *op. cit.*

[24]Scott, *op. cit.*

[25]A. A. Michelson and E. W. Morley, *Am. J. Sci. 34,* 333 (1887).

[26]Kittel, et. al., *op. cit.*, pp. 332-336. [Figs. 2-1 through 2-10 are also from this source.]

[27]Robert R. Leighton, *Principles of Modern Physics* (New York: McGraw-Hill Book Company, Inc., 1959), p. 4.

[28]Kennedy and Thorndike, *Phys. Rev.*, **42**, 400 (1932).

[29]Leighton, *op. cit.*, pp. 4, 5.

[30]Tom Van Flandern, Univ. of Maryland and Meta Research, Open Questions in Relativistic Physics, edited by Franco Selleri (Montreal: Apeiron, 1998), pp. 81-90, as quoted by "What the Global Positioning System Tells Us About Relativity, Meta Research, http://www.metaresearch.org/cosmology/gps-relativity.asp

[31]Leighton, *op. cit.*, pp. 1-4.

[32]*Funk & Wagnall's New Encyclopedia* (New York: Funk & Wagnalls, Inc., 1972), Volume 20, p. 205.

[33]*Ibid.*

[34]*Ibid.*

[35]*Ibid.*

[36]Van Flandern, *op. cit.*

[37]C. MØLLER, *The Theory of Relativity* (Oxford: Clarendon Press, date unknown to author).

[38]J. C. Hafele and R. E. Keating, *Science 177*, 166, 167 (1972).

[39]J. C. Hafele and R. E. Keating, *Science 177*, 168 (1972).

[40]To show this, try the following thought experiment: Position a space station far from any stars and at rest with the fixed stars. From that station simultaneously launch three rockets in the same direction which carry precise atomic clocks at the rotational centers of the rockets. Have all three rockets accelerate at a constant rate at 1 g (= 9.8 m/sec^2). Have rocket A accelerate for 400 seconds, turn around by rotating 180 degrees in one second (rotationally accelerating at a constant rate for 1/2 second and 90 degrees, then rotationally decelerating at a constant rate for 1/2 second and 90 degrees), linearly accelerate for 800 seconds, turn around again in one second and accelerate for 400 hundred seconds, and just before soft-landing on spaceship, make 6 additional 180 degree from start to stop turns in six seconds for a total flight time of

1608 seconds. Have rocket B accelerate for 200 seconds, turn around in one second and accelerate for 400 seconds, turn around in one second and accelerate for 400 seconds, turn around in one second and accelerate for 400 seconds, turn around in one second and accelerate for 200 seconds, and just before soft-landing on spaceship, make 4 additional 180 degree from start to stop turns in 4 seconds for a total flight time of 1608 seconds. Have rocket C accelerate for 100 seconds, turn around in one second and accelerate for 200 seconds, turn around in one second and accelerate for 200 seconds, turn around in one second and accelerate for 200 seconds, turn around in one second and accelerate for 200 seconds, turn around in one second and accelerate for 200 seconds, turn around in one second and accelerate for 200 seconds, turn around in one second and accelerate for 200 seconds, turn around in one second and accelerate for 100 seconds, soft-landing on the spaceship after flight of 1608 seconds.

Compare the atomic clocks of rockets A, B, and C, the weightless space station clock, and another clock on a non-rotating planet with gravitational acceleration at the surface = 1 g = 9.8 m/sec^2, where a vehicle carrying the clock on the surface of the planet inverts itself in one second eight times during the course of 1600 seconds. The space station's clock will have run the fastest. Use that clock as the reference clock. The other clocks will have different time differences from the reference clock. The clock in the gravitational field on the planet will be about 10 nanoseconds slower than the space station clock due to general relativity and the gravitational field. Clock A will be slow by 454 nanoseconds. (The rotation of the clocks will itself yield 0 time dilation (to first order) and the time dilation during the rotations due to rocket velocity is a mere 0.34 nanoseconds for Clock A, and insignificant in all cases.) Clock B will be slow by 114 nanoseconds. Clock

C will be slow by 28 nanoseconds.

The time differences for the rocket clocks can be ascertained as follows:

Let t_0', t_0, $x = 0$. Then, from Eq. (2-2),

$$t' = t\left(1 - \frac{V^2}{c^2}\right)^{-1/2} \quad or \quad t' \approx t\left(1 + \frac{V^2}{c^2}\right)^{1/2}.$$

The time difference at constant velocities $T_{V=K}$ approximately equals

$$T_{V=K} \approx t\left(\left(1 + \frac{V^2}{c^2}\right)^{1/2} - 1\right),$$

which for velocities much less than c

$$\approx t\left(\frac{V^2}{2c^2}\right).$$

The time difference for time intervals where $V = at$ equals

$$T_{V=at} = \int_0^t \left(\frac{V^2}{2c^2}\right) dt$$

$$\approx \int_0^t \frac{a^2}{2c^2} t^2 dt$$

$$\approx \left[\frac{a^2}{2c^2} \frac{t^3}{3}\right]_0^t$$

$$\approx \frac{a^2 t^3}{6c^2}.$$

The time difference between a clock at rest in space and a clock in a gravitational field is proportional to t. But the time difference between a clock at rest in space and an accelerating clock is proportional to t^3. We see there *is* an experiment that can be performed to differentiate between force due to gravity and force due to acceleration. Therefore the absolute principle of equivalence is *false*. Einstein's General Theory of Relativity is undermined.

It is no more necessary to actually perform this thought experiment than Einstein's thought experiment of an observer in a box not being able to tell if he were experiencing gravity or acceleration. The results of the above experiment are consistent with the traditional usage of Einstein's theory of relativity. The problem has not been that no experiment could be performed to disprove the principle of equivalence, but that no one has been thought of until now, which seems strange seeing how straightforward the above experiment is, and how naturally it would come to mind as an experimental test.

[41] Van Flandern, *op. cit.*

[42] *Ibid.*

[43] K. A. Johnson, *Sci. Am. 241:1*, 115 (1979).

[44] L. P. Fulcher, J. Rafelski, and A. Klein, *Sci. Am. 241:6*, 150-159 (1979).

[45] H. C. Dudley, *Ind. Res.*, 43, 44 (Nov. 15 1974).

[46]A. Einstein, as quoted by Kittel, et. al., *op. cit.*, p. 336.

Problem Set 2

1. Did Maxwell view his laws with or without an aether?

2. How many forces can you now explain with an electrical aether? [If you finish this volume you will be able to explain them all with an electrical aether.]

3. Did Michelson-Morley prove there is no aether?

4. What did the Michelson-Morley experiment prove?

5. Does Einstein's Special Theory of Relativity actually require the assumption of the absolute principle of relativity? On what two postulates is it actually based?

6. Can any experiment be devised to show absolute motion through space? What devices would be required?

7. Could any experiment show the difference between gravity and constant acceleration? What's required?

8. What can account for immense pressure in a vacuum? What could account for the ability of the vacuum to ionize?

9. The tug-of-war between particle and wave models of light was in the last century resolved by the recognition of particle-wave duality. How would you resolve the paradox between the need for an absolute reference frame with an aether and the need for relativity (in SR without an aether)?

Chapter 3

SPECIAL QUASI-RELATIVITY

The author was unaware of Lorentzian Relativity (LR), until January, 2001. However, in 1964, as a high school student at age 17, the author concluded that there was an aether after all, and Einstein's Special Theory of Relativity (SR) was wrong. In 1977, the author derived what he called "Quasi-Relativity in an Aether, which is similar to LR. That original derivation follows below with minor editorial changes.

A. Postulates

Science is based on postulates—unproven assumptions. Yet they can be well chosen based on experimental evidence. Einstein's Special Relativity is based on two postulates: 1) the principle of relativity (from which is gleaned the covariance of physical laws); and 2) the constancy of c (the speed of light) to all observers. The following aether model of quasi-relativity is based on postulates also. Aether quasi-relativity postulates are not Einstein's postulates. (The covariance of physical laws and the constancy of c are derived results in aether quasi-relativity.) The postulates for aether quasi-relativity are more fundamental in nature:

Special Quasi-Relativity in an Aether Postulates

1. There is an absolute space mediated by an electrical aether.

2. For every whole particle relativistic action there is an equal and opposite reaction.

52

The first postulate may be expanded to say there is a turbulent absolute space mediated by an electrical aether affected by matter and forces in the Universe. The aether may be electrically polarized, ionized, magnetically aligned, and accelerated relative to the distant stars. The best evidence is the aether consists of a sea of imaginary mass integral spin bosons.

The second postulate (parallel to Newton's third law) is here taken more broadly. Lenz's law with magnetic fields and electrical currents ("the induced emf is always in such a direction as to produce a current and resulting flux change that counteracts the original flux change responsible for the induction"[1]) might be considered a phenomenon explained by a more broad interpretation of Newton's third law. Nature resists the collapse of a magnetic field by the induction of a current just equal to that which would compensate for the loss of the magnetic field. Aether relativity depends on a similar reaction to an action affecting the permittivity of free space, length, and clock speed of an object due to motion of that object in an aether in a "classical mechanics" system.

B. Special Quasi-Relativity Fundamentals

Now let us commence the derivation of an aether model of special quasi-relativity in which relativistic effects occur for motions relative to an absolute space defined by the velocity of the aether. Postulates 1 and 2 will be used in this derivation.

Assume there is an electrical aether (Postulate 1). Consider the electrodynamics of a moving body in a "classical mechanics" type aether system. When a

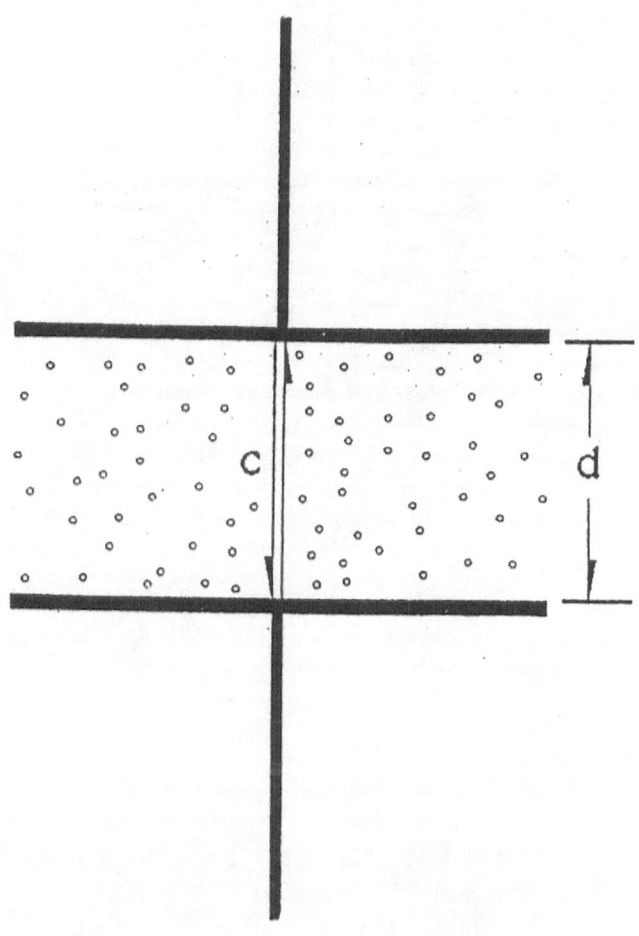

Fig. 3-1. The little circles represent aether particles at "rest" relative to a capacitor of fixed parallel plates. Through this aether at "rest" an electrical signal travels back and forth across the gap in the capacitor at speed c.

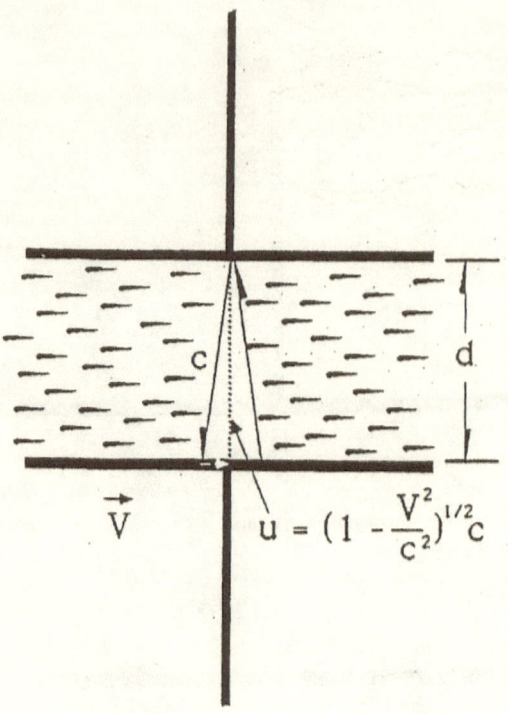

$$u = (1 - \frac{V^2}{c^2})^{1/2} c$$

Fig. 3-2.   The streaked dots represent aether particles traveling at speed V parallel to the plates of the capacitor to the left. The other way of looking at it is that the capacitor is traveling at speed V relative to the aether to the right. The electrical signal actually follows a diagonal path between the plates relative to the aether when going straight across the gap relative to the plates.   The true velocity of the electrical signal relative to the plates in a Galilean frame is

$$u = \left(1 - \frac{V^2}{c^2}\right)^{1/2} c.$$

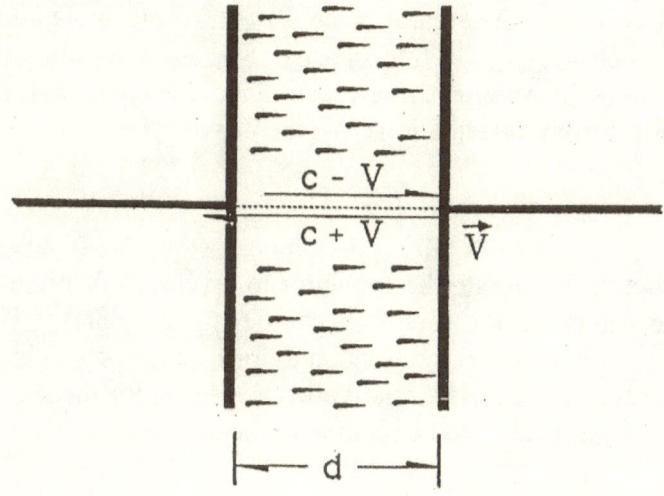

Fig. 3-3. When the capacitor is rotated perpendicular to the aether motion, the electric signal travels at velocity c - V in one direction and c + V in the opposite direction in the Galilean frame.

capacitor is at rest relative to the aether, the electrical signal travels back and forth between the plates of the capacitor at speed c. (See Fig. 3-1.) The separation distance d between the plates can be calculated from a measured time interval τ required for light or an electrical signal to make a round trip across the gap between the plates. The capacitance of the capacitor can be calculated from the equation

$$C = \kappa \varepsilon_0 S / d, \qquad (3\text{-}1)$$

where S is the plate area, κ is the dielectric constant, and ε_0 is the permittivity of free space. Since in our thought problem the capacitor is in a vacuum and κ is equal to 1, let us omit it from the equation. We then have

$$C = \varepsilon_0 S / d . \qquad (3\text{-}2)$$

Now let us accelerate the capacitor to a velocity V relative to the aether, in the direction parallel to the plates. The electric signal traveling back and forth across the plates cannot travel at speed c relative to the plates, for the signal can only travel at speed c relative to the aether.

The signal would actually travel in a diagonal path through the aether, when going straight across the gap, relative to the capacitor. (See Fig. 3-2.) The velocity of the electrical signal relative to the capacitor plates is

$$u = c\left(1 - V^2 / c^2\right)^{1/2}, \qquad (3\text{-}3)$$

as can be seen from Fig. 3-2 and the Pythagorean Theorem. The interval of time required for the round trip for the electrical signal is longer than the τ value for the capacitor at rest. The time required for the electrical signal to make a round trip in the transverse direction at relative aether velocity V is

$$\tau_t' = \tau / \left(1 - V^2 / c^2\right)^{1/2}. \qquad (3\text{-}4)$$

Let us rotate the capacitor so that the relative aether velocity V is in a direction perpendicular to the plates. The time for a round trip signal of light or electricity across the capacitor gap is again changed. The signal goes at speed c + V relative to the capacitor in one direction, and c - V relative to the capacitor in the opposite direction. (See Fig. 3-3.) The interval of time required for the round trip for the electrical signal is longer in this orientation than for the capacitor at rest or for the capacitor traveling parallel to the plates. Whereas

$$\tau = 2d / c \qquad (3\text{-}5)$$

for the capacitor at rest, and

$$\tau_t' = 2d / c\left(1 - V^2 / c^2\right)^{1/2} = \tau / \left(1 - V^2 / c^2\right)^{1/2} \quad (3\text{-}6)$$

for the round trip time of flight for the transverse direction relative to the aether, the round trip time of flight for the longitudinal direction relative to the aether is

$$\tau_L' = \tau_1 + \tau_2, \qquad (3\text{-}7)$$

where

$$\tau_1 = d / (c - V) \qquad (3\text{-}8)$$

and

$$\tau_2 = d / (c + V). \qquad (3\text{-}9)$$

Adding these one way transit times and simplifying terms, we obtain the following for the round trip longitudinal time:

$$\tau_L' = d/(c-V) + d/(c+V) \qquad (3\text{-}10)$$

$$= [d(c+V) + d(c-V)] / [(c-V)(c+V)]$$

$$= 2dc/(c^2 - V^2)$$

$$= 2d/c(1 - V^2/c^2) \qquad (3\text{-}11)$$

$$= \tau/(1 - V^2/c^2). \qquad (3\text{-}12)$$

If we define

$$\gamma \equiv 1/(1 - V^2/c^2)^{1/2}, \qquad (3\text{-}13)$$

as is commonly done, we see that the transverse time of flight τ_t' is

$$\tau_t' = \gamma\tau, \qquad (3\text{-}14)$$

whereas the longitudinal time of flight τ_L' is

$$\tau_L' = \gamma^2\tau. \qquad (3\text{-}15)$$

This increased time of flight of the electric signal in the transverse and longitudinal directions puts stresses upon the classical system. For example: Let us try to treat the capacitor classically, and see what happens. The physical dimensions remain unchanged in the transverse velocity problem. The increased time of flight for the electric signal for both the transverse and longitudinal directions would make the plates of the capacitor appear electrically farther apart than when the capacitor was at rest. The capacitance of the capacitor would therefore decrease to

$$C_t' = C / \gamma, \tag{3-16}$$

and

$$C_L' = C / \gamma^2, \tag{3-17}$$

for a moving capacitor. In this classical mechanics type calculation, this change in capacitance could not arise from changes in S or d. Therefore the effective permittivity of free space must be changed.

$$C_t' = C / \gamma = (\varepsilon_0 / \gamma)(S / d) = \varepsilon_t S / d, \tag{3-18}$$

$$\varepsilon_t = \varepsilon_0 / \gamma. \tag{3-19}$$

$$C_L' = C / \gamma^2 = (\varepsilon_0 / \gamma^2)(S / d) = \varepsilon_L S / d, \tag{3-20}$$

$$\varepsilon_L = \varepsilon_0 / \gamma^2. \tag{3-21}$$

The decrease in the permittivity of free space, as perceived by the moving particle in the "classical" calculated system, is serious business. If the permittivity of free space changed simply as above, the electrical force

$$F = \frac{q_1 q_2}{4 \pi \varepsilon_0 r^2} 1_r \tag{3-22}$$

holding electrons in orbit around the atomic nuclei would be increased. The energy states of the atoms in the moving system would be changed. We realize that this is a physical action. The moving object would be seriously affected. The Universe must compensate for this action by an equal and opposite reaction (Postulate 2). As in Lenz's law of magnetic fields and electrical currents (a broad Newton's third law), the Universe must mandate physical changes to the moving matter to ensure that the permittivity of free space remains constant.

But before seeing how this might be accomplished, consider some other stresses that are forced upon a "classically" calculated system when an object moves through the aether. The time of flight of the signal across the moving capacitor takes longer in actual time (aether rest frame time). If the moving observer used the speed of light in his frame to estimate the distance between the plates, he would notice that it takes more clock pulses (by the factor of γ) for the round trip flight in the transverse direction. The transverse dimension is unchanged classically. How would the moving observer have to account for the increased time for light to traverse the round trip the same transverse distance. He could conclude that his clock is running *faster* by the factor γ. But the time of flight in the longitudinal direction is increased by the factor γ^2. The change of clock speed only accounts for one γ factor. The moving observer is forced to conclude that there has occurred a length *expansion* in the x direction by the factor γ.

So now, with these calculations, we see that when we treat the moving system as "classically" as possible, stresses are put upon the system such as the *reduction of the permittivity of free space* by two different factors in the transverse and longitudinal directions, apparent *speeding up of the clocks*, and *length expansion*. These are actions. The Universe must respond to these actions with equal and opposite reactions (Postulate 2).

We postulate, therefore, that nature reacts to these velocity stresses by corresponding dimensional strains— that it reacts to these actions by equal and opposite reactions. The system tries to restore the permittivity of free space as viewed in the moving frame to the rest permittivity, the clock speed to the rest clock speed, and the length to the rest length. When the length values in the moving system are adjusted so that they appear to be the rest lengths when viewed in the moving frame, then there is real length contraction along the direction of motion as seen by the stationary observer in the aether frame. Likewise, when the clock speeds in the moving system are adjusted so that they appear to be the rest clock speeds when viewed in the moving frame, then there is real slowing of the moving clocks as seen by the stationary observer in the aether frame. This is just as it is in Special Relativity—and in the proper amounts.

Important Derived Results

1. $L_{x'} = \dfrac{L_0}{\gamma}.$ (3-23)

2. $t' = \gamma\tau.$ (3-24)

The first two important derived results show length contraction and time dilation for moving objects and clocks as seen by an observer in the aether rest frame.

These physical changes give the result that the speed of light appears constant to all observers. It takes longer for a light signal to traverse the round trip distance in the transverse direction—longer by the factor γ. But the moving observer doesn't know that, because his clock has slowed down by the factor γ, so that it appears to take the same number of clock pulses for the round trip. Also it takes longer for a light or an electric signal to make the round trip in the longitudinal direction by the factor γ^2. But the moving observer doesn't know that, because his clock has slowed down (accounting for one γ factor) and his distance between the plates has been physically shortened by a γ factor (accounting for the other γ factor). Thus the moving observer counts the same number of clock pulses for the round trip in the longitudinal direction as he would at rest.

If the speed of light is constant to all observers in the transverse direction as well as the longitudinal direction at all relative velocities, then it will be the same to all observers at all relative velocities in all directions.

Important Derived Result

3. The speed of light is constant to all observers with constant linear motion.

Since the increased clock speed is renormalized down to the rest clock speed, and the expanded length is renormalized down in the moving frame to the rest length, then rest clock speed and rest length are constant to all observers with constant linear motion.

Important Derived Result

4. Rest clock speed and rest length are constant to all observers.

If the speed of light is constant to all observers, and the permittivity of free space ε_0 is constant to all observers (as proven in the capacitor thought experiment), then the permeability of free space is likewise constant to all observers, because

$$\varepsilon_0\mu_0 = 1/c^2,\qquad\qquad\qquad(3\text{-}25)$$

where μ_0 is defined to be $4\pi \times 10^{-7}$ H m^{-1}.

Important Derived Result

5. The permittivity of free space ε_0 and the magnetic permeability of free space μ_0 are constant to all observers.

If the speed of light c, the dielectric constant ε_0, the magnetic permeability of free space μ_0, the rest mass m_0, the rest clock speed, and the rest length L_0 are constant to all observers, then of necessity all known force laws must be covariant at all relative constant speeds.

Important Derived Result

6. Physical laws are covariant with respect to uniform motion at any relative constant speed.

The two necessary conditions for the calculation of the Lorentz transformations and Einstein's Special Theory of Relativity, the covariance of physical laws and the constancy of the speed of light to all observers, are derived results in aether relativity. Notice this is obtained without Einstein's first postulate that it is impossible by any means to detect absolute motion in space—the principle of relativity. The covariance of physical laws is possible without the principle of relativity.

In the preceding arguments nature renormalizes rest lengths and rest clock speeds of moving objects so that they are constant to all observers. In the process there is length contraction and time dilation for moving objects as seen by a stationary observer. Of course the above calculated renormalizations will not occur in large lump sums. The renormalizations of length, time will occur by infinitesimal little by littles as the moving observer accelerates. For example

$$L' = \frac{L_0}{\gamma}; \quad d\gamma = \frac{-L_0 dL'}{L'^2}. \tag{3-26}$$

All of these variables are available for measurement in the local aether system.

C. Derivation of Transformations

In Einstein's Special Theory of Relativity, the covariance of physical laws and the constancy of c are first assumed, then the Lorentz transformations calculated, then length contraction and time dilation calculated from the Lorentz transformations. Aether relativity turns the order of these around. Length contraction and time dilation are first derived, then the covariance of physical laws and the constancy of c are derived. Now from these the Lorentz transformations can be derived by the historical calculations. A published source is therefore here quoted for such.

We now look for a length and time transformation which will make the speed of light independent of the motion of the source or receiver. Let the given reference frame S in which the source is at rest be the unprimed frame. The positions and times measured by an observer in this frame will be denoted by the unprimed symbols x, y, z, t. If a light source is at the origin of the frame S and a front is emitted at t = 0, the familiar equation of a spherical wave front is

$$x^2 + y^2 + z^2 = c^2 t^2. \tag{3-27}$$

The Equation (3-27) describes a spherical surface whose radius expands at the speed c.

Let our moving reference frame S' be the primed frame. Positions and times measured by an observer in this frame are denoted by the primed symbols x', y', z', t'. For our convenience we suppose that the zero of t' coincides with the zero of t, and that the origin of x', y', z' coincides with the position of the light source in S at this zero of time. Then to an observer moving with S' the equation of the spherical wave front must be

$$x'^2 + y'^2 + z'^2 = c^2 t'^2. \tag{3-28}$$

Here we have taken the same value c for the speed of light as in the rest frame S.

Suppose that the frame S' is moving in the +x direction with the constant velocity V with respect to the rest frame S. The Galilean transformation of lengths and times connects measurements in the two frames according to the equations

$$x' = x - Vt; \quad y' = y; \quad z' = z; \quad t' = t. \tag{3-29}$$

If we continue by substituting (3-29) in (3-28) we obtain directly that

$$x^2 - 2xVt + V^2 t^2 + y^2 + z^2 = c^2 t^2, \tag{3-30}$$

which equation is certainly not in agreement with Equation (3-27). Thus the Galilean transformation of lengths and times fails. If the principle of the constancy of the speed of light is valid, there should exist *some* transformation which reduces to the Galilean transformation for $V/c \rightarrow 0$ and which transforms $x'^2 + y'^2 + z'^2 = c^2t'^2$ into $x^2 + y^2 + z^2 = c^2t^2$.

We now suspect that the new transformation must be trivial for y' and z', because the terms in y'^2 and z'^2 in (3-28) transform into y^2 and z^2 in (3-27) without anything extra. We need a transformation of lengths and times which is *linear* in x and t, because we want to get a sphere which expands at a uniform rate. Therefore it is no use trying $x' = x^{1/2}t^{1/2}$ or $x' = \sin x$ or other functions. It is also clear from (3-30) that we cannot leave the $t' = t$ transformation unchanged if we want to cancel out the undesired terms - $2xVt + V^2t^2$ because *something certainly must be added* to cancel these terms.

Therefore, let us try next a transformation of the form

$$x' = x - Vt; \quad y' = y; \quad z' = z; \quad t' = t + fx, \quad (3\text{-}31)$$

where f is a constant yet to be determined. Then Equation (3-28) becomes

$$x^2 - 2xVt + V^2t^2 + y^2 + z^2 \qquad (3\text{-}32)$$
$$= c^2t^2 + 2c^2 ftx + c^2 f^2 x^2.$$

Now notice that the terms xt cancel if we set

$$f = -\frac{V}{c^2}, \text{ or } t' = t - \frac{Vx}{c^2}. \qquad (3\text{-}33)$$

With this value of f, Eqn. (3-32) can be written as

$$x^2\left(1 - \frac{V^2}{c^2}\right) + y^2 + z^2 = c^2t^2\left(1 - \frac{V^2}{c^2}\right). \qquad (3\text{-}34)$$

This is much closer to Equation (3-27), but there still remains an unwanted scale factor $(1 - V^2/c^2)$ multiplying x^2 and t^2.

Now we can dispose of the scale factor by taking the transformation to be[2]

Important Derived Result

7. $$x' = \frac{x - Vt}{\left(1 - V^2/c^2\right)^{1/2}}; \quad y' = y;$$ (3-35)

$$z' = z; \quad t' = \frac{t - \left(V/c^2\right)x}{\left(1 - V^2/c^2\right)^{1/2}}.$$

This is the famous *Lorentz transformation*.[3] It is linear in x and t; it reduces to the historic Galilean transformation for $V/c \rightarrow 0$; when substituted in Eqn. (3-28) it gives

$$x^2 + y^2 + z^2 = c^2 t^2,$$ (3-36)

exactly as required. That is,

$$x'^2 + y'^2 + z'^2 = c^2 t'^2$$ (3-37)

is *invariant* under a Lorentz transformation of lengths and times. The form of the equation describing the wave front of light is the same in all frames moving with uniform relative velocity.[4]

In calculations it is sometimes convenient to make use of some standard notation. We now introduce the notation

$$\beta \equiv V/c.$$ (3-38)

That is, β (Greek letter beta) is the velocity measured in a natural system of units in which $c = 1$. It is also convenient to introduce γ (Greek letter gamma}:

$$\gamma \equiv 1/\left(1 - \beta^2\right)^{1/2} \qquad (3\text{-}39)$$

$$\equiv \frac{1}{\left(1 - V^2/c^2\right)^{1/2}} \cdot$$

Note now that $\gamma \geq 1$. In extreme relativistic problems $1 - \beta \ll 1$; it is then useful to note that $1 - \beta^2 = (1 - \beta)(1 + \beta) \approx 2(1 - \beta)$. The Lorentz transformation Equation (3-35) becomes

$$x' = \gamma\left(x - \beta ct\right); \quad y' = y; \qquad (3\text{-}40)$$

$$z' = z; \quad t' = \gamma\left(t - \frac{\beta x}{c}\right),$$

and the inverse transformation is easily seen to be

$$x = \gamma\left(x' + \beta ct'\right); \quad y = y'; \qquad (3\text{-}41)$$

$$z = z'; \quad t = \gamma\left(t' + \frac{\beta x'}{c}\right).$$

This is left now to the reader....[5]

The next logical step in the derivation of special quasi-relativity is the calculation of velocity transformations. Since there is a covariance of physical laws and a constancy of c in quasi-relativity as well as in the traditional special relativity, standard treatments of the velocity transformations can be cited and applied to the development of velocity transformations in quasi-relativity.

Now suppose the S' reference frame moves with uniform velocity $V1_x$ relative to the S reference frame, and a particle moves with uniform velocity components v_x', v_y', v_z' relative to the S' frame. What then are the velocity components v_x, v_y, v_z of the particle relative to the rest S frame?

From Equation (3-41) we have, with $\beta = V/c$,

$$x = \gamma x' + \gamma \beta c t'; \quad t = \gamma t' + \frac{\gamma \beta x'}{c}, \qquad (3\text{-}42)$$

whence

$$dx = \gamma dx' + \gamma \beta c dt'; \qquad (3\text{-}43)$$

$$dt = \gamma dt' + \frac{\gamma \beta dx'}{c}.$$

Thus

$$v_x = \frac{dx}{dt} = \frac{\gamma dx' + \gamma \beta c dt'}{\gamma dt' + \gamma \beta dx'/c} \qquad (3\text{-}44)$$

$$= \frac{v_x' + \beta c}{1 + v_x' \beta / c},$$

or

$$v_x = \frac{v_x' + V}{1 + v_x' V / c^2}. \qquad (3\text{-}45)$$

This result may be compared with the Galilean transformation result $v_x = v_x' + V$. Similarly, because y = y' and z = z',

$$v_y = \frac{dy}{dt} = \frac{dy'}{\gamma dt' + \gamma \beta dx'/c} \qquad (3\text{-}46)$$

$$= \frac{v_y'}{1 + v_x'V/c^2}\left(1 - \frac{V^2}{c^2}\right)^{1/2},$$

and

$$v_z = \frac{v_z'}{1 + v_x'V/c^2}\left(1 - \frac{V^2}{c^2}\right)^{1/2}. \qquad (3\text{-}47)$$

The inverse transformations follow from Equation (3-40), or by solving (3-45), (3-46), (3-47) for the primed velocity components:

$$v_x' = \frac{v_x - V}{1 - v_x V/c^2}; \qquad (3\text{-}48)$$

$$v_y' = \frac{v_y}{1 - v_x V/c^2}\left(1 - \frac{V^2}{c^2}\right)^{1/2}; \qquad (3\text{-}49)$$

$$v_z' = \frac{v_z}{1 - v_x V/c^2}\left(1 - \frac{V^2}{c^2}\right)^{1/2}. \qquad (3\text{-}50)$$

(Endnote 6.)

Important Derived Results

8. Velocity transformations (see Eqs. (3-45 through 3-47).

9. Inverse velocity transformations (see Eqs. (3-48 through 3-50).

D. Relative Mass Increase and the Equivalence of Mass and Energy

With the derivation of the Lorentz transformations and the velocity transformations, we are at last prepared to derive the equivalence of mass and energy from our first principles without assuming any part of the conclusions. Since in aether quasi-relativity there is the covariance of physical laws and the constancy of c, just as in Einstein's Special Theory of Relativity, the necessary calculations to accomplish this task are the same as historic calculations, which can be alluded to in lieu of independent derivations.

Now we want to find a definition of the ordinary momentum **p** which reduces to M**v**, where M is the rest mass,[7] for $v/c \ll 1$, and which assures momentum conservation in collisions regardless of the velocities of the particles relative to the reference frame [see Figures 3-4 to 3-10]. We will subsequently find the appropriate definition by consideration of a particular collision. We first show by example that the Newtonian (nonrelativistic) ordinary momentum M**v** is not conserved in collisions involving *relativistic* velocities.

Consider now the accompanying figures (Figure 3-4 and Figure 3-5) which describe a collision between particles of *equal mass*. For our first case we choose a reference frame S such that the particles approach each other with equal and opposite velocities: the y velocity component of particle 1 is $-v_y$ before the collision and $+v_y$

after the collision. In this chosen and illustrated reference frame the center of mass is at rest. The total y component of the momentum (the total from sphere 1 and sphere 2) must be zero by symmetry, both before and after the collision. This will be true no matter what physical definition we use for momentum, provided that it has opposite signs for $\pm v_y$. We therefore encounter no trouble here (whether or not the expression is correct) with the Newtonian definition $\mathbf{p} = M\mathbf{v}$: the change in p_y of particle 1 is $+2Mv_y$, and the change in p_y of particle 2 is $-2Mv_y$, so that the total change in the y component of the Newtonian momentum is zero.

The primed reference frame S' (in Figure 3-6) moves with the particular velocity $\mathbf{V} = v_x \mathbf{1}_x$ with respect to the unprimed reference fame S (also in Figure 3-6); here v_x is the x component of the velocity of particle 2 before the collision. We know from the relation for the relativistic addition of velocities in Equation (3-52) that a particle velocity component seen as v_y by an observer at rest in S will be seen as

$$v_y' = \frac{v_y}{1 - v_x V / c^2}\left(1 - \frac{V^2}{c^2}\right)^{1/2} \qquad (3\text{-}51)$$

by an observer at rest in S'. This result in Equation (3-51) has the consequence that the magnitude of the y components of the velocities of particles 1 and 2 are not equal in S', even though they were equal in S. For we see from the Equation (3-51) with $V = v_x$ (of the particle 2) that we have for the initial y velocity components as viewed in S':

$$-v_y'(1) = -\frac{v_y}{1 + v_x^2 / c^2}\left(1 - \frac{v_x^2}{c^2}\right)^{1/2} ; \qquad (3\text{-}52)$$

$$v_y'(2) = \frac{v_y}{1 - v_x^2/c^2}\left(1 - \frac{v_x^2}{c^2}\right)^{1/2}.\qquad(3\text{-}53)$$

These velocity components of particles (1) and (2) are now not equal in magnitude. Because the x components of velocity of particles (1) and (2) are not equal in S (they are the negatives of each other), it follows from Equation (3-51) that the y components of velocity of particles (1) and (2) have different magnitudes as seen from any reference frame moving relative to S in the x direction. Thus the momentum change $2Mv_y'$ of particle (2) is not equal in magnitude to the momentum change $2Mv_y'$ of particle (1). Of a consequence we see that a definition in which the momentum is directly proportional to velocity cannot assure momentum conservation in all reference systems. Therefore, either momentum conservation is incompatible with Lorentz invariance (which is unacceptable to us) or there exists another definition of momentum such that momentum conservation is valid in all systems with constant relative velocities.

We now look for such a definition of momentum which is Lorentz invariant. That definition must be such that the y component of the momentum of a particle is independent of the x component of the velocity of the reference frame in which the collision is observed. If we can find such a definition, then conservation of the y component of momentum in one reference frame ensures its conservation in all reference frames. We know that under the Lorentz transformations the displacement Δy in the y direction is the same for all reference frames. But the time Δt to go the distance Δy depends upon the reference frame, and thus the velocity component $v_y = \Delta y/\Delta t$ depends on the reference frame. Instead of laboratory clocks to measure Δt,

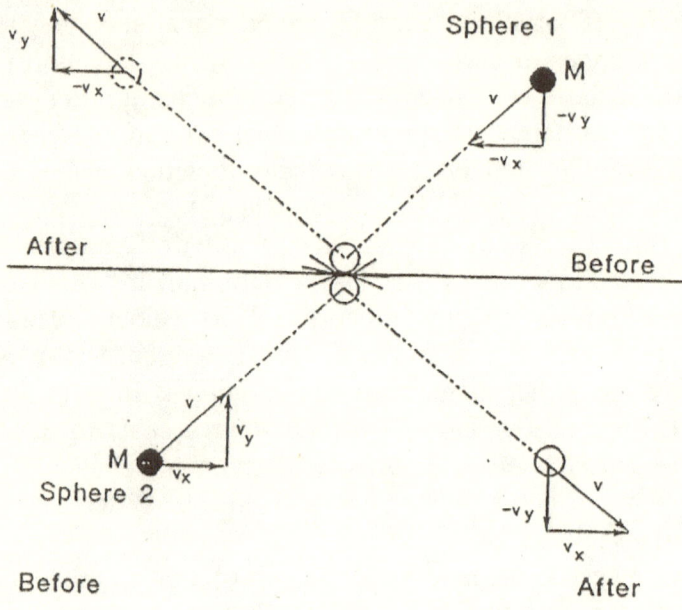

Fig. 3-4*.   A collision between two spheres of mass M, taking place in the xy plane. The velocities in the x an y directions before and after collision are shown.

*Figures 3-4 to 3-10 are adapted from Charles Kittel, *et. al.*, *Mechanics*: Berkeley Physics Course–Volume 1 (New York: McGraw-Hill, 1965), pp. 382-385.

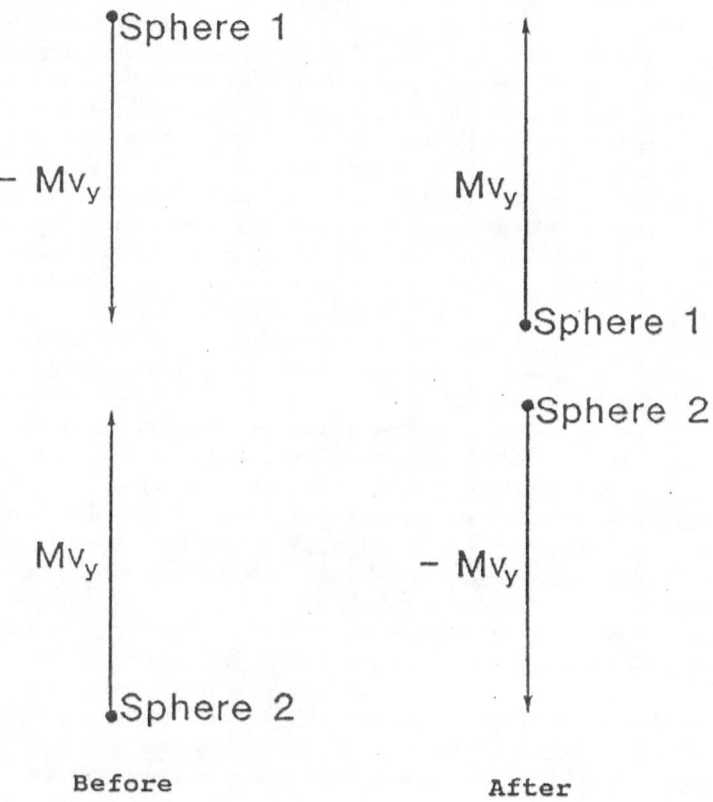

Fig. 3-5. The individual nonrelativistic momenta in the y direction are shown. The *total* momentum in the y direction is zero before *and* after the collision.

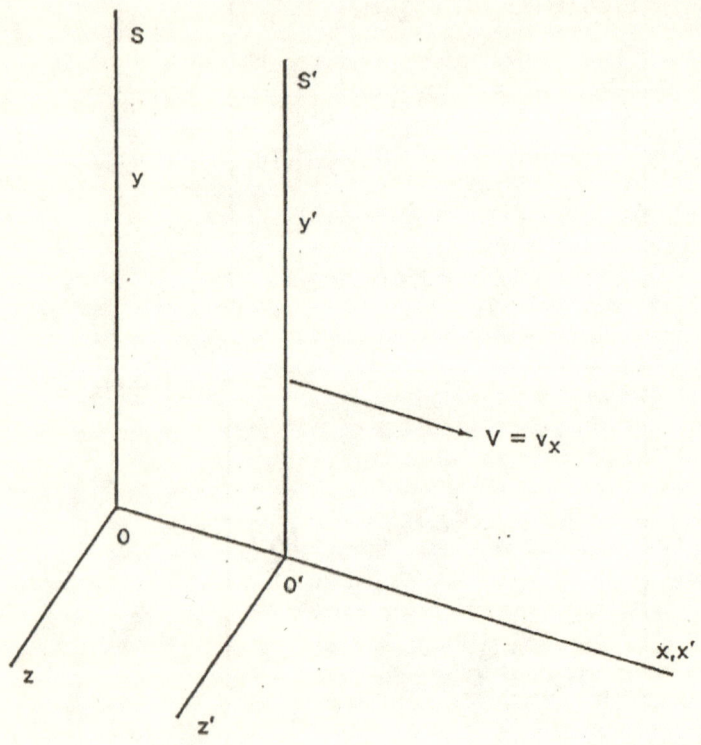

Fig. 3-6. We have viewed the collision in frame S. What if we view it in frame S', which has the particular velocity $V = v_x$ with respect to S, as shown?

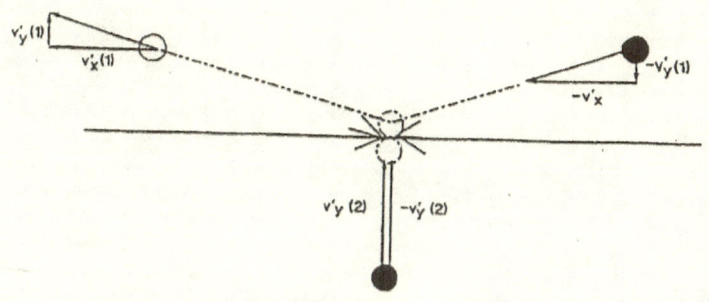

Fig. 3-7. In S', we find

$$v_x'(1) = -\frac{2V}{1+V^2/c^2}; \quad v_x'(2) = 0;$$

$$v_y'(1) = \frac{v_y}{1+V^2/c^2}\left(1-\frac{V^2}{c^2}\right)^{1/2};$$

and

$$v_y'(2) = \frac{v_y}{\left(1-V^2/c^2\right)^{1/2}} > v_y'(1).$$

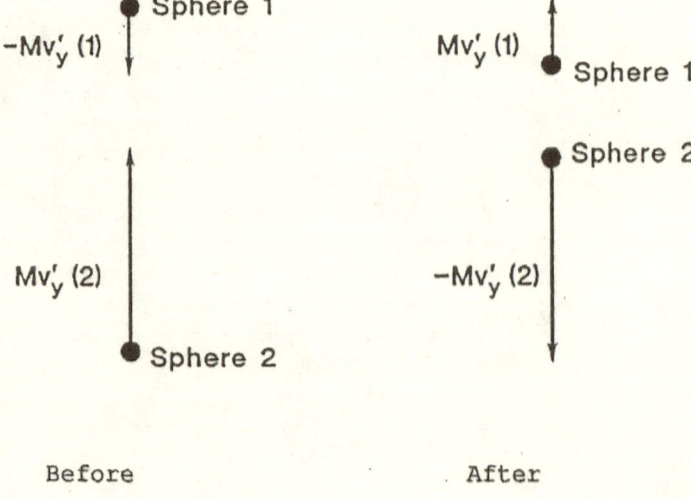

Fig. 3-8.   In the new frame S', the nonrelativistic momentum is *not* the same in the y' direction before and after the collision.   There is a net increase in the y component of the nonrelativistic momentum.

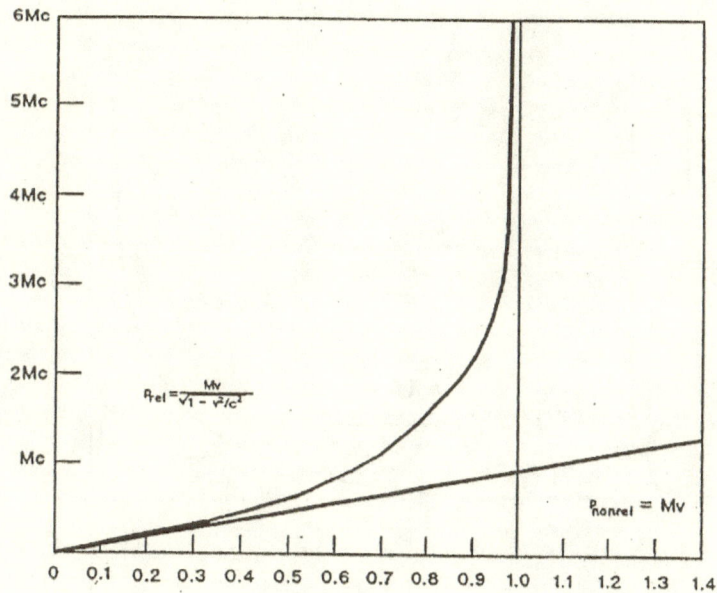

Fig. 3-9. So that momentum conservation will hold in *all* frames, we redefine **p** as follows: For a particle with velocity **v** and rest mass M,

$$p = \frac{Mv}{\sqrt{1 - v^2 / c^2}}.$$

The magnitudes of both the relativistic momentum and the nonrelativistic momentum are plotted in the graph.

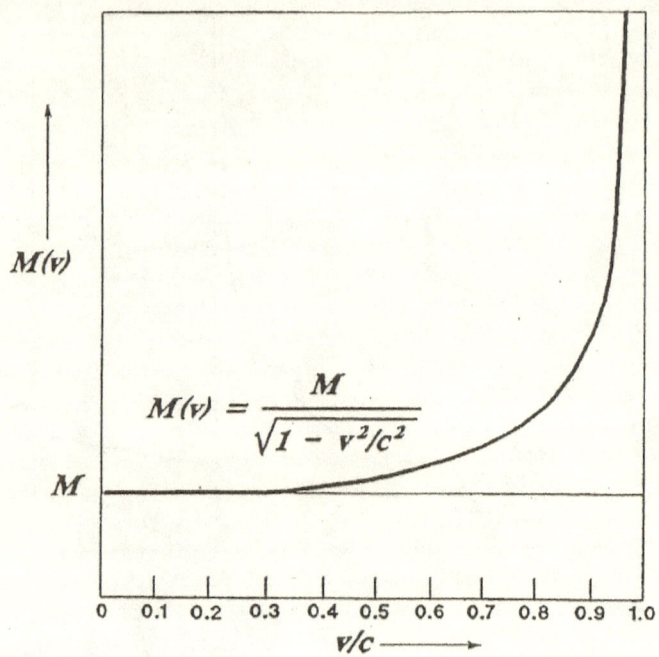

Fig. 3-10. The new definition of momentum leads to this behavior of the mass:

$$M(v) = \frac{M}{\sqrt{1 - V^2 / c^2}}.$$

we can refer to a clock carried on the particle. This clock naturally measures the proper time interval $\Delta\tau$ of the particle. *All observers in every frame will agree on the value of* $\Delta\tau$. Thus the quantity $\Delta y/\Delta\tau$ is the same in all the reference frames.

We know that Δt and $\Delta\tau$ differ by the time-dilation factor Equation (3-24), so we have

$$\Delta\tau = \Delta t\left(1 - \frac{v^2}{c^2}\right)^{1/2}, \tag{3-54}$$

whence

$$\frac{\Delta y}{\Delta\tau} = \frac{\Delta y}{\Delta t}\cdot\frac{\Delta t}{\Delta\tau} = \frac{\Delta y}{\Delta t}\frac{1}{\left(1 - v^2/c^2\right)^{1/2}}. \tag{3-55}$$

We see from Equation (3-55) that the y component of $\mathbf{v}/(1 - v^2/c^2)^{1/2}$ will be the same in all reference frames which differ only in their x component of velocity. If we now *define* the relativistic momentum by

$$p \equiv Mv\left(1 - v^2/c^2\right)^{1/2}, \tag{3-56}$$

then conservation of the y component of momentum is valid in any other inertial reference frame which differs from the rest frame by a constant velocity in the x direction. Now note that we may write

$$p = Mc\beta\gamma, \tag{3-57}$$

from the definitions $\beta = v/c$ and $\gamma = (1 - v^2/c^2)^{-1/2}$ introduced [previously].

For the reader's and authors' convenience we have used axes arranged in a symmetrical way so that in this collision there is no change in the x component of velocity of either particle. It follows that p_x as defined by Equation (3-56) is automatically conserved. Thus with the definition Equation (3-56) we have complete relativistic momentum conservation in the two-body collision of identical particles. The student may himself or herself show that even if particle (2) has a different mass from particle (1), the above argument holds, so that we have a relativistic law of momentum conservation. For v/c $\ll$ 1 the definition of momentum reduces to the non-relativistic result $\mathbf{p} = M\mathbf{v}$ (an application of the Correspondence Principle). It is an experimental fact that the momentum as defined by Equation (3-56) is conserved in all collision processes.

We may write the relativistic momentum Equation (3-56) as

$$\mathbf{p} = M(v)\mathbf{v}, \tag{3-58}$$

so that we may interpret

$$M(v) \equiv M / \left(\left(1 - v^2 / c^2 \right)^{1/2} \right) = M\gamma \tag{3-59}$$

as the relativistic mass of the particle of rest mass M when in motion with the speed v. We define the rest mass as the mass for v → 0. In turn, as v → c, the relativistic mass M(v) → ∞. The relativistic increase of mass has been verified in various electron deflection experiments already performed; it is also verified implicitly in the operation of every high-energy particle accelerator.[8]

We have now derived the relative mass increase relation without first assuming the equivalence of mass and energy. We have done so from the Lorentz transformations, which were derived from the covariance of physical laws and the constancy of c, which in quasi-relativity are derived from two postulates: 1) there is an absolute space mediated by an electrical aether, and 2) for every whole particle relativistic action there is an equal and opposite reaction.

Important Derived Result

10. $M = \gamma M_0$ (3-60)

Let us pause to make some historical notes. In 1895, G. F. FitzGerald advanced a length contraction as an explanation of the null result of the 1887 Michelson-Morley experiment.

$$L_{x'} = \frac{L_0}{\gamma}. \qquad (3\text{-}61)$$

A. Lorentz independently discovered the length contraction (named the Lorentz-FitzGerald Contraction in their honor), and expanded the considerations to include a relation for relative mass increase.

$$M = \gamma M_0. \qquad (3\text{-}62)$$

This followed the length contraction, because the masses of particles are inversely proportional to the volumes of the particles

$$M = \frac{k}{v}, \qquad (3\text{-}63)$$

[where k is some proportionality constant, and v is particle volume], and the length contraction reduces their volumes. By 1900, mass measurements on speeding particles showed that the Lorentz equations were followed exactly.[9] This insight was years before Einstein's Special Relativity.

The equivalence of mass and energy can now be derived from our postulates through the study of relativistic energy.

We first consider the relativistic energy from a formal viewpoint.   From Equation (3-60) the square of the relativistic momentum can be written as

$$p^2 = M^2 c^2 \beta^2 \gamma^2. \qquad (3\text{-}64)$$

The identity

$$\frac{1}{1 - v^2/c^2} - \frac{v^2/c^2}{1 - v^2/c^2} = 1, \qquad (3\text{-}65)$$

or

$$\gamma^2 - \beta^2 \gamma^2 = 1, \qquad (3\text{-}66)$$

is a ready-made Lorentz invariant, because the number 1 is a constant. On multiplying by $M^2 c^4$, we have

$$M^2 c^4 \left(\gamma^2 - \beta^2 \gamma^2 \right) = M^2 c^4, \qquad (3\text{-}67)$$

or

$$M^2 c^4 \gamma^2 - p^2 c^2 = M^2 c^4, \qquad (3\text{-}68)$$

using Equation (3-64).   Because the *rest mass* M is a constant, we know that $M^2 c^4$ is a constant and therefore a Lorentz invariant as required. But what physical quantity is $M^2 c^4 \gamma^2$? Its role in Equation (3-68) strongly suggests that it must be an important physical quantity, for when $p^2 c^2$ is subtracted from it, we have a number ($M^2 c^4$) which is invariant under Lorentz transformations.
    Let us consider the quantity

$$Mc^2 \gamma = Mc^2 \frac{1}{\left(1 - \beta^2\right)^{1/2}} . \qquad (3\text{-}69)$$

For $\beta \ll 1$, this becomes

$$Mc^2\gamma \approx Mc^2(1 + \tfrac{1}{2}\beta^2 + \ldots) \qquad (3\text{-}70)$$
$$\approx Mc^2 + \tfrac{1}{2}Mv^2 + \ldots .$$

The second term $\tfrac{1}{2}Mv^2$ we recognize as the kinetic energy K in the non-relativistic limit. Now suppose we next define the *total relativistic energy* E of a free particle by the equation

$$E \equiv Mc^2 \gamma \equiv \frac{Mc^2}{\left(1 - v^2 / c^2\right)^{1/2}} . \qquad (3\text{-}71)$$

Then Equation (3-68) tells us that

$$E^2 - p^2 c^2 = M^2 c^4 , \qquad (3\text{-}72)$$

which is a Lorentz invariant. If we transform from one reference frame to another, with $p \to p'$ and $E \to E'$, then the invariance of Equation (3-72) means that

$$E'^2 - p'^2 c^2 = E^2 - p^2 c^2 = M^2 c^4 . \qquad (3\text{-}73)$$

This is what we mean when we say that Equation (3-72) is a Lorentz invariant. We emphasize here that M denotes the rest mass of the particle and is a number invariant under a Lorentz transformation.[10]

We now have the mathematical tools to see that if $v = 0$, then $p = 0$ and $E^2 = M^2c^4$ by Eq. (3-72). Thus $E = Mc^2$ for a particle at rest. This energy must be the energy due to the mass of the particle. Also $E = M(v)c^2$ for an object at any relative constant velocity.

Important Derived Result

11. $$E = Mc^2. \tag{3-74}$$

E. Transformation of Momentum and Energy

Before concluding this chapter, we shall allude to the derivation of some more important transformations. Again, these derivations are valid in special quasi-relativity as well as in special relativity because both models adhere to the covariance of physical laws and the constancy of the speed of light, and the calculations are derivations from those two principles.

We have seen from Equation (3-57) and Equation (3-59) that

$$p_x = M\frac{dx}{d\tau}; \quad p_y = M\frac{dy}{d\tau}; \quad p_z = M\frac{dz}{d\tau}. \tag{3-75}$$

It follows from Equation (3-57) and Equation (3-71) that we can write E as

$$E = Mc^2\frac{dt}{d\tau}, \tag{3-76}$$

using $dt/d\tau = (1 - \beta^2)^{-1/2}$. Because M and τ are Lorentz invariants, it follows from Equation (3-75) and Equation (3-76) that p_x, p_y, p_z, and E/c^2 must transform under a Lorentz transformation exactly as x, y, z, and t transform. Because we know how the latter transform, the result of

Equation (3-77) follows simply. Using the transformations given previously, we have *the transformation relations* for momentum and energy:

$$p_x'= \gamma\left(p_x - \frac{\beta E}{c} \right); \quad p_y'= p_y;$$

(3-77)

$$p_z'= p_z; \quad E'= \gamma\left(E - p_x c\beta \right).$$

The inverse transformations of momentum and energy follow on changing $-\beta$ into $+\beta$ and interchanging primed and unprimed quantities:

$$p_x = \gamma\left(p_x'+ \frac{\beta E'}{c} \right); \quad p_y = p_y';$$

(3-78)

$$p_z = p_z'; \quad E = \gamma\left(E'+ p_x'c\beta \right).$$

We can determine the velocity of the particle from its momentum and energy, using Equation (3-75) and Equation (3-76):

$$v_x = \frac{dx}{dt} = \frac{dx}{d\tau}\cdot\frac{d\tau}{dt}$$

(3-79)

$$\frac{p_x}{M}\cdot\frac{Mc^2}{E} = \frac{c^2 p_x}{E},$$

or

$$\mathbf{p} = \mathbf{v}E/c^2.$$

(3-80)

(Endnote 11.)

The rate of change of momentum in general is given by

$$\frac{dp}{dt} = M\frac{d}{dt}\frac{v}{\left(1 - v^2 / c^2\right)^{1/2}},$$ (3-81)

where M is the rest mass of the particle.   In another constant speed frame S' we are concerned with

$$\frac{dp'}{dt'}.$$

We are here trying to avoid the use of the phrase relativistic force (because d**p**/dt does not behave as part of a four-vector).

Suppose that the particle is instantaneously at rest in the inertial frame S', which moves with velocity v**1**$_x$ with respect to S.   By the previously derived Lorentz transformation laws for **p** and E, we know that

$$\Delta p_y = \Delta p_y'; \quad \Delta p_z = \Delta p_z'.$$ (3-82)

The time interval Δt' in S' is just the proper time interval Δτ; hence

$$\Delta t' = \Delta \tau = \left(1 - \frac{v^2}{c^2}\right)^{1/2} \Delta t.$$ (3-83)

It follows that

$$\frac{\Delta p_y}{\Delta t} = \left(1 - \frac{v^2}{c^2}\right)^{1/2} \frac{\Delta p_y'}{\Delta t'}$$ (3-84)

$$= \left(1 - \frac{v^2}{c^2}\right)^{1/2} \frac{\Delta p_y'}{\Delta \tau},$$

and similarly for the quantity $\Delta p_z / \Delta t$. Thus

$$\frac{dp_y}{dt} = \left(1 - \frac{v^2}{c^2}\right)^{1/2} \frac{dp_y'}{d\tau}; \qquad (3\text{-}85)$$

$$\frac{dp_z}{dt} = \left(1 - \frac{v^2}{c^2}\right)^{1/2} \frac{dp_z'}{d\tau}.$$

The x components transform differently. We know from Equation (3-78) that

$$p_x = \frac{p_x' + vE'/c^2}{\left(1 - v^2/c^2\right)^{1/2}}, \qquad (3\text{-}86)$$

or

$$\Delta p_x = \frac{\Delta p_x' + vE'/c^2}{\left(1 - v^2/c^2\right)^{1/2}}. \qquad (3\text{-}87)$$

Now

$$E' = \left(M^2 c^4 + p'^2 c^2\right)^{1/2}, \qquad (3\text{-}88)$$

so that

$$\Delta E' = \frac{p_x' \Delta p_x' c^2}{\left(M^2 c^4 + p'^2 c^2\right)^{1/2}}. \qquad (3\text{-}89)$$

But in the inertial frame S' we know that $p_x' = 0$ at the instant of time considered; thus Equation (3-87) becomes

$$\Delta p_x \cong \frac{\Delta p_x'}{\left(1 - v^2 / c^2\right)^{1/2}}, \qquad (3-90)$$

and, using Equation (3-83),

$$\frac{\Delta p_x}{\Delta t} \cong \frac{\Delta p_x'}{\Delta t'} = \frac{\Delta p_x'}{\Delta \tau}. \qquad (3-91)$$

In the limit $\Delta t \to 0$ we have

$$\frac{dp_x}{dt} = \frac{dp_x'}{d\tau}. \qquad (\textit{Endnote} \, 12.) \qquad (3-92)$$

F. Differences Between Relativity and Quasi-Relativity

This concludes the derivation of special relativistic transformations which are the same formulas as special quasi-relativistic transformations. Note that all the above results were derived from only two postulates without further assumptions: 1) there is an absolute space mediated by an electrical aether; and 2) for every whole particle relativistic action there is an equal and opposite reaction.

Since there is covariance of physical laws and the constancy of the speed of light both in Einstein's Special Theory of Relativity and in quasi-relativity, and since the formulas of the derived transformations are identical in both systems, one might think that the theories are identical. Not so. The assumptions are different, the conclusions are different, and the interpretation of the formulas is different. Einstein assumed that there was no absolute space or aether and that one uniformly moving reference frame was as good as another as a base for calculating velocity transformations. Quasi-relativity does not assume this. It assumes that there is an absolute space and an aether and that the above transformations are all calculated relative to the rest frame of that aether.

In Einstein's Special Theory of Relativity, relative mass increase, time dilation, and length contraction are only *apparent*—observational phenomena due to rotation in complex space-time. In quasi-relativity, mass increase, time dilation, and length contraction are *real*—caused by motion relative to an aether. The resultant constancy of the speed of light in quasi-relativity is only *apparent*, not *real* as in Einstein's special relativity. Quasi-relativity reverses what is considered *real* or *actual* lengths and times and what are *apparent* lengths and times. The mass, clock speed, and length of a moving observer as seen by himself only appear to be the rest mass, rest clock speed, and rest length. They are not actually rest quantities, and invariant, as in Einstein's theory.

Transformations in quasi-relativity are different than in special relativity. To transform position, time, momentum, or energy (mass) to another frame in quasi-relativity, one must first make an inverse transformation to the aether rest frame according to the direction and magnitude of the observer's velocity relative to the aether, then make an appropriate rotation of the x axis to the direction of the velocity of the target frame relative to the aether, then make a transformation according to the magnitude of the velocity of the target relative to the aether. Every correct transformation should be a triple transformation like this. Such a triple transformation [1] an inverse transformation of increased clock speed, decreased mass, and length expansion when going from the observer to the aether rest frame; 2) an appropriate rotation; and 3) a transformation of decreased clock speed, increased mass, and length contraction when going from the aether rest frame to the other arbitrary frame, where all transformations go by the clock in the aether frame] is not in general equal to either a single transformation or a single inverse transformation. Here is where Einstein gets into trouble with the clock paradox. Not every uniformly moving frame is equivalent. Relative velocities between observer and target can vary from 0 to ± 2V, with time and energy (mass) remaining constant. To see this, consider two examples:

Suppose there is an arbitrary x, y, z axis coordinate frame at rest in the aether. Suppose there are three particles of equal mass in the aether rest frame. Suppose particle 1 travels at velocity +V along the x axis in that frame, and particle 2 travels at velocity -V along the x axis in that frame. Particle 3 is at rest at the origin of the coordinates. Einstein would interpret the Lorentz transformations such that the mass of particle 2, as observed by particle 1, would be increased by the gamma factor, where

$$v^2 = (2V)^2 = 4V^2. \tag{3-93}$$

The clock speed would be slowed by the same gamma factor. Also Einstein would interpret the Lorentz transformations such that the mass of particle 1 as observed by particle 2 would be increased by the gamma factor and its clock slowed by the gamma factor as observed by particle 2. In quasi-relativity, the gamma factor for mass increase, time dilation, and length contraction works only in the aether rest frame. In quasi-relativity, particles 1 and 2 would have the same gamma factors in the aether rest frame. Thus their masses would be the same in the aether frame. In the frames of particles 1 and 2, the mass of particles 1 and 2 are also equal. A reference particle 3 of equal mass in the rest frame, and at rest in the ether at the origin of our coordinates, would seem to have faster clock speed and reduced mass by the factor gamma compared to particle 1, where $v^2 = V^2$. Particle 2 has relative mass increase and slowing of the clocks also by the factor gamma, where $v^2 = V^2$. These gamma factors of speeding and slowing the clocks cancel out so that the clock rate of particle 2 appears to be the same as the clock rate of particle 1 to an observer traveling with particle 1. Particles 1 and 2 will appear to have the same mass and clock speed in any uniformly moving frame.

The same sort of problem could be developed with particle 1 traveling along the x axis and particle 2 traveling along the y axis and particle 3 at rest at the origin of the coordinates. In this case the velocity of particle 2 relative to particle 1 will be different than in the first problem, according to the inverse Lorentz transformations. Einstein would predict relative mass increase and time dilation for particle 2 relative to particle 1 and for particle 1 relative to particle 2. But quasi-relativity predicts the masses and clock speeds will be equal in any uniformly moving frame.

We see that we need more than relative velocity to compute mass increase, time dilation, and length contraction. We need to know also the aether velocity relative to the observer. We need to make three transformations from any arbitrary frame to any other arbitrary frame: 1) an inverse transformation of increased clock speed, decreased mass, and length expansion when going from the observer to the aether rest frame; 2) an appropriate rotation; and 3) a transformation of decreased clock speed, increased mass, and length contraction when going from the aether rest frame to the other arbitrary frame, where all transformations go by the clock in the aether frame.

Why has the necessity for three transformations in making relativistic calculations not been detected before? When compared to the speed of light, absolute velocity of the earth relative to the aether is small. Small differences in aether velocity for the observer on earth make little difference when calculating the relativistic effects of particles going close to the speed of light, such as accelerated particles. Thus we have not noticed variations from the expected according to the Lorentz transformations when calculating the relativistic effects of fast particles.

That is because the earth is nearly an aether rest frame for such problems. To measure a difference between special relativity and special quasi-relativity, we need to be able to detect relativistic effects of velocities on the order of the speed of the earth's rotation. Our precision in measuring mass is not sufficient to measure relative mass increase at such velocities. Our precision in measuring length is not sufficient to measure length contraction at such velocities. The precision of our atomic clocks, however is sufficient to detect time dilation at such velocities. Therefore an experiment involving atomic clocks could show, for example, that jets flying westward have faster clocks than clocks on earth, which are faster than clocks in jets flying eastward. This experiment was performed with these results by Hafele and Keating[13, 14], as was previously mentioned. The Hafele-Keating experiment shows that clocks traveling at lower velocities relative to absolute space run faster as in quasi-relativity, not slower or indeterminate, as in Einstein's Special Relativity.

The difference between special quasi-relativity and special relativity is not just a technicality. Special relativity will not work in uniting relativity with particle physics and deriving a unified field theory. But special quasi-relativity will. Special quasi-relativity combines better with general quasi-relativity also—in a much more intuitively obvious way than the complex Riemannian geometry and tensor algebra necessary in Einstein's General Theory of Relativity. We turn our attention now to general relativity and general quasi-relativity.

---

[1]H. F. E. Lenz, *Ann. der Phys.*, **31**, 483 (1834) as quoted by William Taussig Scott, *The Physics of Electricity and Magnetism,* Second Edition (New York: John Wiley & Sons, Inc., 1966), p. 339.

[2]Charles Kittel, Walter D. Knight, Malvin A. Ruderman, *Mechanics: Berkeley Physics Course--Volume 1*

(New York: McGraw-Hill Book Company, 1965), pp.346-348.

[3]This transformation has a long history. It was first used by J. Larmor to explain the null result of the Michelson-Morley experiment, in his *Aether and matter*, pp. 174-176 (Cambridge University Press, New York, 1900). Larmor claims accuracy only to order v^2/c^2; in fact, his results are exact.

[4]*Ibid.*, p. 349.

[5]*Ibid.*, pp. 349, 350.

[6]*Ibid.*, pp. 350, 351.

[7]The rest mass is defined as the inertial mass in the nonrelativistic limit $v/c \ll 1$.

[8]*Ibid.*, pp. 382-386.

[9]Isaac Asimov, *Asimov's Biographical Encyclopedia of Science and Technology*, Second Revised Edition (Garden City, New York: Doubleday & Company, Inc., 1982), [821] **FITZGERALD**, George Francis; [839] **LORENTZ**, Hendrik Antoon, pp. 530, 531, 543, 544.

[10]Charles Kittel, *op. cit.*, pp. 386-388.

[11]*Ibid.*, p. 388.

[12]*Ibid.*, pp. 400, 401.

[13]J. C. Hafele and R. E. Keating, *Science 177*, 166, 167 (1972).

[14]J. C. Hafele and R. E. Keating, *Science 177*, 168 (1972).

Problem Set 3

1. In a "classical" aether reference frame, what does velocity relative to the aether do to the permittivity of free space, to the speed of clocks, to length, and to mass in the rest frame of the moving observer? Why do we not observe these results in real life?

2. What does an observer moving relative to the aether observe for the permittivity of free space, speed of clocks, length, and mass in his own rest frame? Why are these different from those calculated in Problem 1?

3. Using the above reasoning, explain why there is slowing of clocks, length contraction, and mass increase of a moving object as seen by an observer at rest.

4. Show how the speed of light is constant to all observers in an aether system.

5. Show why the laws of nature are covariant in an aether system.

6. What does the Lorentz transformation mean in Einstein's aetherless system of relativity? What does the Lorentz transformation mean in the author's aether system of quasi-relativity?

7. How much are the clocks slowed for a cosmic particle traveling 0.99 c relative to the earth laboratory frame? [Assume the aether velocity relative to the earth laboratory frame is small relative to the 0.99 c.]

8. Observer A is at rest in the aether. Observer B travels away from Observer A at 0.9 c. Observer C travels away from Observer A also at 0.9 c, but at an angle of 30° from the line AB. What is the clock rate of C relative to B according to Einstein's etherless theory of Special Relativity? What is the clock rate of C relative to B according to Quasi-Relativity?

9. Does Einstein's Special Relativity have a clock paradox? Does the author's Special Quasi-Relativity have a clock paradox?

Chapter 4

GENERAL QUASI-RELATIVITY

General quasi-relativity, like general relativity, is largely about gravity and inertia. A full understanding of gravity and inertia requires an aether unified field theory. While a unified field theory is developed in subsequent chapters, the author has discovered a unified field theory depends on the structure of matter, which in turn depends on a correct model of special and general quasi-relativity. Fortunately the reasoning need not be wholly circular. Simple, reasonable assumptions can be made from which a solid general theory of quasi-relativity can be deduced. Then when we get to deriving a unified field theory, the reasonableness of our postulates will be seen with additional clarifications.

General quasi-relativity requires an additional postulate over the two postulates of special quasi-relativity.

General Quasi-Relativity Postulate

3. All non-zero mass particles (including imaginary mass particles) accelerate at the same rate at the same potential in a gravitational field.

A comparison is now in order between Einstein's postulates for special and general relativity and the author's postulates for special and general quasi-relativity. Einstein listed three postulates for special and general relativity: 1) the principle of relativity (from which was gleaned the covariance of physical laws); 2) the constancy of c; and 3) the equivalence principle. The close of chapter two disproved both the principle of relativity and the principle of equivalence in the absolute. Both are relatively true, or partially true. But consider now the number of Einstein's postulates relative to the author's postulates. The author's second postulate is Newton's third law; the author's third postulate is Galileo's law of falling objects, used by Newton in classical physics. Newton's third law and Galileo's law are both laws of physics that would be implicit in the covariance of physical laws in Einstein's system. If Einstein would have counted those laws as additional postulates in his model, and if he were to count the assumption that there is no aether (which was an additional assumption beyond his postulates) as an additional postulate, then Einstein would have had at least six postulates for his system. In the same degree of explicitness, the author has only three postulates. The author's system does more with fewer a priori assumptions. In quasi-relativity, covariance of physical laws and the constancy of c are derived results. Also such equivalence between the force of gravity and the force of inertia that really exists is a derived result in quasi-relativity.

From postulate 3 we expect an integral spin boson light weight aether particle and a canon ball to fall at the same rate in a vacuum. If a gravitational body is at rest relative to the fixed stars, we expect the graviton aether particles to fall from at rest at infinity to the escape velocity v_e at any given radius r outside of a spherically symmetric gravitational body (where r is measured from the center of the body). We expect a gravitational body to accelerate the aether particles in the negative vertical direction—inward radially relative to the center of the gravitational body. The aether particles should drift through the gravitational body and be decelerated on the opposite side of the body.

Consider a gravitational body traveling at high velocity relative to an aether. There are two effects: a special relativistic effect upon the clocks, lengths, and masses, common to every particle in the system, and a general relativistic effect due solely to the acceleration of the aether particles relative to the gravitational body. We have already seen from special quasi-relativity in an aether that the lengths, times, and masses in the system will be so adjusted that rest mass, rest length, and rest time will be conserved for every observer. Thus the special relativistic effect, as seen by a distant observer at rest in the aether, will transform away in the system of the moving gravitational body. An observer in the moving system will not be able to detect any aether velocity except that due solely to the acceleration of the aether by the gravitational body (except through costly rocket-clock experiments). Therefore, conveniently, no matter how fast the aether is traveling in the system due to translational motion of the system, the aether can be treated in the system as at rest except for the acceleration of the aether by the gravitational body.

The gravitational effect upon the aether, then, can be calculated by assuming that the aether is at rest at infinity relative to the gravitational body. The aether is then accelerated inward at the rate any massive object would be, and decelerated out the other side of the body. The acceleration of the aether by the planet would be such that at any elevation the aether should be traveling at the escape velocity ($-v_e$) toward the planet and ($+v_e$) away from it. For weak gravitational fields (nonrelativistic aether velocities), the escape velocity for any given elevation on or outside the surface of a planet or sun is

$$v_e = \left(\frac{2GM}{r} \right)^{1/2} ,$$
                                                                    (4-1)

where M is the mass of the large gravitational body from which the escape velocity v_e is measured, r is the distance from the point in space at which v_e is measured to the center of the large gravitational body, and G is the universal gravitational constant,

$6.67259(85) \times 10^{-11} \, m^3 kg^{-1} s^{-2}$.

A. Gravitational Slowing of Atomic Clocks

Now we are ready for our first problem in aether general relativity. The gravitational slowing of atomic clocks occurs because the gravitational body accelerates the aether to the escape velocity v_e relative to a stationary observer near a gravitational body. The clock of the observer is slowed because of the relative aether velocity through it (as in special quasi-relativity in an aether). The gravitational slowing of atomic clocks at a fixed elevation on or above a gravitational body is simply found by using the escape velocity as the aether velocity relative to the observer in the time dilation relation already derived for special relativity or special quasi-relativity. (Set x = 0 in the Lorentz transformation for time, equation (3-38), in differential form, thus finding time dilation alone.)

$$dt' = \left(1 - \left(V^2 / c^2\right)\right)^{-1/2} d\tau \qquad (4\text{-}2)$$

(from special relativity). Substituting the escape velocity v_e for V, we get

$$dt' = \left(1 - 2GM / rc^2\right)^{-1/2} d\tau. \qquad (4\text{-}3)$$

The gravitational potential energy of a particle of mass m near a large gravitational body of mass M is given by the relation

$$U(r) = -GMm / r \qquad (4\text{-}4)$$

(from elementary physics). The gravitational potential φ is found by dividing the gravitational potential energy U(r) by m. Thus

$$\varphi = -GM/r \qquad (4\text{-}5)$$

(from elementary physics). Rewriting equation (4-3) in terms of φ, we get:

$$dt' = \left(1 + \left(2\varphi/c^2\right)\right)^{-1/2} d\tau. \qquad (4\text{-}6)$$

This equation, derived easily through aether quasi-relativity, is the same as the general relativity relation for the gravitational slowing of clocks. The differential $d\tau$ is the differential time interval in a stationary frame fixed at elevation r, and is the so-called "proper time" of that frame. The differential dt' is the "coordinate-time" interval of gravitation-free space. The equation expressed in this form is valid anywhere in a weak gravitational field, and not just outside a spherically symmetric mass distribution.

It is of interest at this point to observe that, historically speaking, Einstein developed his Special Theory of Relativity upon the principle of relativity before he developed the General Theory of Relativity upon the principle of equivalence. Had the reverse been true a very interesting result might have occurred. Without the prior notion that absolute motion through space is unmeasurable, the principle that the force due to gravity is indistinguishable from the inertial force due to acceleration might have led to the idea that a gravitational body actually does accelerate the surrounding space (or aether). The gravitational acceleration of the aether would have been then the most natural hypothesis.

B. Gravitational Red Shift

Intimately related to the gravitational slowing of clocks is the gravitational red shift. The theory behind the red shift is this: A spectral emission line from an atom or molecule in a gravitational field is emitted at the same frequency (as measured by it's proper time) as the same emission line of an atom emitted in a gravitation-free frame (as measured in the proper time of a gravitation free frame). However, the proper time of the gravitational field frame is slowed relative to the coordinate time of gravitation-free space. Thus the emission frequency of the light is actually reddened at the time of emission as measured in coordinate time. It is theorized that the light maintains the same frequency as measured in coordinate time throughout its flight, but seems to change to the observer in that the proper time (that the observer measures the frequency by) may be faster or slower depending on whether the observer is in a weaker or stronger gravitational field than the spectral line emitting atom. The photon does not lose energy in escaping a gravitational field (as measured in the coordinate frame. The energy loss is in being emitted at a lower frequency in coordinate time, and being observed differently by observers with different proper times.

The calculation of this effect is easily done with the formula we derived from our aether model for the proper time of a gravitational frame in terms of the gravitational potential and coordinate time. The derivation follows closely that in *Introduction to General Relativity*, by Adler, Basin, and Schiffer, which is alluded to below:

Consider now, for example, a light wave emitted on the sun and received on the earth. Also, let the gravitational potential at the surface of the sun be φ_s. By the way, proper-time intervals are related to coordinate-time intervals [in the weak gravitational field approximation] by the equation

$$d\tau_s = \left(1 + \frac{2\varphi_s}{c^2}\right)^{1/2} dt. \qquad (4\text{-}7)$$

Similarly, on the earth [a weak field case], proper-time intervals are related to coordinate-time intervals by

$$d\tau_e = \left(1 + \frac{2\varphi_e}{c^2}\right)^{1/2} dt, \qquad (4\text{-}8)$$

where φ_e is the value of the gravitational potential on the earth. Let us suppose now n waves of frequency v_0 are emitted in proper time $\Delta\tau_s$ from an atom on the sun. Then we have

$$n = v_0 \Delta\tau_s. \qquad (4\text{-}9)$$

On the earth one certainly receives n waves, but the frequency and time duration of the wave train have changed from the sun. Using a frequency-duration relation for the earth analogous to Equation (4-9) for the sun,

$$n = v_e \Delta\tau_e, \qquad (4\text{-}10)$$

we obtain the following, since n is a constant,

$$v_0 \Delta\tau_s = v_e \Delta\tau_e. \qquad (4\text{-}11)$$

Thus

$$v_e = v_0 \frac{\Delta\tau_s}{\Delta\tau_e}. \qquad (4\text{-}12)$$

From Equation (4-7) the coordinate-time duration of the wave corresponding to $\Delta\tau_s$ is

$$\Delta t = \frac{\Delta\tau_s}{\sqrt{1+2\varphi_s/c^2}}. \qquad (4\text{-}13)$$

We suppose that the coordinate-time duration of the wave Δt is the *same* on the earth as on the sun. Equation (4-8) then gives

$$\Delta t = \frac{\Delta\tau_e}{\sqrt{1+2\varphi_e/c^2}}. \qquad (4\text{-}14)$$

By virtue of this and Equation (4-13), we then have

$$\frac{\Delta\tau_s}{\Delta\tau_e} = \left(\frac{1+2\varphi_s/c^2}{1+2\varphi_e/c^2}\right)^{1/2}. \qquad (4\text{-}15)$$

Substitution of this into Equation (4-12) gives

$$v_e = v_0\frac{\Delta\tau_s}{\Delta\tau_e} = v_0\left(\frac{1+2\varphi_s/c^2}{1+2\varphi_e/c^2}\right)^{1/2}. \qquad (4\text{-}16)$$

Now expanding to first order in the small quantities φ_s/c^2 and φ_e/c^2, we obtain

$$\frac{v_e-v_0}{v_0} = \frac{\varphi_s-\varphi_e}{c^2}, \qquad (4\text{-}17)$$

or in briefer notation,

$$\frac{\Delta \nu}{\nu_0} = \frac{\Delta \varphi}{c^2}. \qquad (4\text{-}18)$$

Now since the sun is at a large negative potential relative to the earth, we see that $\Delta \varphi$ is negative. Thus the frequency of light emitted on the sun *decreases* as it leaves the sun, and when it is received on earth, we see a shift toward the red end of the spectrum. Another way of looking at it is as though the atoms of the sun vibrated in slow motion when we viewed them from the earth. Now, of course, there is nothing special about using the earth and sun as the two points considered, and we can just as well use two points at different heights on the earth if our measurement is precise enough to detect the correspondingly small shift in light frequency.[1]

We see that the aether quasi-relativity red shift result is just the same as the general relativity red shift inasmuch as the same derivation could be used in both models to go from the proper-time-to-coordinate-time relationship (4-6), (4-7), and (4-8) to the red shift formula (4-17) and (4-18). The proper-time-to-coordinate-time relationship (4-6) was derived independently in the aether model and in general relativity. The aether derivation of (4-6) was simple. One had only to insert the escape velocity into the special quasi-relativity relation for time dilation.

C. Other Gravitational Relativistic Effects

Since the gravitational slowing of clocks in the aether model arises from an actual velocity of the clock relative to the aether, and is in that sense special quasi-relativistic, we expect by the aether model that the other quasi-relativistic effects—inertial mass increase and length contraction will also occur in the gravitational frame. Gravitational mass increase may be obtained by substituting the velocity of the aether relative to the observer (the escape velocity for the case of a stationary frame at fixed elevation) in the special relativistic equations for relative mass increase. Thus we have

$$m' = \left(1 - V_e^2 / c^2\right)^{-1/2} m = \left(1 - 2GM / rc^2\right)^{-1/2} m$$

$$= \left(1 + 2\varphi / c^2\right)^{-1/2} m. \tag{4-19}$$

Gravitational alteration of the measure of distance in the radial direction in a gravitational field may be obtained by setting t = 0 in the x' transformation equation (3-40), converting to differential notation, setting V equal to the escape velocity v_e, and substituting dr for dx.

$$dr' = \left(1 - V_e^2 / c^2\right)^{-1/2} dr = \left(1 - 2GM / rc^2\right)^{-1/2} dr$$

$$= \left(1 + 2\varphi / c^2\right)^{-1/2} dr. \tag{4-20}$$

D. Aether Derivation of the Schwarzschild Line Element

In Euclidean space, the distance between two points in Cartesian coordinates is

$$\Delta s^2 = \Delta x^2 + \Delta y^2 + \Delta z^2. \tag{4-21}$$

A differential line element would be

$$ds^2 = dx^2 + dy^2 + dz^2. \tag{4-22}$$

In special relativity we are interested in knowing the four-dimensional distance between two events in space-time. Since the time dimension part squared must have the opposite sign as the square of the space part, and since the time-like distance is normally longer than the space-like distance for events which we consider, the four-dimensional distance is usually written as

$$\Delta s^2 = c^2 \Delta t^2 - \Delta x^2 - \Delta y^2 - \Delta z^2. \tag{4-23}$$

A differential line element would be

$$ds^2 = c^2 dt^2 - dx^2 - dy^2 - dz^2. \tag{4-24}$$

This line element is called the Lorentz line element, and is here expressed in Cartesian form. The spherical coordinate form will be more interesting to us.

Let us convert this Lorentz line element into spherical coordinates. There are different conventions for the use of the symbols φ and θ in describing the angles in the spherical coordinates.   To be in harmony with the convention of Adler, Bazin, and Schiffer in *Introduction to General Relativity*,[2] let us use φ for the longitude and θ for the azimuth. The x, y, and z transformations to spherical coordinates are then

$$x = r \cos\varphi \sin\theta; \quad y = r \sin\varphi \sin\theta; and \quad z = r \cos\theta. \tag{4-25}$$

To find the spherical form of the Lorentz line element, we find the differentials dx, dy, and dz, and then find their squares dx^2, dy^2, and dz^2, all in terms of r, φ, and θ. We substitute these results in equation (4-24), collect terms, and simplify the result by utilization of the fact that $\sin^2\theta + \cos^2\theta = 1$ and $\sin^2\varphi + \cos^2\varphi = 1$. This is a prodigious, though simple calculation which will not be done here. The result is

$$ds^2 = c^2 dt^2 - dr^2 \left(d\theta^2 + \sin^2\theta \, d\varphi^2 \right). \tag{4-26}$$

This is the Lorentz line element in spherical coordinates from special relativity for gravity-free space.

This equation can easily be converted into the Schwarzschild line element by recognizing that this formula is valid for the proper time and proper distance of a frame. Since in general relativity we often denote proper time, etc., by primes and reserve unprimed times and lengths for the coordinate times and lengths, let us rewrite (4-26) with primes.

$$ds^2 = c^2 dt'^2 - dr'^2 \left(d\theta'^2 + \sin^2 \theta' d\varphi'^2 \right). \qquad (4\text{-}27)$$

Our aether model has given us a relation for the proper time for a gravity frame in terms of coordinate time ((4-3), (4-6), and (4-7)). The relation for differential vertical proper lengths in a gravity frame is given by (4-20). Since the motion of the aether particles is in the radial direction, the transverse proper lengths are equal to the transverse coordinate lengths, and may be expressed as

$$r' d(\theta' - \pi / 2) = r d(\theta - \pi / 2) \, or \, r' d\theta' = r d\theta; \;\; r' d\varphi' = r d\varphi.$$
$$(4\text{-}28)$$

In general, in strong gravitational fields, $\theta' \neq \theta$. But in the problems we will be working in the equatorial plane, $\theta' = \theta = \pi/2$. By substituting equations (4-6), (4-20), and (4-28) into equation (4-27), we may express the line element in a gravity field in terms of coordinate time and coordinate distance.

$$ds^2 = c^2 \left(1 - \frac{2GM}{rc^2} \right) dt^2 - \qquad (4\text{-}29)$$

$$\frac{dr^2}{1-\dfrac{2GM}{rc^2}} - r^2\left(d\theta^2 + \sin^2\theta'\,d\varphi^2\right).$$

In the equatorial plane $\theta' = \theta = \pi/2$. Also in weak gravitational fields $\theta' \approx \theta$. Under those conditions, the aether derivation agrees with the traditional Schwarzschild line element (4-30).

$$ds^2 = c^2\left(1 - \frac{2GM}{rc^2}\right)dt^2 - \qquad\qquad (4\text{-}30)$$

$$\frac{dr^2}{1-\dfrac{2GM}{rc^2}} - r^2\left(d\theta^2 + \sin^2\theta\,d\varphi^2\right).$$

The aether derivation agrees with the Schwarzschild line element in the equatorial plane and in weak gravitational fields, which covers all the general relativistic effects calculated in this chapter. The original Schwarzschild line element may be in error off the equatorial plane in strong gravitational fields. This may yield errors in some singularity studies and may even affect the shape of the sun near the poles.

Equation (4-30) is the Schwarzschild line element. We derived it easily from special quasi-relativity in an aether merely by substituting into the spherical form of the Lorentz line element the proper times and lengths for an aether velocity equal to the escape velocity radially—the very average velocity we expect for an aether attracted by a gravitational body. Notice that the aether derivation did not require any complicated tensors or gravitational field equations.

With the Schwarzschild line element we can calculate most of the known tests for general relativity. The perihelic shift of Mercury, the bending of starlight in a gravitational field, and radar ranging of planets can be calculated given the Schwarzschild line element. These calculations will be demonstrated in Parts E through I. These calculations are more difficult than most in this book. Therefore the reader can skip parts E, F, and H and I if so desired. But be sure to read part G.

The aether model can account for other geometries also than the spherical mass distribution in Schwarzschild's solution.   Dipole, quadripole, and higher order mass distributions will effect the shape of the aether velocity field and will give rise to different line elements than the Schwarzschild line element. The dynamic aether model is a versatile theory.

E. Calculus of Variation Methods

In order to calculate the perihelic shift of Mercury and the deflection of light grazing the sun from the result we have obtained, the Schwarzschild line element, calculus of variations methods are ordinarily used. Since this branch of mathematics may not be prior common knowledge to some of the readers of this book, a brief review of relevant facets of this branch of calculus will here be included. We shall draw largely from *An Introduction to the Calculus of Variations*, by Fox.[3]

Weak variations

Let our first problem in the calculus of variations be a relatively simple one. Now let

$$I = \int_a^b F(x, y, dy/dx)dx, \qquad (4\text{-}31)$$

where I in this system is a convenient symbol for the integral and F denotes a given functional form. In our problem, the functional relation between y and x is not known and the problem consists in finding this relation so that I is a maximum or a minimum. In other words, given F, find the path of integration for F for which I is a maximum or a minimum. We here confine ourselves to the case where y is a single-valued function of x in the interval (a,b).

We commence our problem by finding the stationary values of I and then proceed to develop tests which enable us to discriminate between the cases when I is a maximum or a minimum or is neither.

Evidently in our problem the arc of integration must be of such a nature that the integral in Equation (4-31) can be determined; such an arc is known as an *admissible* arc. Subsequent analysis of the problem requires us to assume that F(x,y,p) possesses partial derivatives with respect to the variables x, y, and p of at least the fourth order in an interval which includes the points x = a and x = b. This will justify our employment in our problem of the mean-value theorem for functions of several variables. We simplify the problem appreciably by assuming that a and b, the limits of integration, are prescribed in this case. In addition, in our problem, although the functional relation between y and x is not yet known, we assume that the values of y corresponding to x = a and x = b, say α and β respectively, are also prescribed. Geometrically speaking, in this calculus of variations problem, the integral I must be taken along a plane curve from the given point A, coordinates (a,α), to the given point B, coordinates (b,β), as is shown in Fig. 4-1.

Our problem then resolves itself into that of finding the admissible curve or curves joining A and B for which I is stationary.

Let

$$y = s(x) \tag{4-32}$$

be the equation of the admissible curve for which I is stationary and [see Fig. 4-1] let APB be the curve whose equation is Equation (4-32). In calculus of variation problems we encounter, the symbol s when used to denote a functional form will always refer to the stationary case; those investigations in which s is used to denote the length of arc of a curve will be sufficiently self-explanatory to avoid the possibility of confusion. Let AQB, in Fig. 4-1, be another admissible curve joining A and B and let its equation be

$$y = s(x) + \varepsilon t(x) \tag{4-33}$$

where ε we take to be an arbitrary constant independent of x and y and t(x), in our problem, denotes any arbitrary function of x which is independent of ε. In this science, with this restriction on t(x) the ordinate y is said to be subjected to weak variations.

In Fig. 4-1 the points P and Q have the same abscissa x, and $PQ = \varepsilon t(x)$. Since the curve in Equation (4-33) also passes through the points A and B we must have

$$t(a) = t(b) = 0. \tag{4-34}$$

Denoting differentiations by dashes or primes, we have from Equation (4-33)

$$dy / dx = s'(x) + \varepsilon t'(x). \tag{4-35}$$

Hence, in our problem, for weak variations, as ε tends to zero Q tends to P and simultaneously the slope of AQB at Q tends to that of APB at P.

Let the value of the integral in Equation (4-31), when taken along the curve APB, for which it is stationary, be

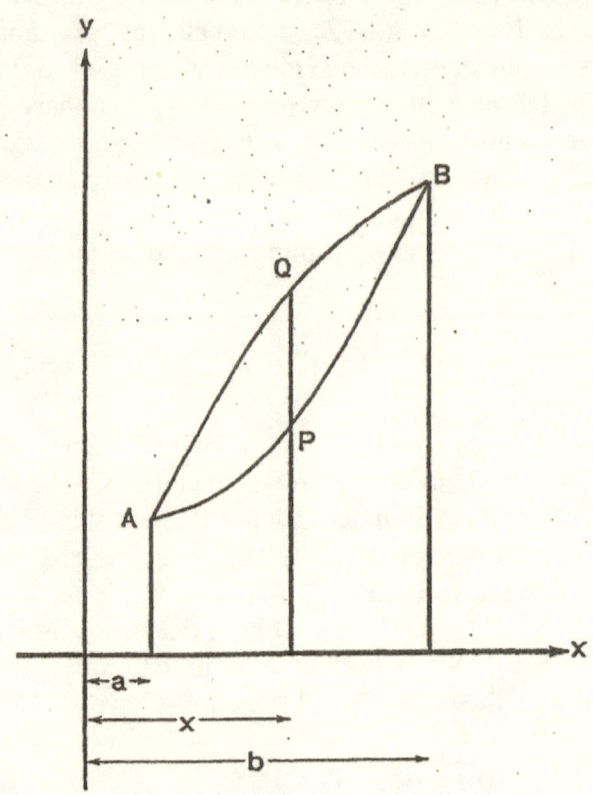

Fig. 4-1. Admissible curves APB and AQB joining A and
B.

denoted by I_s and when taken along the neighbouring curve AQB be denoted by $I_s + \delta I_s$. Then

$$I_s = \int_a^b F(x,s,s')dx \qquad (4\text{-}36)$$

and

$$I_s + \delta I_s = \int_a^b F(x,s+\varepsilon t,s'+\varepsilon t')dx, \qquad (4\text{-}37)$$

where s, s', t, and t' are abbreviations for s(x), s'(x), t(x), and t'(x) respectively.

The assumption that F(x,y,p) possesses continuous partial derivatives justifies an application of the mean-value theorem for functions of several variables. If the derivatives are continuous up to at least the third order we have

$$F(x,s+\varepsilon t,s'+\varepsilon t') = F(x,s,s') + \qquad (4\text{-}38)$$

$$\varepsilon\left(t\frac{\partial F}{\partial s} + t'\frac{\partial F}{\partial s'}\right) +$$

$$\frac{\varepsilon^2}{2!}\left(t^2\frac{\partial^2 F}{\partial s^2} + 2tt'\frac{\partial^2 F}{\partial s\partial s'} + t'^2\frac{\partial^2 F}{\partial s'^2}\right) +$$

$$O(\varepsilon^3),$$

$$where\ \frac{\partial F}{\partial s}\ denotes\ \frac{\partial F(x,s,s')}{\partial s},$$

$$\frac{\partial F}{\partial s'} \ denotes \ \frac{\partial F(x,s,s')}{\partial s'}, etc.$$

From (4-36) and (4-37) we finally have

$$\delta I_s = \varepsilon \int_a^b \left(t \frac{\partial F}{\partial s} + t' \frac{\partial F}{\partial s'} \right) dx + \qquad (4\text{-}39)$$

$$\frac{\varepsilon^2}{2!} \int_a^b \left(t^2 \frac{\partial^2 F}{\partial s^2} + 2tt' \frac{\partial^2 F}{\partial s \partial s'} + t'^2 \frac{\partial^2 F}{\partial s'^2} \right) dx +$$

$$O(\varepsilon^3).$$

Denoting the coefficient of ε by I_1 and that of ε^2 by I_2, the quantities εI_1 and $\varepsilon^2 I_2$ are sometimes referred to as the 'first variation' and 'second variation' respectively.

Evidently if I_s is a maximum then δI_s must be negative for all sufficiently small values of ε, whether positive or negative.   Hence sufficient conditions for a maximum are $I_1 = 0$ and $I_2 < 0$. Similarly for a minimum value of I_s it is sufficient to have $I_1 = 0$ and $I_2 > 0$.

The Eulerian characteristic equation

The equation $I_1 = 0$ is easily modified to a more convenient form. Integrating by parts we have

$$\int_a^b t' \frac{\partial F}{\partial s'} dx = \left(t \frac{\partial F}{\partial s'} \right)_{x=b} - \left(t \frac{\partial F}{\partial s'} \right)_{x=a} - \qquad (4\text{-}40)$$

$$\int_a^b t \frac{d}{dx}\left(\frac{\partial F}{\partial s'}\right) dx.$$

In the term $\partial F/\partial s'(= \partial F(x,s,s')/\partial s')$, the variables x and s are treated as constants and only the s' terms are differentiated. In the term $d/dx\ (\partial F/\partial s')$, s and s' must be treated as functions of x after the partial differentiation with respect to s' and before the differentiation with respect to x.

Now it has been stipulated that $t(a) = t(b) = 0$, equation (4-34), and so the first two terms on the right-hand side of (4-40) vanish. The equation $I_1 = 0$ then readily reduces to

$$\int_a^b t(x)\left[\frac{\partial F}{\partial s} - \frac{d}{dx}\left(\frac{\partial F}{\partial s'}\right)\right] dx = 0. \qquad (4\text{-}41)$$

So far no use has been made of the arbitrariness of the function $t(x)$. . . . If $t(x)$ is an arbitrary function of x, then (4-41) can be satisfied if and only if

$$\frac{\partial F}{\partial s} - \frac{d}{dx}\left(\frac{\partial F}{\partial s'}\right) = 0, \qquad (4\text{-}42)$$

for all values of x between a and b.[4]

Fox then proceeds to give a proof that equation (4-41) holds if and only if equation (4-42) is satisfied. However, since this point is fairly intuitively obvious, we shall omit his lengthy proof.

The discovery of (4-42) by Euler in 1744 inaugurated the calculus of variations in its modern form. It is a differential equation of the second order known as the characteristic equation or as Euler's equation. Its solution is the equation of a curve known as a characteristic curve or more generally as an extremal.

In application of (4-42) we shall always replace s(x) by the more convenient variable y. The results obtained may then be summed up as follows:

THEOREM 1. The integral

$$\int_a^b F(x,y,y')dx,$$

whose end points are fixed, is stationary for weak variations if y satisfies the differential equation

$$\frac{\partial F}{\partial y} - \frac{d}{dx}\left(\frac{\partial F}{\partial y'}\right) = 0. \qquad (4\text{-}43)$$

In chapter three of his book, Fox proves the more generalized case of Theorem 1 for a function F with n arbitrary parameters, q_1, q_2, q_3, . . . ,q_n. The result is his Theorem 7.[5]

THEOREM 7. Let the values of t_0 and t_1 and the functional form of F be given. Then the integral

$$\int_{f_0}^{f_1} F(q_1, q_2, ..., q_n; \dot{q}_1, \dot{q}_2, ..., \dot{q}_n; t)dt,$$

where the q's are arbitrary functions of t, is stationary for weak variations when the q's satisfy the n equations

$$\frac{\partial F}{\partial q_m} - \frac{d}{dt}\left(\frac{\partial F}{\partial \dot{q}_m}\right) = 0 \quad (m = 1,2,...,n). \qquad (4\text{-}44)$$

The last theorem we shall use in the calculation of the perihelic shift of Mercury and the calculation of the angle of deflection of starlight grazing the sun.

F. Perihelic Shift of Mercury

Two sections earlier we derived the Schwarzschild line element, equation (4-30). Now with some tools from the calculus of variations and the understanding that the motion of a body in a gravitational field follows a four-dimensional geodesic line, we can solve the general relativistic Kepler problem for the orbit of a planet in a gravitational field. With a few interpolated comments for ease of comprehension, we can follow the solution of the problem as per Adler, Bazin, and Schiffer.[6]

As a shorthand notation, and to be in agreement with the notation of Adler, et. al., let us define

$$m \equiv \frac{GM}{c^2}. \qquad (4\text{-}45)$$

We may now rewrite equation (4-30) as

$$ds^2 = \left(1 - \frac{2m}{r}\right)c^2 dt^2 - \qquad (4\text{-}46)$$

$$\left(1 - \frac{2m}{r}\right)^{-1} dr^2 - r^2\left(d\theta^2 + \sin^2\theta d\varphi^2\right).$$

The motion of a body in a gravitational field follows a four-dimensional geodesic line. Hence, to find the orbit of a planet, we need the Euler-Lagrange equations for the following variational problem

$$\delta \int ds = 0, \tag{4-47}$$

where ds is given by the Scharzschild line element (4-46). This is obviously equivalent to the problem

$$\delta \int (1) ds = 0. \tag{4-48}$$

By dividing the Schwarzschild line element (4-46) by ds^2 we can arrive at the following identity

$$1 = \left(1 - \frac{2m}{r}\right) c^2 \frac{dt^2}{ds^2} - \tag{4-49}$$

$$\left(1 - \frac{2m}{r}\right)^{-1} \frac{dr^2}{ds^2} - r^2 \left(\frac{d\theta^2}{ds^2} + \sin^2 \theta \frac{d\varphi^2}{ds^2}\right).$$

If we take a dot to indicate differentiation with s, we may rewrite (4-49) as

$$1 = \left(1 - \frac{2m}{r}\right) c^2 \dot{t}^2 - \tag{4-50}$$

$$\left(1 - \frac{2m}{r}\right)^{-1} \dot{r}^2 - r^2 \left(\dot{\theta}^2 + \sin^2 \theta \dot{\varphi}^2\right).$$

We may substitute the identity (4-50) in the place of the (1) in equation (4-48) to obtain the equivalent variational problem to that of equation (4-47),

$$\delta \int \left[\left(1 - \frac{2m}{r}\right) c^2 \dot{t}^2 - \left(1 - \frac{2m}{r}\right)^{-1} \dot{r}^2 - r^2 \left(\dot{\theta}^2 + \sin^2 \theta \dot{\varphi}^2\right) \right] ds = 0.$$

(4-51)

We may now use Fox's Theorem 7 to find the Euler-Lagrange equations for θ, φ, and t associated with this variational problem. The variables we just named correspond to q_1, q_2, and q_3, and the arc length s corresponds to the independent variable t in the theorem. The function F in the theorem corresponds to the right side of equation (4-50). The Euler-Lagrange equation for θ, then, is

$$\frac{\partial F}{\partial \theta} - \frac{d}{ds}\left(\frac{\partial F}{\partial \dot{\theta}}\right) = 0.$$

(4-52)

Doing the partial differentiation we obtain

$$- 2r^2 \sin\theta \cos\theta \dot{\varphi}^2 - \frac{d}{ds}\left(- 2r^2 \dot{\theta}\right) = 0.$$

(4-53)

Dividing the equation by 2 and rearranging, we obtain the Euler-Lagrange equation for θ:

$$\frac{d}{ds}\left(r^2 \dot{\theta}\right) = r^2 \sin\theta \cos\theta \dot{\varphi}^2.$$

(4-54)

Similarly for φ we obtain

$$\frac{\partial F}{\partial \varphi} - \frac{d}{ds}\left(\frac{\partial F}{\partial \dot{\varphi}}\right) = 0.$$

(4-55)

$$0 - \frac{d}{ds}\left(-2r^2 \sin^2 \theta \dot{\varphi}\right) = 0. \qquad (4\text{-}56)$$

Again, dividing the equation by 2 and rearranging, we have

$$\frac{d}{ds}\left(r^2 \sin^2 \theta \dot{\varphi}\right) = 0. \qquad (4\text{-}57)$$

The equation for t is

$$\frac{\partial F}{\partial \dot{t}} - \frac{d}{ds}\left(\frac{\partial F}{\partial \dot{t}}\right) = 0. \qquad (4\text{-}58)$$

$$0 - \frac{d}{ds}\left(2c^2\left[1 - \frac{2m}{r}\right]\dot{t}\right) = 0. \qquad (4\text{-}59)$$

Dividing the equation by $-2c^2$, we obtain the Euler-Lagrange equation for t:

$$\frac{d}{ds}\left[\left(1 - \frac{2m}{r}\right)\dot{t}\right] = 0. \qquad (4\text{-}60)$$

Note that we have not included the Euler-Lagrange equation for r; it is more convenient to . . . [use equation (4-50) obtained by dividing the Schwarzschild line element by ds^2 as the fourth differential equation along with equations (4-54), (4-57), and (4-58)].

Using the above four differential equations for t, r, θ, and φ as functions of s, it is possible to obtain and solve the equations of a planetary orbit. In classical mechanics the orbit of a body in a central force field lies in a plane. We can show that the same holds true in the present theory. By an appropriate orientation of the axes we can make

$$\theta = \pi / 2 \text{ and } \dot{\theta} = 0$$

at some initial s. Then, from (4-54), it follow that, for all s,

$$\theta = \frac{\pi}{2}, \tag{4-61}$$

since the initial conditions determine a unique solution of (4-54), and (4-61) is surely such a solution. Substitution of $\theta = \pi/2$ in (4-57) immediately:

$$r^2 \dot{\varphi} = h = const. \tag{4-62}$$

Equation (4-60) integrates to

$$\left(1 - \frac{2m}{r}\right)\dot{t} = l = const. \tag{4-63}$$

Substituting the results (4-61), (4-62), and (4-63) into (4-50), we obtain the following differential equation for r(s):

$$1 = \left(1 - \frac{2m}{r}\right)^{-1} c^2 l^2 - \tag{4-64}$$

$$\left(1 - \frac{2m}{r}\right)^{-1} \dot{r}^2 - \frac{h^2}{r^2}.$$

As in the classical Kepler problem, one can simplify matters by considering r as a function of φ instead of s.   Denoting differentiation with respect to φ by a prime, we then have

$$r' = \frac{dr}{d\varphi} = \frac{\dot{r}}{\dot{\varphi}}.$$

(4-65)

From (4-62) and (4-65) we obtain

$$\dot{r} = \dot{\varphi}r' = \frac{h}{r^2}r'.$$

(4-66)

The differential equation for r(φ) is then obtainable from (4-64):

$$\left(1 - \frac{2m}{r}\right) = c^2 l^2 - \frac{h^2}{r^4}r'^2 - \frac{h^2}{r^2}\left(1 - \frac{2m}{r}\right).$$

(4-67)

. . . Following once more the example of the classical Kepler problem, we substitute for the dependent variable

$$r = \frac{1}{u},$$

(4-68)

which implies

$$r' = -\frac{u'}{u^2}.$$

(4-69)

Using these relations, we can convert (4-67) to a differential equation for u(φ):

$$(1 - 2mu) = c^2 l^2 - h^2 u'^2 - h^2 u^2 (1 - 2mu).$$

(4-70)

This reduces to

$$u'^2 = \left(\frac{c^2l^2 - 1}{h^2}\right) + \frac{2m}{h^2}u - u^2 + 2mu^3....\quad(4\text{-}71)$$

To make the problem more transparent and to establish a closer connection with the classical Kepler problem (which involves a second-order differential equation), we shall convert the first-order equation (4-71) to a second-order equation by differentiation with respect to φ. We obtain

$$2u'u'' = \frac{2m}{h^2}u' - 2uu' + 6mu^2u'.\quad(4\text{-}72)$$

One possible solution is then obtained by setting the common factor u' equal to zero:

$$u' = 0 \quad u = const \quad r = const.\quad(4\text{-}73)$$

Thus circular motion occurs in relativity theory just as in classical theory. . . . The other possible solution, which is much more interesting, will result from canceling the common factor u' from (4-72):

$$u'' + u = \frac{m}{h^2} + 3mu^2.\quad(4\text{-}74)$$

This last equation is quite similar in structure to the orbit equation of the classical Kepler problem....

For sake of brevity, we shall omit Adler's, et al, derivation of orbit equation of the classical Kepler problem. The result they obtained is

$$u'' + u = \frac{GM}{H^2} \qquad (4\text{-}75)$$

[where we have substituted our symbol G for the gravitation constant in place of their symbol κ]

where H is twice the constant areal velocity:

$$H = r^2 \frac{d\varphi}{dt} = const. \qquad (4\text{-}76)$$

The analogous term in the relativistic equation (4-74) is m/h^2, which, by virtue of (4-45) and (4-62), is explicitly given by

$$\frac{m}{h^2} = \frac{GM}{c^2 r^4 \left(\dfrac{d\varphi}{ds}\right)^2} = \frac{GM}{c^2 r^4 \left(\dfrac{d\varphi}{dt}\right)^2 \left(\dfrac{dt}{ds}\right)^2}. \qquad (4\text{-}77)$$

Furthermore, we know . . . that, for slowly moving bodies in weak gravitational fields, $(dt/ds)^2$ is approximately $1/c^2$; substituting this in (4-77), we obtain an approximate form for m/h^2:

$$\frac{m}{h^2} \cong \frac{GM}{r^4 (d\varphi/dt)^2} = \frac{GM}{H^2}. \qquad (4\text{-}78)$$

Thus we see that the relativistic equation (4-74) differs from the classical equation (4-73) through the addition of the quadratic term $3mu^2$ and has a slightly different constant term m/h^2. One might furthermore expect the term $3mu^2$ to be small relative to the leading constant term; we may easily verify that this is indeed the case by forming the ratio of it and the constant term m/h^2. This ratio is

$3u^2h^2$, which, by virtue of (4-62), is $3r^2\varphi^2 \cong 3[r(d\varphi/dt)]^2 \cdot 1/c^2$. The quantity $r(d\varphi/dt)$ is the lateral velocity of the planet (the velocity perpendicular to r), so the above ratio may be written as $3v^2_{lateral}/c^2$, which is always very small and equal to 7.7×10^{-8} in the case of Mercury. The close similarity between the relativistic equation (4-74) and the classical theory (4-75) is now quite clear. . . .

Let us now investigate the relativistic equation (4-74) with a view to calculating the perihelion shift. We saw above that the term $3mu^2$ represents a small addition to the classical equations, so let us try a perturbation approach. Define

$$A = \frac{m}{h^2} \cong \frac{GM}{H^2} \qquad (4\text{-}79)$$

and the small dimensionless quantity

$$\varepsilon = 3mA \cong \frac{3G^2 M^2}{c^2 H^2}. \qquad (4\text{-}80)$$

The relativistic orbit equation then takes the form

$$u'' + u = A + \frac{\varepsilon u^2}{A}. \qquad (4\text{-}81)$$

To solve this we assume a solution of the form

$$u(\varphi) = u_0(\varphi) + \varepsilon v(\varphi) + O(\varepsilon^2). \qquad (4\text{-}82)$$

Substituting this form for u in the differential equation (4-81), we obtain

$$u_0'' + \varepsilon v'' + u_0 + \varepsilon v = A + \varepsilon u_0^2/A = O(\varepsilon^2). \qquad (4\text{-}83)$$

Equating the zeroth-order terms in ε, we have

$$u_0'' + u_0 = A, \tag{4-84}$$

which is essentially the classical equation (4-75). The solution is easily checked to be

$$u_0 = A + B\cos(\varphi + \delta) \tag{4-85}$$

where B and δ are arbitrary constants. By an appropriate orientation of the axes we may make δ equal to zero, in which case we obtain the familiar equation of an ellipse,

$$u_0 = A + B\cos\varphi. \tag{4-86}$$

Similarly, equating the first-order ε terms in 4-83), we obtain

$$v' + v = \frac{u_0^2}{A} = A + 2B\cos\varphi + \frac{B^2}{A}\cos^2\varphi \tag{4-87}$$

$$= \left(A + \frac{B^2}{2A}\right) + 2B\cos\varphi + \frac{B^2}{2A}\cos 2\varphi.$$

Note that we need only a nonhomogeneous solution to this equation since the zeroth-order solution already contains a term $B\cos\varphi$, which is the general solution to the homogeneous equation. Despite the cumbersome appearance of (4-87) it is readily solved; since it is *linear* in v, we may write v as the sum $v = v_a + v_b + v_c$, where v_a, v_b, and v_c are solutions of the equations

$$v_a'' + v_a = A + \frac{B^2}{2A} \qquad (4\text{-}88)$$

$$v_b'' + v_b = 2B\cos\varphi$$

$$v_c'' + v_c = \frac{B^2}{2A}\cos 2\varphi$$

that is, we superpose the three solutions (4-88) to get (4-87). The nonhomogeneous solutions to (4-88) are easily checked to be

$$v_a = A + \frac{B^2}{2A} \qquad v_b = B\varphi\sin\varphi \qquad (4\text{-}89)$$

$$v_c = -\frac{B^2}{6A}\cos 2\varphi;$$

so a nonhomogeneous solution to (4-87) is

$$v = v_a + v_b + v_c \qquad (4\text{-}90)$$

$$\left(A + \frac{B^2}{2A}\right) + B\varphi\sin\varphi - \frac{B^2}{6A}\cos 2\varphi.$$

Combining this with the zeroth-order solution (4-86), we have the entire solution for the orbit to first order in ε:

$$u = u_0 + \varepsilon v = \left(A + \varepsilon A + \frac{\varepsilon B^2}{2A}\right) + \qquad (4\text{-}91)$$

$$\left(B\cos\varphi - \frac{\varepsilon B^2}{6A}\cos 2\varphi \right) + \varepsilon B\varphi\sin\varphi.$$

Using this solution, we can readily calculate the perihelion shift.   Since only the last term is nonperiodic, it is clear that whatever irregularities occur in the perihelion position must be due to this term.   To clarify further the effect of the non-periodic term, note that, to first order in ε,

$$\cos(\varphi - \varepsilon\varphi) = \cos\varphi\cos\varepsilon\varphi + \sin\varphi\sin\varepsilon\varphi \qquad (4\text{-}92)$$

$$= \cos\varphi + \varepsilon\varphi\sin\varphi,$$

so the solution may be written as

$$u = A + B\cos(\varphi - \varepsilon\varphi) + \qquad (4\text{-}93)$$

$$\varepsilon\left(A + \frac{B^2}{2A} - \frac{B^2}{6A}\cos 2\varphi \right).$$

In this form the effect of the various terms on the orbit is apparent.   The basic elliptical orbit is represented by A + Bcosφ. The effect of the last term is to introduce small *periodic* variations in the radial distance of the planet.   Such effects are difficult to detect, and since they are periodic, they cannot influence the perihelic motion.   However, the εφ which appears in the cosine argument does indeed introduce a nonperiodicity, and since φ can become large, the effect is not negligible. Accordingly let us write (4-93) in the form

$$u = A + B\cos(\varphi - \varepsilon\varphi) + \qquad (4\text{-}94)$$

(periodic terms of order ε).

The perihelion of a planet occurs when r is a minimum or when u = 1/r is a maximum. From (4-94) we see that u is maximum when

$$\varphi(1 - \varepsilon) = 2\pi n \qquad (4\text{-}95)$$

or approximately

$$\varphi = 2\pi n(1 + \varepsilon). \qquad (4\text{-}96)$$

Therefore successive perihelia will occur at intervals of

$$\Delta \varphi = 2\pi(1 + \varepsilon) \qquad (4\text{-}97)$$

instead of 2π as in periodic motion. Thus the perihelion *shift* per revolution is given by

$$\delta\varphi = 2\pi\varepsilon = 2\pi\left(\frac{3G^2 M^2}{c^2 H^2}\right). \qquad (4\text{-}98)$$

For the case of Mercury, Eq. (4-98) gives a total shift of 42.89" per century. This is in excellent agreement with the observational result of 42.6" ± 1.0" which is unaccounted for classically.[7]

The aether model perihelic shift result is the same as that of General Relativity, inasmuch as the same derivation was used to go from the Schwarzschild line element, equation (4-29), to the perihelic shift equation (4-98). However, the aether model used an entirely different method of deriving the Schwarzschild line element than did the General Theory of Relativity. The consideration of a gravitationally accelerated aether gave an intuitively obvious modification of differential lengths and times in the Lorentz line element, which otherwise was not easily anticipated in the General Relativity model.

G. Evidence of Aether Drift Across Planets

Equation (4-98) may be rephrased in terms of the ratio $v^2_{lateral}/c^2$ by recalling equation (4-76) and noting that since planetary orbits are nearly circular,

$$v_{lateral} \approx \left(\frac{GM}{r} \right)^{1/2} = r \frac{d\varphi}{dt}. \qquad (4\text{-}99)$$

Using equations (4-76) and (4-99), we can rewrite equation (4-98) as

$$\delta\varphi = 2\pi\varepsilon = 2\pi \left(\frac{3v^2_{lateral}}{c^2} \right). \qquad (4\text{-}100)$$

Heretofore the numeral 3 in equation (4-100) may have seemed peculiar, giving no further insight than that this is just the way the equation turned out. But in the aether model, the numeral 3 in equation (4-100) has an exciting and simple interpretation, which we shall demonstrate by the following argument:

Suppose we are correct about the existence of an electrical aether. And suppose it is gravitationally accelerated by the sun in the amount we have predicted. Its inward velocity at r is

$$v_{escape} \approx \left(\frac{2GM}{r} \right)^{1/2}. \qquad (4\text{-}101)$$

A planet not moving at all at the distance r from the sun (though that is impossible for more than an instant) would feel an aether wind of velocity (in addition to planetary acceleration) v_{escape} (as calculated relative to the sun) through the planet in the direction of the sun by virtue of the acceleration of the aether by the sun. However, planets do not just sit still in space, they travel with the lateral velocity (for nearly circular orbits)

$$v_{lateral} \approx \left(\frac{GM}{r} \right)^{1/2}. \qquad (4\text{-}102)$$

This means the planet will sense an additional relative aether wind of velocity $v_{lateral}$ coming from the direction in which the planet is going, just as rain that is coming straight down has an apparent horizontal velocity component when one is running. The true relative aether velocity experienced by a planet will have to be the vectorial sum of v_{escape} and $v_{lateral}$. Thus, since v_{escape} and $v_{lateral}$ are at right angles to each other, we have:

$$v_{true}^2 = v_{escape}^2 + v_{lateral}^2. \qquad (4\text{-}103)$$

But from equations (4-101) and (4-102) we see that

$$v_{escape}^2 = 2 v_{lateral}^2. \qquad (4\text{-}104)$$

Thus

$$v_{true}^2 = 2 v_{lateral}^2 + v_{lateral}^2, \qquad (4\text{-}105)$$

or

$$v_{true}^2 = 3v_{lateral}^2.$$ (4-106)

Thus equation (4-100) may be written

$$\delta\varphi = 2\pi\varepsilon = 2\pi\left(\frac{v_{true}^2}{c^2}\right).$$ (4-107)

This result gives us a degree of confidence that there is not only an aether drift across the planet due to the gravitational acceleration of the aether by the sun, but also because of the lateral orbital velocity of the planet, which, by the way, was the very thing that Michelson and Morley set out to detect in their experiment and failed to find. It seems, however, that in actuality, the general relativistic result for the perihelion shift is a demonstration partly (in disguise) that there is an aether velocity across the planet due to its orbital velocity.

H. Bending of Starlight in a Gravitational Field

The derivation of the Schwarzschild line element (4-30) also assures the aether model will give the same value for the angle of deflection of light rays grazing the sun as does the General Relativity Theory. We can find this relation by considering the null geodesic, and adapting a result we obtained in studying the perihelic shift of Mercury.

The choice of a null geodesic for the equation of the path of

$$ds^2 = 0$$ (4-108)

can lose some of its mystery by considering a null geodesic in the Cartesian form of the Lorentz line element:

$$ds^2 = c^2 dt^2 - \left(dx^2 + dy^2 + dz^2\right) = 0 \qquad (4\text{-}109)$$

simply means that

$$\left(dx^2 + dy^2 + dz^2\right) = c^2 dt^2, \qquad (4\text{-}110)$$

which simply says we are considering the case of the equation of the wave front of light, which obviously was our intent. Taking the null geodesic in the Schwarzschild line element, then, should give us the equation of the wave front of light in a distorted gravitational frame.

It is interesting to note that [the equation for a light-ray trajectory] can . . . be deduced from (4-74) by intuitive reasoning. Equation (4-74) describes the orbit or trajectory of a particle in the Schwarzschild field:

$$u'' + u = \frac{m}{h^2} + 3mu^2. \qquad (4\text{-}111)$$

Using the expression for m given by (4-45) and the (exact) expression for m/h² given by (4-77), we can write this as

$$u'' + u = \frac{GM}{c^2 r^4}\left(\frac{ds}{d\varphi}\right)^2 + 3\frac{GMu^2}{c^2}. \qquad (4\text{-}112)$$

This equation for the geodesics follows directly from the variational problem (4-51) and involves no approximation. In order to specialize this to the case of a light ray, we must additionally set ds² = 0.

Since the angular interval dφ will in general be non-zero as the light ray sweeps by the sun, we conclude that, for the limiting case of a null geodesic,

$$\frac{m}{h^2} = \frac{GM}{c^2 r^4} \left(\frac{ds}{d\varphi}\right)^2 = 0. \qquad (4\text{-}113)$$

It follows that the equation of the trajectory is . . . ,

$$u'' + u = 3mu^2 \quad (\text{null geodesic}). \qquad (4\text{-}114)$$

As with the orbit equation (4-74), we can show that the term $3mu^2$ is small relative to the other terms of the equation. To do this, form the ratio of $3mu^2$ to the term u; that is, consider 3mu. Using the definition of the Schwarzschild radius $r_s = 2m$ [= $2GM/c^2$], we may also write this ratio as $(3/2)(r_s/r)$. . . . The Schwarzschild radius of the sun is on the order of a kilometer; thus, for a trajectory outside the sun's surface, the above ratio is evidently very small. This allows us to regard $3mu^2$ as a small perturbation term in Eq. (4-114). Accordingly, let us call

$$3m = \varepsilon \qquad (4\text{-}115)$$

and write the equation of the light-ray trajectory as

$$u'' + u = \varepsilon u^2. \qquad (4\text{-}116)$$

As in . . . [the derivation of the perihelic shift of Mercury], we shall use a standard perturbation approach to treat the above equation; we suppose a solution to (4-116) of the form

$$u = u_0 + \varepsilon v + O(\varepsilon^2), \qquad \varepsilon = 3m. \qquad (4\text{-}117)$$

Substituting this in (4-116), we obtain

$$u_0'' + u_0 + \varepsilon v'' + \varepsilon v = \varepsilon u_0^2 + O(\varepsilon^2). \qquad (4\text{-}118)$$

Equating the zeroth-order terms in ε, we have

$$u_0'' + u_0 = 0. \qquad (4\text{-}119)$$

This has the solution

$$u_0 = A\cos(\varphi + \delta), \qquad (4\text{-}120)$$

which, by an appropriate orientation of axes, may be written without the arbitrary constant δ:

$$u_0 = A\cos\varphi. \qquad (4\text{-}121)$$

In terms of the first-order radius $r = 1/u_0$; this becomes

$$r\cos\varphi = \frac{1}{A}. \qquad (4\text{-}122)$$

Since $r\cos\varphi$ is simply the Cartesian coordinate x, this evidently represents a straight line parallel to the y axis. This is indeed precisely what we should expect: in first approximation the light ray is not deflected at all by the sun's gravitational field. From Eq. (4-122) it is clear that the distance of closest approach to the origin (the sun) is $1/A$, so we shall call this constant r_0 and write the zeroth-order solution as

$$u_0 = \frac{1}{r_0}\cos\varphi. \tag{4-123}$$

Next, equating the first-order ε terms of Equation (4-118), we obtain

$$v' + v = u_0^2 = \frac{1}{r_0^2}\cos^2\varphi \tag{4-124}$$

$$= \frac{1}{2r_0^2}(1 + \cos 2\varphi).$$

To solve this we shall use a trial solution with unknown coefficients:

$$v = \alpha + \beta\cos 2\varphi. \tag{4-125}$$

Now differentiation gives

$$v' = -4\beta\cos 2\varphi, \tag{4-126}$$

so that

$$v' + v = \alpha - 3\beta\cos 2\varphi. \tag{4-127}$$

Comparing this term by term with Equation (4-124), we see that Equation (4-125) will be a solution if

$$\alpha = \frac{1}{2r_0^2}, \quad \beta = -\frac{1}{6r_0^2}. \tag{4-128}$$

Thus a solution of the differential Equation (4-124) is

$$v = \frac{1}{2r_0^2} - \frac{1}{6r_0^2}\cos 2\varphi \tag{4-129}$$

$$= \frac{2}{3r_0^2} - \frac{1}{3r_0^2} \cos^2 \varphi.$$

Using this and the zeroth-order solution in Equation (4-123), we have the full first-order solution to the trajectory Equation (4-116):

$$u = \frac{1}{r_0} \cos\varphi - \frac{\varepsilon}{3r_0^2} \cos^2 \varphi + \frac{2\varepsilon}{3r_0^2}. \qquad (4\text{-}130)$$

Now as we have seen above, the trajectory of a light ray as given by Equation (4-130) is essentially a straight line with a perturbation of order ε. We find that the effect of this perturbation will alter the trajectory to produce a small overall deflection; that is, light approaches the sun along an asymptotic straight line, is deflected by the gravitational field, and recedes again on another asymptotic straight line. The total deflection of the light can be measured observationally for the case of starlight grazing the sun and arriving finally on the earth. Let us therefore see what total deflection is predicted by Equation (4-130) for such a situation.

The asymptotes of the trajectory will clearly correspond to those values of the angle φ for which r becomes infinite or (equivalently) u becomes zero in Equation (4-130). Thus, to find the angles of the asymptotes, we must solve the quadratic equation obtained by setting u = 0 in Equation (4-130)

$$\frac{\varepsilon}{3r_0^2} \cos^2 \varphi - \frac{1}{r_0} \cos\varphi - \frac{2\varepsilon}{3r_0^2} = 0 \qquad (4\text{-}131)$$

or

$$\cos^2 \varphi - \frac{3r_0}{\varepsilon} \cos\varphi - 2 = 0. \qquad (4\text{-}132)$$

Now the solution is

$$\cos\varphi = \frac{3r_0}{2\varepsilon} \pm \left(\frac{9r_0^2}{4\varepsilon^2} + 2\right)^{1/2} \qquad (4\text{-}133)$$

$$= \frac{3r_0}{2\varepsilon}\left[1 \pm \left(1 + \frac{8\varepsilon^2}{9r_0^2}\right)^{1/2}\right].$$

We find that in order that cosφ be less than or equal to 1, we must choose the minus sign. Expansion of Equation (4-133) to first order in ε then gives the result

$$\cos\varphi = -\frac{2\varepsilon}{3r_0} = -\frac{2m}{r_0}. \qquad (4\text{-}134)$$

From this result in Equation (4-134) it is clear that φ is near π/2 or - π/2 (as we might expect), since m/r$_0$ is a small quantity. Therefore we set φ = π/2 + δ, which gives us for one asymptote

$$\sin\delta = \frac{2m}{r_0}. \qquad (4\text{-}135)$$

or approximately

$$\delta = \frac{2m}{r_0}. \qquad (4\text{-}136)$$

Now a similar procedure for the other asymptote (for which φ is taken to be - π/2 - δ) yields the same value for the deflection 2m/r₀. Thus the total deflection of the light ray (the angle between asymptotes) is finally by Equation (4-45)

$$\Delta = \frac{4GM}{c^2 r_0}.$$ (4-137)

For a light ray which just grazes the sun, Equation (4-137) predicts a deflection of 1.75".[8]

I.   Radar Ranging in the Solar System

From results in this chapter it should now be clear that a necessary consequence of the gravitational deflection of electromagnetic waves is a reduction in speed of light in the deflecting field. This is now observable and constitutes a further experimental test of relativistic gravitation.

We find if radar is bounced off a planet when it is near superior conjunction, so that the line of sight passes close to the limb of the Sun, there will be an excess delay introduced in the time for the round trip.[9]

M. G. Bowler calculates this excess delay from tensor gravitational field equations and the gravitational refractive index of space.[10] We shall calculate it initially from the information in the Schwarzschild line element derived from aether General Quasi-Relativity.

Observe the figure 4-2. The distance between the sun and the target planet is r_p. The distance between the sun and the earth is r_e. The constant b is the distance from the sun to, and perpendicular to, a line joining the earth and the target planet. Notice the differentials drawn on that line:

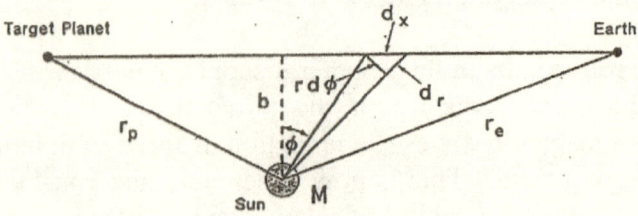

Fig. 4-2. Radar Ranging Planets in the Solar System. Radar from the earth travels near the sun to strike a target planet and return to earth. The radius of earth orbit is r_e. The radius of the planetary orbit is r_p. The perpendicular distance from the sun to the x line is b. The differential $dx^2 = dr^2 + r^2 d\varphi^2$.

The Schwarschild line element Eq. (4-30) shows that in a gravitational field dt^2 is reduced by the factor $(1 - 2GM/rc^2)$, dr^2 is increased by division by $(1 - 2GM/rc^2)$, and the square of the transverse differential lengths, $r^2(d\theta^2 + \sin^2\theta d\varphi^2)$, is unaffected. In this problem we are working in the equatorial plane, so that $\theta = \pi/2$ and $d\theta = 0$. Under such conditions, the square of the transverse differential lengths in the Schwarschild line element reduces to the $r^2d\varphi^2$ of our radar ranging problem.

We first find the velocity squared

$$\frac{dx'^2}{dt'^2} = c^2 = \frac{dr^2}{dt^2\left(1 - \dfrac{2GM}{rc^2}\right)^2} + \frac{r^2d\varphi^2}{dt^2\left(1 - \dfrac{2GM}{rc^2}\right)}. \qquad (4\text{-}139)$$

Next we divide the equation by c^2, multiply the equation by dt^2, take the binomial approximation with terms involving $(1 - 2GM/rc^2)$, and combine terms by taking advantage of the fact that $dx^2 = dr^2 + r^2d\varphi^2$.

$$dt^2 \approx \frac{dx^2}{c^2} + \frac{4GM}{c^4}\frac{dr^2}{r} + \frac{2GM}{c^4}rd\varphi^2. \qquad (4\text{-}140)$$

$$dt \approx \frac{dx}{c}\left[1 + \frac{4GM}{c^2}\frac{dr^2}{rdx^2} + \frac{2GM}{c^2}\frac{rd\varphi^2}{dx^2}\right]^{1/2}. \qquad (4\text{-}141)$$

$$dt \approx \frac{dx}{c} + \frac{2GM}{c^3}\frac{dr^2}{rdx} + \frac{GM}{c^3}\frac{rd\varphi^2}{dx}. \qquad (4\text{-}142)$$

We modify the terms in Eq. (4-142) using four identities:

$$dr^2 = \frac{x^2dx^2}{r^2}; \quad dx^2 = r^2d\theta^2 + dr^2; \qquad (4\text{-}143)$$

$$d\varphi^2 = \frac{dx^2 - dr^2}{r^2}; \quad r^2 = x^2 + b^2.$$

$$dt \approx \frac{dx}{c} + \frac{2GM}{c^3} \frac{x^2 dx}{r^3} + \frac{GM}{c^3} \frac{\left(dx^2 - dr^2\right)}{rdx} \qquad (4\text{-}144)$$

$$\approx \frac{dx}{c} + \frac{2GM}{c^3} \frac{x^2 dx}{r^3} + \frac{GM}{c^3} \frac{dx^2 - \frac{x^2 dx^2}{r^2}}{rdx}$$

$$\approx \frac{dx}{c} + \frac{GM}{c^3} \frac{x^2 dx}{r^3} + \frac{GM}{c^3} \frac{dx}{r}$$

$$\approx \frac{dx}{c} + \frac{GM}{c^3} \frac{dx}{r} \left[1 + \frac{x^2}{r^2} \right]. \qquad (4\text{-}145)$$

Since b is small compared to x and r, $x \approx r$ for most of the light path length. Therefore Eq. (4-145) is approximately

$$dt \approx \frac{dx}{c} + \frac{2GM}{c^3} \frac{dx}{\left(x^2 + b^2\right)^{1/2}}. \qquad (4\text{-}146)$$

Integrating we have

$$t = \frac{x}{c} + \frac{2GM}{c^3} \int \frac{dx}{\left(x^2 + b^2\right)^{1/2}}. \qquad (4\text{-}147)$$

M. G. Bowler arrived at Eq. (4-147) from tensor gravitational equations and the gravitational refractive index in space near the sun. We arrived at Eq. (4-147) independently from information in the Scharzschild line element derived from aether General Quasi-Relativity. But since the two approaches are in agreement, we shall quote Bowler for the rest of the derivation.

The excess delay is the second term. We evaluate it in pieces. The excess delay introduced in going between Earth and the point of closest approach is

$$\frac{2GM}{c^3}\left[\ln\left(x + \sqrt{x^2 + b^2}\right)\right]_0^{\sqrt{r_e^2 - b^2}}$$

and in going between the point of closest approach to the Sun and the planet is

$$\frac{2GM}{c^3}\left[\ln\left(x + \sqrt{x^2 + b^2}\right)\right]_0^{\sqrt{r_p^2 - b^2}}.$$

The one-way journey between Earth and the target planet thus introduces an excess delay

$$\frac{2GM}{c^3}\ln\left(\left[\frac{r_e + \sqrt{r_e^2 - b^2}}{b}\right]\left[\frac{r_p + \sqrt{r_p^2 - b^2}}{b}\right]\right)$$

and the total excess delay in the round trip is

$$t_E = \frac{4GM}{c^3}\ln\left(\left[\frac{r_e + \sqrt{r_e^2 - b^2}}{b}\right]\left[\frac{r_p + \sqrt{r_p^2 - b^2}}{b}\right]\right)$$

with $b \ll r_p, r_e$ we have

$$t_E = \frac{4GM}{c^3}\ln\left[4\frac{r_e r_p}{b^2}\right] \qquad (4\text{-}148)$$

and

$$\frac{\partial a_E}{\partial b} = -\frac{8GM}{bc^3}.$$ (4-149)

If we take Mercury as the target planet with

$$r_e = 1.495 \times 10^{13} \text{ cm},$$ (4-150)

$$r_p = 0.565 \times 10^{13} \text{ cm},$$ (4-151)

$$b = R = 7 \times 10^{10} \text{ cm},$$ (4-152)

and

$$M = 2 \times 10^{33} \text{ gm},$$ (4-153)

we find

$$t_E = 220 \text{ μsec}$$ (4-154)

in a total time of

$$\approx 2\frac{\left(r_e + r_p\right)}{c} \approx 22 \min s,$$ (4-155)

(note that 100 μsec is the time taken for light to travel only 30 km), and

$$\left[\frac{\partial t_E}{\partial b}\right]_R \approx -5.7 \times 10^{-16} \, s\, cm^{-1}.$$ (4-156)

The planets Mercury, Venus and Mars and certain suitable space probes have been used as targets: currently the predictions of general relativity are verified at ≤3% level. We should note that a measurement of the coefficient in Eq. (4-149) removes the need to know planetary orbits with great accuracy (the best way to determine them is by radar ranging) and also sweeps under the rug subtleties relating the real earth based system of coordinates, immersed in the Sun's gravitational field, to the idealized coordinates we have used.[11]

J. Significance of the Four Tests of General Relativity

We have seen how that by the consideration of an actual radial velocity of an aether of amount

$$v_e = \left(\frac{2GM}{r} \right)^{1/2} , \qquad (4\text{-}157)$$

the equations of Special Quasi-Relativity give rise to gravitational effects in dt and dr (Eqns. (4-6) and (4-20), respectively). Equation (4-6) is sufficient to predict the gravitational red shift, and equations (4-6) and (4-20) are sufficient to modify the Lorentz line element (4-26) to the Schwarzschild line element (4-30) for a gravitational field. Equation (4-30), in turn, by means of calculus of variations methods, gives rise to the General Relativistic perihelic shift relation for Mercury (4-98) and the General Relativistic bending of starlight in a gravitational field (4-137) by means of standard General Relativistic calculations. The information contained in the Schwarzschild line element also can calculate the excess delay in radar ranging of the planets. As a net result, we see that the aether model obtains the same values, and accounts for every result of Special and General Relativity!

Considering the previous literature on relativity and an aether, this is a remarkable result. The null result of the Michelson Morley experiment has usually been taken before as proving there is no aether or aether drift. But there in fact can be an aether compatible with the null result of the Michelson-Morley experiment, giving rise to every result of Special and General Relativity.

The fact that the aether model of the universe gives rise to the same experimentally measurable results in General Relativity, means that experimentally speaking, the aether model is on as good footing as Special and General Relativity. Special and General Relativity have the advantage only in seniority—custom of scientific thought. The aether model has the advantage of bringing in a more intuitively obvious physical approach to the relativistic problems. The aether model and its consequences are much more easily mentally visualized, and the initial phases of it are much more easily calculated—nowhere requiring tensor equations. The real advantage of aether quasi-relativity, however, is that it enables a tremendous breakthrough in the derivation of the structure of elementary particles and in a unified field theory. To such applications we will now turn in subsequent chapters.

---

[1]Ronald Adler, Maurice Bazin, and Menahem Schiffer, *Introduction to General Relativity* (New York: McGraw-Hill Book Company, 1965), pp. 125-127.

[2]*Ibid.*, pp. 179-187, 189-194.

[3]Charles Fox,*An Introduction to the Calculus of Variations* (London: Oxford University Press, 1963), pp. 3-8.

[4]*Ibid.*

[5]*Ibid.*, pp. 62, 63.

[6]Adler, Bazin, and Schiffer, *op. cit.*, pp. 179-187.

[7]*Ibid.*

[8]*Ibid.*, pp. 189-193.

[9]M. G. Bowler, *Gravitation and Relativity* (Oxford: Pergamon Press, 1976), p. 59.

[10]*Ibid.*, pp. 57-60.

[11]*Ibid.*, pp. 59, 60.

Problem Set 4

1. How do we know that aether particles (near-zero mass bosons) at rest at infinity will reach the velocity squared V^2 = 2GM/r near a gravitational mass M?

2. According to the gravitational red shift theory presented on pp. 102-3, would photons be stopped in leaving a black hole or only red shifted to zero frequency?

3. Calculate the red shift of light shown up a 10,000 foot mine shaft.

4. Calculate the blue shift of light shown down a 10,000 foot mine shaft.

5. If length is contracted and time dilated with relative motion to the aether, why is dt^2 multiplied by $(1 - 2GM/rc^2)$ and dr^2 is divided by $(1 - 2GM/rc^2)$ in the Schwarzschild line element?

6. Why is $r^2(d\theta^2 + \sin^2\theta d\varphi^2)$ in the Schwarzschild line element not multiplied or divided by $(1 - 2GM/rc^2)$?

7. Calculate the perihelic shift of earth. [$R_{earth\ orbit}$ = 1.496 x 10^8 km; M_{sun} = 1.991 x 10^{30} kg.]

8. Calculate the perihelic shift of a satellite in a near circular orbit one million kilometers from the center of the sun.

9. Calculate the total deflection of a light ray grazing the surface of the earth. [R_e = 6371 km; M_e = 5.975 x 10^{24} kg.]

10. What would happen to a light ray grazing a massive object at the Swarzschild radius for that object [R_s = $2GM/c^2$]?

11. Does light go faster or slower in a gravitational field?

12. Is an aether compatible with general relativity? What velocity of the aether is key in converting Special Quasi-Relativity into General Quasi-Relativity?

Chapter 5

GRAVITY AND INERTIA

A. Powerful Insights

The reader may have followed along thus far in the text, seeing how an aether model can account for special and general relativistic effects, without seeing the staggering and simple consequences of this model. If special relativistic effects are observational phenomena only, without an aether, as in Einstein's theory, then there is nothing that can be done, for example, about unwanted relative mass increase. Nothing can be done about purely observational phenomena, if there is no physical mechanism that links the phenomena. Relative mass increase means kinetic energy. And kinetic energy means work done upon the object in accelerating it. And work means force times distance. There will be an inertial force accompanying the relative mass increase. If special relativity is just an observational phenomena of rotation in complex space-time, then there is nothing that can be done about inertial forces except to live with them. That means there is a limit to the maneuverability of jet aircraft, for there is a limit to human endurance to G-forces.

But we have seen there may indeed be a physical explanation of relativistic effects—an aether. The relative mass increase may not come about merely by the relative velocity to some other object in space or even to some distant fixed stars, as in Mach's Principle, but by the relative velocity to the aether—to the intimately local aether at that—not some distant aether. In our aether model, relative mass increase comes about due to the change in the velocity of the aether through the particles. In order to affect the mass of the particle, this aether speed change has to be in the dimensions of the particle.

152

There is one powerful and simple thought question. What if some means were devised whereby the aether particles were accelerated around a man-made craft, instead of permitted to zing through the craft, as has always before occurred with man-made vehicles. Let the aether particles on the outside be guided around the craft much as air flows around a car or airplane, and let the aether particles within the craft ride with the vehicle at the velocity of the craft, as air does in a car or airplane. What would happen to the relative mass increase then? Well, according to the aether model, relative mass increase only occurs when the aether travels *through* a massive particle, redefining the speed of light *within* the particle. The particle has no way of sensing the speed of the aether some distance away from itself. Therefore, if the aether particles in the craft ride with the vehicle, and the aether particles outside the craft are deflected around the vehicle, then the craft experiences no relative mass increase—regardless of the speed of travel. If there is no relative mass increase, then there is no increase in kinetic energy. If there is no increase in kinetic energy, then there is no work done in accelerating the craft. And if there is no work done in accelerating the craft, then there is no inertial G-force experienced by the occupants upon acceleration and deceleration—regardless of the rate of acceleration. And if there is no relative mass increase, then there is no speed of light barrier. As long as the aether particles could be accelerated out of the way fast enough to provide passage for the craft, the craft could go at any arbitrarily large speed. (As it turns out, streamline acceleration of the exterior aether particles may be impossible. But reflection of the aether from the surface of the craft to achieve inertialess travel may be possible.) Travel to distant planets in the solar system may take only minutes instead of months and years as at present. And trips to distant solar systems and galaxies would then become humanly feasible. By the way, the clocks would not be slowed in inertia-less craft. The vehicles could operate on standard coordinate time. And there would be no large kinetic energy investment in space travel. The aether model of the universe would liberate man from the confines of this solar system.

While inertia may be a local aether phenomenon, gravitational acceleration may also be a local aether phenomenon. Einstein's equivalence principle is relatively, though not absolutely, true. Except for clock differences, there are no apparent differences between the force of gravity and the inertial force of acceleration. That is because both may be due to changes of aether velocities either in space or in time. We recall that we depended upon the aether being accelerated by gravitational bodies in order to account for general relativistic effects. This means that there would not only be a length contraction in the radial direction near a gravitational body, and a slowing of atomic clocks in a gravitational field, there must also be a relative mass increase for mass in a gravitational field. And we have seen from special relativity that with relative mass increase there is an increase in kinetic energy and an inertial force. Thus we expect an inertia-like force in a gravitational field, due to the acceleration of the aether.

B. Centrifugal Force

Consider the so-called fictitious centrifugal force of inertia in a rotating reference frame. Newton noticed that absolute rotational motion can be detected by a water in a bucket type experiment. General Relativity and even Mach's Principle shed little light on the mechanics of such a phenomenon. Those scientific theories only offer a mathematical description of the phenomena—comparable to kinematics, as opposed to dynamics. But the aether model gives a simple physical, dynamical explanation of the inertial force in a rotating reference frame. In the aether model, there is aether distributed in the space in which the moving frame rotates. The aether is not disturbed by the spin of objects in the frame. The objects plow through the aether (or rather the aether zings through the objects) without the objects deflecting the aether. But the relative aether speed through objects in the rotating frame alters the speed of light in the particles of each object, thus causing relative mass increase depending upon distance from the center of rotation and the angular velocity. At the center of rotation, velocity relative to the aether is zero (in the plane of rotation), and the velocity relative to the aether (in the plane of rotation) increases linearly with distance from the center of rotation. This means there is a gradient of the square of the aether velocity relative to objects in the rotating frame. This we believe is the cause of the inertial centrifugal force felt by objects in a rotating frame of reference.

In centripetal acceleration,

$$a = v^2 / r, where\, v = \omega r\, and\, v << c. \tag{5-1}$$

Thus

$$a = \omega^2 r. \tag{5-2}$$

Now,

$$v^2 = \omega^2 r^2, \tag{5-3}$$

so that

$$\nabla\left(v^2\right) = 2\omega^2 r. \tag{5-4}$$

Thus

$$1/2\nabla\left(v^2\right) = \omega^2 r = a, and\, F = ma. \tag{5-5}$$

Del(v^2) is directed in the direction of increasing r, as is also the felt centrifugal force. We see that

$$F = m \; \tfrac{1}{2} \nabla v^2 \qquad\qquad (5\text{-}6)$$

gives the correct value for centrifugal force in terms of the velocity and velocity gradient of the aether.

C. Gravitational Acceleration

If an object feels an inertial force because of acceleration relative to the aether, or a gradient of the square of the aether speed (as in centripetal acceleration), then an object in a gravitationally accelerated aether fluid must also feel an inertial force due to the gradient of the square of the aether speed caused by the gravitational acceleration of the aether. We cannot expect an object to differentiate identical aether velocity gradients which are caused by different phenomena.

A body held at rest (in the coordinate frame) in a gravitational field experiences a gravitational force. The aether particles are accelerated in time in their travel in the vertical direction through the object. Yet the object senses no time derivative aether acceleration at any point in the object. There would be, however, an aether velocity gradient through the object. The gradient is a spatial derivative of the field evaluated at a point. Thus the gradient is a value that is a limit as the spatial difference goes to zero. Thus, to test the aether v^2 gradient in a gravitational field, we can take the approximation of a constant acceleration, such as is common in calculations for gravitation acceleration near the surface of the earth. In such calculations, the force is taken to be

$$F = -mg, \quad (5\text{-}7)$$

Where g is approximately a constant.

In the simple gravitational force problem, let us allow the aether velocity at some point P in space and T in time be v (as the aether is accelerated near a gravitational body). Let the acceleration of the aether be -g at that point. At some brief time Δt later, the aether will be at a distance $v\Delta t$ below P. But since the particle is undergoing acceleration, the new velocity will be $v + g\Delta t$ at the new time $T + \Delta t$ and the new position $v\Delta t$ below P. Let us now calculate the value of $\nabla(v^2)$. In order to keep dr positive, we subtract the later velocity squared from the earlier velocity squared.

$$\nabla\left(v^2\right) = \frac{\lim\limits_{\Delta t \to 0} v^2 - \left(v + g\Delta t\right)^2}{v\Delta t} \quad (5\text{-}8)$$

$$= \frac{\lim\limits_{\Delta t \to 0} -2vg\Delta t - g^2\left(\Delta t\right)^2}{v\Delta t} = -2g1_r.$$

Let us use the same aether force equation we derived in the last section for centrifugal force, namely:

$$F = m\left(1/2\nabla\left(v^2\right)\right). \quad (5\text{-}9)$$

For an object in a gravitational field the (inertial) force felt is

$$F = m\left(1/2\nabla\left(v^2\right)\right) = m\left(1/2\left(-2g1_r\right)\right) = -mg1_r, \quad (5\text{-}10)$$

which is the value we expected.

We can look at the gravitational aether velocity gradient another way. In Chapter 4 we learned that each point in a gravitational field has an escape velocity associated with it, namely

$$v_e = \left(\frac{2GM}{r}\right)^{1/2}.$$ (5-11)

It is the change from point to point of this escape velocity (the inward velocity of the aether particles) that in our ether model causes a felt gravitational (inertial) force by an observer in that gravitational field. Let us now take the gradient of the square of that escape velocity as above:

$$\nabla\left(v_e^2\right) = \nabla\left(\frac{2GM}{r}\right) = 2GM\left(-1r^{-2}\right)1_r$$ (5-12)

$$= -\frac{2GM}{r^2}1_r.$$

Let us now apply the aether inertial force formula we first derived for centrifugal force:

$$F = m\left(1/2\nabla\left(v^2\right)\right).$$ (5-13)

Applying Eqn. (5-13) to our current problem we obtain

$$F = m\left(1/2\nabla\left(v_e^2\right)\right) = 1/2\,m\left(-\frac{2GM}{r^2}\right)1_r$$ (5-14)

$$= -\frac{GMm}{r^2}1_r,$$

where the final expression in the above equation is the familiar Classical formula for the gravitational force valid in all non-relativistic cases (weak gravitational fields or velocities $\ll c$, which is the same.)

Notice that our equation of the inertial force gives the expected value of the gravitational force. We must conclude that the gravitational force itself is an inertial force mediated only by the aether, and has no direct action-at-a-distance force complementing it. A gravitational body may accelerate the aether, and the aether then accelerates the objects in a gravitational field. The mechanics of this aether particle acceleration may be simple. Each massive body is accelerated by an aether velocity gradient (or rather gradient of v^2) of the aether particles it feels across its particles. But aether particles such as gravitons are in this model themselves non-zero massive particles. They may simply obey the inertial law that every other massive particles do, and be accelerated by the aether velocity gradient of the aether particles that go through them. The geometry is such that such aether accelerations toward a massive particle or group of particles are perpetually self sustaining as long as, and in the proportion that, the aether particles closest to the massive source are accelerated.

D. Ordinary Acceleration

In the simple case of linear acceleration of an object relative to a zero gradient aether velocity field, we would have

$$F = m\frac{dv}{dt}.$$ (5-15)

This is just as we would expect from Classical physics. In Classical physics the applied force equals $F = ma$, and the inertial force is $F = -ma$, when the acceleration is calculated relative to the rest frame. But in our equation, we are calculating the aether velocity relative to the observer, so the inertial relation bears the positive sign.

E. Formal Aether Model Expression for Gravitational and Inertial Force

Equation (5-15) is different than Eqn. (5-13), yet the two equations are not contradictory. One contains a spatial derivative of aether velocity, and the other contains a time derivative of aether velocity. We may now combine these two equations to obtain a unification of gravity and inertia in nonrelativistic conditions.

$$F = m\left[\frac{dv}{dt} + 1/2\nabla\left(v^2\right)\right]. \qquad (5\text{-}16)$$

This equation holds for subrelativistic aether velocities (v ≪ c) and weak gravitational fields (which are the same thing). The v in this equation is not the velocity of an object relative to the rest frame of the aether. It is just the reverse—the aether velocity relative to the observer. Also, accordingly, the force **F** is not the force applied to accelerate the object, but the reverse, the inertial force felt by the object.   The equation gives the inertial or gravitational force felt by an observer in a changing velocity field of the aether particles.

F. Vertical Free Fall

Before we move on to more important things, let us give a couple of more simple tests to Eqn. (5-16). Let us consider first the case of a vertical free fall of an object in a gravitational field. In this case neither the term (dv/dt) nor ($\frac{1}{2}\nabla(v^2)$) individually vanish. But they add to zero.

First, let us calculate the term (dv/dt). The whole object travels at the velocity of the center point P. At T = 0 the point P is at elevation r = R. We set the object at rest at time T = 0 in the coordinate frame. Thus its velocity at time zero in the coordinate frame is v = 0. The aether velocity in the coordinate frame at point P at time T = 0 is

$$V = -\left(\frac{2GM}{R}\right)^{1/2}.$$ (5-17)

The aether velocity relative to the point P in the object, at time T = 0 is

$$u = V - v = -\left(\frac{2GM}{R}\right)^{1/2}.$$ (5-18)

At time T' = 0 + Δt, the point P in the object has fallen to a point

$$r \approx R - 1/2\, a(\Delta t)^2, \text{where}\, a = -\frac{GM}{r^2},$$ (5-19)

as can be seen by

$$F = -\frac{GMm}{r^2}1_r = ma.$$ (5-20)

The new v' at T' = Δt is

$$v' \approx -\frac{GM}{R^2}\Delta t,$$ (5-21)

whereas the new coordinate aether velocity

$$V' = -\left(\frac{2GM}{R - \dfrac{GM}{2R^2}(\Delta t)^2}\right)^{1/2}.$$

(5-22)

The new aether velocity relative to the point P is

$$u' \approx -\left(\frac{2GM}{R - \dfrac{GM}{2R^2}(\Delta t)^2}\right)^{1/2} + \frac{GM}{R^2}\Delta t.$$

(5-23)

Now let us take the time derivative of the relative aether velocity vector.

$$\frac{du}{dt} = \lim_{\Delta t \to 0} \frac{u' - u}{\Delta t}$$

(5-24)

$$\approx \lim_{\Delta t \to 0} \frac{-\left(\dfrac{2GM}{R - \dfrac{GM}{2R}(\Delta t)^2}\right)^{1/2} + \dfrac{GM}{R^2}\Delta t + \left(\dfrac{2GM}{R}\right)^{1/2}}{\Delta t} 1_r$$

$$\approx \lim_{\Delta t \to 0} \frac{\left[\left(\dfrac{2GM}{R}\right)^{1/2} - \left(\dfrac{2GM}{R}\right)^{1/2}\right]}{\Delta t} 1_r + \frac{GM}{R^2} 1_r$$

$$= \lim_{\Delta t \to 0} \frac{0}{\Delta t} 1_r + \frac{GM}{R^2} 1_r$$

(5-25)

$$= \frac{GM}{R^2} 1_r.$$ (5-26)

Now let us calculate the gradient of v^2. At time zero, the position of point P in the object is at location $r = R$. At that position the aether velocity relative to the coordinate frame is

$$V = -\left(\frac{2GM}{R}\right)^{1/2} 1_r;$$ (5-27)

and we let the object velocity be v relative to the coordinate frame. At point P, at the fixed point in time, the net aether velocity relative to the object at point P is

$$u = -\left(\frac{2GM}{R}\right)^{1/2} - v.$$ (5-28)

$$u^2 = \frac{2GM}{R} + 2v\left(\frac{2GM}{R}\right)^{1/2} + v^2.$$ (5-29)

We now take the gradient of u^2.

$$\nabla\left(u^2\right) = \lim_{\Delta r \to 0} \frac{\left(\frac{2GM}{R} + 2v\left(\frac{2GM}{R}\right)^{1/2} + v^2\right)}{\Delta r} 1_r -$$ (5-30)

$$\lim_{\Delta r \to 0} \frac{\left(\frac{2GM}{R-\Delta r} + 2v\left(\frac{2GM}{R-\Delta r}\right)^{1/2} + v^2\right)}{\Delta r} 1_r$$

$$= \lim_{\Delta r \to 0} \frac{\left(\dfrac{2GM}{R} - \dfrac{2GM}{R - \Delta r} \right)}{\Delta r} 1_r +$$

$$\lim_{\Delta r \to 0} \frac{\left(2v \left[\dfrac{2GM}{R} \right]^{1/2} - 2v \left[\dfrac{2GM}{R - \Delta r} \right]^{1/2} \right)}{\Delta r} 1_r$$

$$= \lim_{\Delta r \to 0} \frac{2GM(R - \Delta r) - 2GMR}{R(R - \Delta r)\Delta r} 1_r +$$

$$\lim_{\Delta r \to 0} 2v \sqrt{\frac{2GM}{R}} \left(1 - \sqrt{\frac{1}{1 - \dfrac{\Delta r}{R}}} \right) 1_r$$

$$= \lim_{\Delta r \to 0} \frac{-2GM\Delta r}{R(R - \Delta r)\Delta r} 1_r + 0 1_r$$

$$= \lim_{\Delta r \to 0} \frac{-2GM}{R(R - \Delta r)} 1_r = \frac{-2GM}{R^2} 1_r. \tag{5-34}$$

We now combine the results we obtained for the time derivative of the relative aether velocity vector and the gradient of the aether velocity squared, by Eqn. (5-16), to obtain the felt inertial force upon a vertically falling object in a gravitational field. The result is

$$F = m \left[\frac{du}{dt} + 1/2 \nabla \left(u^2 \right) \right] \tag{5-35}$$

$$= m\left[\frac{GM}{R^2}1_r + 1/2\left(\frac{-2GM}{R^2}1_r\right)\right]$$

$$= 0. \tag{5-36}$$

The inertial force vanishes in the case of a free falling object falling in the vertical direction in a gravitational field.

G. Orbital Fall

Now let us try the case of the free falling object in a circular orbit about a gravitational body. In this case (if we consider the object making one rotation every orbit), the time derivative of the relative ether velocity vector vanishes. The gradient of v^2 also vanishes, but for an unusual reason. As we learned in the section "Evidence of Aether Drift Across Planets," Chapter 4, Section G., the square of the inward accelerated aether velocity is twice the square of the lateral orbital aether velocity. Thus the true relative aether velocity to an orbiting satellite is the vector sum of two velocity vectors at right angles to each other.

In order for the gradient of v^2 to vanish for an orbiting object, the magnitude of v_{net} must remain constant for a small change in the position (in the radial direction) of a point in an object in a fixed circular orbit. If the point P is at distance $r = R$ from the center of the gravitational body, and point Q is at the distance $r = R + \Delta r$ in the same orbiting object along the same radial, then the vertical aether velocity component is less at Q than at P, but the horizontal velocity component is more at Q than at P because the angular velocity is the same for both P and Q, and Q is farther from the origin than P. In order for the gradient of v^2 to vanish for this case, the following relation must hold true:

$$\left(v_e + dv_e\right)^2 + \left(v_l + dv_l\right)^2 = v_e^2 + v_l^2. \qquad (5\text{-}37)$$

Thus the following equation must also hold true:

$$2v_e dv_e + \left(dv_e\right)^2 + 2v_l dv_l + \left(dv_l\right)^2 = 0. \qquad (5\text{-}38)$$

We can neglect the squares of the differentials and divide the equation by 2. We see that in order for the gradient of v^2 to vanish for this case, the following relation must hold true:

$$v_e dv_e = -v_l dv_l. \qquad (5\text{-}39)$$

At point P, the inward aether velocity component is

$$v_e = \sqrt{2GM}r^{-1/2}. \qquad (5\text{-}40)$$

$$dv_e = -1/2\sqrt{2GM}r^{-3/2}dr. \qquad (5\text{-}41)$$

$$v_l = \sqrt{GM}r^{-1/2}. \qquad (5\text{-}42)$$

The differential dv_l cannot be calculated directly like dv_e, because the point Q is not orbiting independently of the point P, but is rigidly bound to point P and orbits at the angular velocity at which point P orbits. We find dv_l by the following method:

$$v_l + dv_l = v_l \frac{(r + dr)}{r}, \qquad (5\text{-}43)$$

$$dv_l = v_l \frac{(r + dr)}{r} - v_l = v_l \frac{dr}{r}$$

$$\sqrt{GM} r^{-1/2} \frac{dr}{r} = \sqrt{GM} r^{-3/2} dr. \qquad (5\text{-}44)$$

Now we inquire,

$$Does\, v_e dv_e = -v_l dv_l\,? \qquad (5\text{-}45)$$

$$Does\, \sqrt{2GM} r^{-1/2}\left(-1/2\sqrt{2GM} r^{-3/2} dr\right) \qquad (5\text{-}46)$$

$$= -\sqrt{GM} r^{-1/2}\left(\sqrt{GM} r^{-3/2} dr\right)?\; \text{ Yes.}$$

$$-\frac{GM}{r^2} dr \equiv -\frac{GM}{r^2} dr, \qquad (5\text{-}47)$$

and the point is proved. The length of the aether velocity vector remains unchanged with change in radial position in an orbiting object in a circular orbit around a gravitational body, though the direction of the aether velocity vector changes. Thus grad v^2 vanishes, and since the time derivative of the aether velocity vector also vanishes, the inertial force vanishes in a free falling body in a circular orbit about a gravitational body. The equation of inertia

$$F = m\left[\frac{dv}{dt} + 1/2\,\nabla\left(v^2\right)\right], \qquad (5\text{-}48)$$

where **v** is the aether velocity relative to the observer, appears to hold true for all cases for subrelativistic mechanics.

H. Relativistic Aether Model Force Equation for Gravity and Inertia

It is now desirous that we should expand and refine our synthesis of gravitational and inertial force to apply to all velocities and velocity gradients of the aether particles relative to the observer. In equation (5-48) we recognize the first term as the time derivative of the nonrelativistic momentum. In relativity the mass as well as the velocity can vary in the momentum. So we generalize this term as the time derivative of the three vector relativistic momentum. The second term in equation (5-48) is the three vector spatial gradient of the kinetic energy (for nonrelativistic velocities, but we can generalize to the relativistic kinetic energy). But the kinetic energy is the t component of the four-vector momentum P_4. We can therefore translate equation (5-48) into the following valid at all velocities:

$$F = \left[\frac{\partial}{\partial t} P_3 + \nabla P_t \right]. \qquad 5\text{-}49)$$

I. Summary and Forward Look

In the next chapters we will be able to continue our study of gravity and inertia and the other forces. We will study their effects in the smallest structures of matter--electrinos and echons. In the author's model, electrinos are the smallest particles in the Universe, and echons are the lowest level orbital structures made out of electrinos. We will find how gravity and inertia are united to the other forces.

We derived also in this chapter two important results which will be useful in future work.

Important Derived Results

$$F = m\left[\frac{dv}{dt} + 1/2\nabla\left(v^2\right)\right], \quad where\, v << c. \qquad (5\text{-}50)$$

$$F = \left[\frac{\partial}{\partial t}P_3 + \nabla P_t\right]. \qquad (5\text{-}51)$$

Problem Set 5

1. In the aether model, what is responsible for relative mass increase?

2. What would happen if aether exterior to a craft were kept exterior to the craft and aether interior to the craft were kept interior to the craft as the craft accelerated?

3. Give a reason why centrifugal force should not be considered a "fictitious" force in the aether model.

4. Give an evidence that gravity is an inertial force mediated only by the aether, and has no direct action-at-a-distance force complementing it.

5. What could accelerate aether particles if a gradient of aether velocity squared accelerates other massive particles?

6. What formulae in this chapter especially unite gravity and inertia?

7. In terms of Eq. (5-16), why is the felt inertial force on a vertical free-falling object 0?

8.   In terms of Eq. (5-16), why is the felt inertial force on a circularly orbiting object 0?

Chapter 6

ELECTRINOS AND ECHONS

I. Getting Started

A. Introduction

For centuries man has sought the smallest particles of matter. First atoms were thought to be the smallest subdivisions of matter. Then electrons, protons, and neutrons were discovered—subatomic particles. Scores of additional subatomic "elementary" particles were discovered. Then quarks were theorized to construct many of these pieces of matter.[1] Some now ask, "are there pieces of quarks—or preons?"[2] "Are there an infinite number of subdivisions of matter?" "Or is there some smallest limit which we will eventually reach in our theoretical and experimental physics?"

The successful scientific approach for centuries has been to move down a step at a time from the macroscopic to the microscopic. We are now dealing with very small pieces of matter. It is difficult with experiment to probe very much deeper into the structure of matter. Therefore the author here temporarily takes a different approach. The author attempts to start at the very bottom and to work up, to derive a particle system from first principles, and then to see if the products of those derivations match the elementary particles observed in the real world.

After years of trying to derive such a model, the author is convinced that deriving a particle system from first principles is impossible with Einstein's Special and General Theories of Relativity. That problem has been said to be "as difficult as uniting fire and ice."[3] The feat is possible, however, with Quasi-Relativity in an Aether. One must be able to assign a radial velocity to space about an elementary particle to unlock the riddle of elementary particles and the unified field theory. The particle structure derivation here, then, will follow the preliminary work in quasi-relativity discussed in Chapters 3 and 4.

Not everything one would like to know about electrinos and echons will be derived in this chapter. But enough will be derived to show their structure and a number of interesting mathematical relationships between their parts.

B. Postulates

In addition to the three postulates of Special and General Quasi-Relativity in an Aether, several postulates are necessary to derive the structure of elementary particles and unite the forces. The postulates will be listed here and discussed where appropriate in the text and derivation.

Particle and Field Postulates

4. Parsimony Principle: The Universe is constructed according to the simplest design possible to account for the many varied natural phenomena.

5. A spherically or cylindrically symmetric smooth charge distribution cannot have detectable spin.

6. In every particle other than nothing, electrinos have a velocity component along one direction equal to or greater than c. Electrinos may also have velocity components or zero in perpendicular directions.

7. Total momentum **P**, observable angular momentum $J\hbar$, and total ordinary energy **W** are conserved in every natural non-accelerating frame and reaction.

8. The observable angular momentum (spin) of the parameterized particle system is

$$s_p = \left[r_p \, x \left(m_{f_{s_p}} c_t \right) \right] \qquad (6\text{-}1a)$$

$$= \sum_i \left[f_p \hbar / 2 \right]_i 1_J ,$$

in the aether rest frame, where the variables are defined and illustrated in the text.

Definitions of Postulates' Variables and Terms

In postulate 6, c is the ordinary speed of light.

In postulate 7, the word "natural" is employed because the postulate holds true for natural particles—those made of unitons, semions, and quartons and their anti-particles in matter, light, and gravitons.   Octons, however, are supernatural particles, which, while they may conserve order energy (positive and negative energy in the creation of particles), they do not conserve entropy energy (ordinary energy **W**—the absolute value, term for term, of order energy in the equations).

In postulate 8, s_p is the observable spin of the overall particle or particles studied. $s_p = J\hbar$. Σ_i means sum over all the sub-particles being considered. r_p means the radius vector of the overall particle contemplated. x means the vector cross product. m_{fp} means the mass of an electrino i making up the overall particle p considered.   For an electron it's the mass of a semion.   For a photon it's the mass of a uniton.   In the relativistic frame, it is the strong mass of the electrino (to be derived).   In the nonrelativistic frame, it is a portion of the overall particle considered (such as half of the electron). c_t means the tangential (perpendicular to the radius) or orbital vector velocity at the speed of light. f_p equals the fraction of a whole particle for electrino i in overall particle p, or the fusion state of the electrino. f_p are fractions of charge, and come in +1, -½, +¼, -⅛, 0, +⅛, -¼, +½, and -1 only. The first four are matter, and the last four are antimatter. $\hbar$ equals Planck's constant h divided by 2π. $\mathbf{1}_J$ is a unit vector in the direction of the angular momentum $J\hbar$.

Equation (6-1a) holds true in the hyperoptic speed frame as well as the suboptic speed frame. In the hyperoptic speed frame there may be imaginary radii, minus imaginary mass, and positive or negative v_t. The cross product is real and obeys the right hand side of the equation. Particle systems are mass singularities. Because the observer cannot see across to the back side of a mass singularity, but can observe only the front side at the event horizon, $\mathbf{s}_p$ is the product of r_p, m_{fp}, and c_t.

Equation (6-1a) holds true in the aether rest frame. That is as if an observer rode with the aether particles from at rest at infinity to the surface of the particle and made the observation as he/she passed the surface of the particle on the way in, or rode with the aether particles from at rest at infinity through the particle, and observed at the surface of the particle on the way out. The aether rest frame is not the customary, natural, or convenient frame from which to observe and calculate. It is more natural to select the echon relative rest frame. That frame is at rest with the center of mass of the particle. In that frame the aether travels with velocity $\mathbf{v}$ radially. Calculating the angular momentum or spin in this frame requires the addition of terms as shown in Equation (6-1b).

$$s_p = \left[r_p \, x \left(m_{f_p} V_p \right) \right] \qquad\qquad (6\text{-}1b)$$

$$= \sum_i \left[f_p \frac{1}{2} \hbar k_p \right]_i 1_J$$

V_p is different than c_t in Equation (6-1a). For the electron, it is the vectorial sum of c_t and v_p, which are the orbital velocity of the electrino and the radial velocity of the aether at the surface of particle p, respectively. k_p is added. $k_p = (1 + v_p^2/c^2)^{1/2}$ and is on the order of $1 + 10^{-45}$ for electrons and differs negligibly from 1 for conceivable v_p. v_p is v for particle p, the magnitude of v_p. Equation (6-1a) is a good approximation for equation (6-1b).

The Parsimony Principle is a long-standing physical view of the Universe. It has been employed in many theories. This principle is the foundation for this derivation of particle and field structure. Some of the conservation laws in Postulate 7 have long been known and used in physics. The rest of the postulates are unfamiliar ones introduced by the author.

C. Charge Distribution

Many physicists wonder about the charge distribution of spinning particles, but few ask themselves the necessary sequence of basic questions to resolve the matter, such as: What should be the shape of the charge distribution of a spinning particle? What if the charge distribution were spherically symmetric? What physical feature on such a smooth spinning charge distribution should signal that there is a motion of the charge relative to space and thus there should be a magnetic field? What besides a magnetic field could prove the charge is in motion? The answer to the last two questions is "nothing". Therefore, a spherical charge distribution in the microscopic world of elementary particles shouldn't have detectable spin or a magnetic field (Postulate 5). But elementary particles like electrons have magnetic moments. Therefore they must not have a spherically symmetric spinning charge distribution.

Then what could be the charge distribution of spinning elementary particles like electrons? The charge would have to come in discrete "lumps." There could not be just one lump of charge orbiting about nothing, for the particle must have a stable center of mass. Therefore there must be two or more discrete "lumps" of charge orbiting about a common center in a simple spinning elementary particle— two, three, four, a dozen, or a million "lumps" in the particle. But how many "lumps" should there be? Let us now employ the Parsimony Postulate. What is the simplest configuration? Two "lumps" or sub-particles of charge orbiting about a common center.

What should be the shape of the "lumps"? Again we appeal to the Parsimony Postulate. The Parsimony Postulate requires that we select the very simplest shape possible for the "lumps." A bumpy lump would be very complex mathematically to describe. A cube or tetrahedron would be simpler. What would be the very simplest shape to describe mathematically? It would be an extremely thin spherical shell. We conclude therefore that these sub-particles should be spherically symmetric and have no spin other than that occurring in their orbit about each other as part of the spinning "elementary" particle. The author believes two of these sub-particles orbit about each other for positive and negative electrons, muons, and tauons. Since the existence of muons and tauons is an added complexity to the scheme, the Parsimony Postulate would also require that at least one deeper complexity in the system exists in the structure of those particles.

What would be the simplest charge distribution for a zero-spin particle? A very small thin spherical shell symmetric charge not orbiting. Such a nearly point charge, however, would be very strong massive by itself (see Chapter 7), and may not long remain isolated without other charges orbiting about it. The author believes such a spherically symmetric nearly point charge is a constituent and the core particle for complex particles such as protons and neutrons.

Mesons have zero spin and much less mass than this. How could that be? Let us employ the Parsimony Postulate again.   What is the next simplest possible charge distribution for a zero-spin particle? Two fourth charges can orbit one way, two more fourth charges can orbit the same way in a plane, but in a separate orbit. Then the two orbits of fourth charges can orbit the opposite way about each other. Each sub-particle would have one fourth the charge of the elementary particle. We could call these four sub-particles "quartons." The author believes pi-, K-, and D-mesons are simple quarton systems.

D. Nomenclature

We could call the two sub-particles in electrons, muons, and tauons "semions." The whole nearly point charges we could call "unitons." We could call quartons, semions, and unitons all "electrinos." The author believes every particle discovered and undiscovered can be composed of octons (halves of quartons), quartons, semions, and unitons. This is a model different than quarks (which divide charges into one thirds and two thirds).

An electron would be composed of two semions—two half charges revolving around each other.   But, at all energies attained and attainable in the foreseeable future, the electron will always appear as a whole charge. The semions are trapped in very deep binding energy wells, with currently no known way of blasting them apart. We have no way of making free electrinos, and have never observed free electrinos.

A pi-meson, or pion, would be composed of four quartons—as described above—two pairs of quartons orbiting oppositely to the inner orbits. Again, a pion is a whole particle. Currently detectable elementary particles are either, or they are composed of, whole particles. Quartons always come in fours as whole particles.

This whole particle concept is important. Therefore let us give it names. Let us call the simplest whole particle made up of parts an "echon" and a whole particle made up of only one part (a uniton) a "yachon". The etymology behind the coining of these words is as follows: In Hebrew "yachid" means one, only, sole, single, but one. The Hebrew word "'echad" means one, unity, a whole made up of parts. Electrons and pions, etc., are wholes made up of parts. And they are particles. Particles are traditionally named with the suffix "on". Therefore the author calls whole elementary particles (four quartons or two semions) "echons" and unitons "yachons". The study of balancing equations of echons and yachons in decay schemes or other particle reactions can be called chonomics.

E. A Glimpse of Possible Higher Particle Systems

A neutron may be a uniton in imaginary space and an electron in imaginary space orbiting about each other in real space, held together by the electric force. A Λ particle and a Ξ^0 particle may just be a neutron with the semions in higher energy states, making more massive particles. A proton may be a neutron, as described above, with a pion orbiting about it, held to the neutron by the strong nuclear force. A Σ^+ particle may be just a proton with the electron semions in higher energy states, making the particle more massive. A neutrino may be just an electron in imaginary space and a pion in imaginary space orbiting about each other in real space, held together by the electric force. Electron neutrinos are in ground state. Muon neutrinos and tauon neutrinos are just neutrinos with higher energy states, making the particles more strong massive. And in all the above cases, the antiparticles of all the above particles are just the opposite charges in the same charge orbital structures.

II. Deriving Electrinos

A. Starting at the Bottom

As explained in the introduction to the last section, the author derives his model by starting at the very bottom and working up. We must begin, then, with the very smallest particles, which we calculate to be on the order of i 10^{-35} m in radius. Our model is a boson (an aether) model. The very smallest particles would accelerate the aether to very high velocities. In fact, there would be nothing stopping the tiniest particles from accelerating the aether to the speed of light c. The aether velocity through the tiniest particles would be no more than c, but would be c exactly. We will therefore use the aether speed c in our calculations and let that parameter dictate the sizes and masses of the tiniest particles.

B. The Fine Structure Constant

If any particles are fine structures, certainly the tiniest particles in the Universe are fine structures. It is known that the charge varies with distance because of quantum effects.[4] In fine structures the electric force

$$F = \frac{qq'1_r}{4\pi\varepsilon_0\alpha r^2} \qquad (6\text{-}2)$$

is[5] 137.035 999 679(94) or 1/α times stronger than we would expect from the Coulomb force equation

$$F = \frac{qq'1_r}{4\pi\varepsilon_0 r^2} \qquad (6\text{-}3)$$

But that is not all. The strong electric force, employing the fine structure constant, acts in imaginary radii in echons. The imaginary radii squared in the denominator of the force equation reverses the force. We would expect that the electric force in such a system could be different in the imaginary short-range frame than in the real long-range frame. This difference is calculated at the close of this chapter by equating the strong electric force with the strong gravitational force. In the real long-range frame, the Coulomb electric force may be that in Eq. (6-3). But inside echons, and with electrinos, the electric force may be

$$F = \frac{qq'1_r}{4\pi\varepsilon_0\alpha(ir)^2}, \qquad (6\text{-}4)$$

where α is the fine structure constant, and $1/\alpha = 137.035\ 999\ 679(94)$. All echons and electrinos are fine structures.

Scientists have long known that

$$\alpha = \frac{e^2}{4\pi\varepsilon_0\hbar c}, \qquad (6\text{-}5)$$

but what is the numerical value of α has been the subject of continued theorizing and experimentation. In this chapter, however, the fine structure constant can now be derived from first principles in this model of a unified Universe. That derivation shall be done at the close of this chapter. We will first use the fine structure constant in particle-force equations to derive the smallest charge structures.

C. Aether Velocities through Particles

From Postulate 6, each electrino's velocity relative to the aether particles is $\geq$c. The simplest non-zero mass system that we can have that obeys this postulate is an electrical gravitational thin spherical shell body that accelerates the aether particles (bosons) to speed c inward at the surface of the particle and speed c outward at the same surface for aether particles that have fallen all the way through the gravitational body and are on their way out. In each case the particle surface travels at speed c relative to the aether particles, though in opposite directions. The aether particles travel at ± c relative to the gravitational body in this simplest case.

So much of the calculations of the structure and composition of particles have their foundation in characteristic equations of the aether velocity relative to the particles. For electrinos at rest other than the gravitational aether velocity (a frame the author calls the "relative rest frame" for the electrinos), the characteristic equation is

$$v^2 = -c^2. \tag{6-6}$$

For electrinos in the inner relativistic frame, v^2 is negative (where mass is minus imaginary, and radius is imaginary). In the outer non-relativistic frame and macroscopic frame, v^2 is positive (where mass and radius are positive real).

We wish now to calculate masses in the relative rest frame of the electrinos from our characteristic equations.

D. The Smallest Charge Structures

The electrinos are essentially point charges, but they cannot be quite point charges, or their electric self-energy would be infinite. Electrinos attract each other by strong electric and strong gravitational forces (which are equivalent). (See V. Synthesizing and Generalizing the Model—Fields (last of this Chapter) and Chapter 7.) They attract themselves. Why do they not fuse down to zero radius? Since, if they did, their self-energy would be infinite, they fuse down to a given radius, where their self-energy is finite—of the appropriate value. The electrinos travel at c relative to the aether, in the simplest case, but not faster than c. The inward attraction of the aether due to gravity cannot be greater than that the gravitational aether speed will equal c for an isolated uniton:

$$v^2 = \frac{Gm}{r} = \frac{2Gm_q}{r} = -c^2. \tag{6-7}$$

Now let us calculate the electric self-energy of such a particle. Since the size of the particle is very small, we must use the fine structure constant α in this calculation. Classical electrostatics gives us[6] the capacitance of a spherical surface of radius r:

$$C = 4\pi\varepsilon_0 r. \tag{6-8}$$

For fine structures we multiply that by α:

$$C = 4\pi\varepsilon_0 \alpha r. \tag{6-9}$$

The work required to add a charge q to a capacitor with capacitance C is

$$W = 1/2 \, q^2 / C. \tag{6-10}$$

Therefore the potential energy of a spherical capacitor in a fine structure, i.e. the energy of its electrostatic field, is

$$E_{pot} = \frac{q^2}{8\pi\varepsilon_0 \alpha r_p} = m_q c^2, \tag{6-11}$$

where m_q is the mass associated with the electric field. For the charge mass of the uniton, the energy U is just that of Eq. (6-11). We see that

$$c^2 = \frac{-2Gm_q}{r_p} = \frac{U}{m_q} = \frac{e^2}{8\pi\varepsilon_0 \alpha r_p m_q}. \tag{6-12}$$

Continuing the calculations:

$$m_q^2 = \frac{e^2}{-8\pi\varepsilon_0 \alpha 2G}, \tag{6-13}$$

$$m_q = -i\left(\frac{e^2}{4\pi\varepsilon_0 \alpha 4G}\right)^{1/2} = -i\left(\frac{\hbar c}{4G}\right)^{1/2} = M_q, \tag{6-14}$$

where

$$\alpha = \frac{e^2}{4\pi\varepsilon_0 \hbar c}. \tag{6-15}$$

The charge mass M_q is a very large minus imaginary value in the electrino relative rest frame. We continue calculating masses in this frame.

E.   Kinetic Mass and Total Mass in the Electrino Relative Rest Frame

In addition to the electric self potential mass, the uniton or any electrino has kinetic mass in the electrino relative rest frame. The aether field is not static. Every portion of charge in the uniton is traveling at speed c relative to the aether. Therefore m_q or M_q has kinetic energy also. If m_q had a small velocity relative to the aether, its kinetic energy

E_{kin} would be $\frac{1}{2}m_q v^2$. But since $v^2 = -c^2$, we take the relativistic form $E_{kin} = -m_q c^2$. The total fundamental mass of the uniton is $M_q + (-M_q) = 0$. The total absolute value mass is $|M_q| + |-M_q| = 2M_q = M_0$, which is the imaginary Planck mass, composed simply of the following constants:

$$M_0 = -i \left(\frac{\hbar c}{G} \right)^{1/2}. \qquad (6\text{-}16)$$

Numerically it is

$$M_0 \approx -i\, 2.176\ 44(11) \times 10^{-08}\, kg.\,[5] \quad (6\text{-}17)$$

$$R_0 = \frac{2GM_q}{-c^2} = i \left(\frac{\hbar G}{c^3} \right)^{1/2} \approx i\, 1.616\ 252(81) \times 10^{-35}\, m. \quad (6\text{-}18)$$

The physical size of the uniton is very small—essentially a point charge. But it is imaginary in radius. The mass in the relative rest frame is very large on particle scales. But it is minus imaginary. Essentially the radius of the uniton has been relativistically contracted and the mass relativistically increased. The circumferences of the uniton, however, are not in the direction of aether motion. We might think they are not contracted. But they are. The circumference is 2π times the relativistic imaginary radius. The relativistic particles are relativistic throughout.

Semions and quartons (1/2 and 1/4 charges) are also of interest to us. From equations parallel to equations (6-7) through (6-18) we see that, while the fundamental masses of the particles are all zero, the absolute value masses and radii are:

$$m_{quarton} = 1/4\ m_{uniton} \approx -i\, 5.441\ 1(03) \times 10^{-09}\, kg. \qquad (6\text{-}19)$$

$$r_{quarton} = 1/4 \ r_{uniton} \approx i \ 4.040 \ 63(21) \ x \ 10^{-36} m. \qquad (6\text{-}20)$$

$$m_{semion} = 1/2 \ m_{uniton} \approx -i \ 1.088 \ 22(06) \ x \ 10^{-08} \ kg. \qquad (6\text{-}21)$$

$$r_{semion} = 1/2 \ r_{uniton} \approx i \ 8.081 \ 26(41) \ x \ 10^{-36} m. \qquad (6\text{-}22)$$

F. Electrino Rest Frames

In an electrino frame where the aether velocity is truly zero at the surface of the electrino, the mass of the electrino would be zero, but the radius of the electrino would have to be infinite. This is not a natural "rest frame" for the electrino. A natural "rest frame" for an electrino is where there is no x-, y-, or z-motion for the electrino system in the frame, but there is speed of light aether motion radially, both inward and outward, at the surface of the electrino. For the most part, the aether motion with an electrino is generated within a short distance of the electrino. The aether turbulance just follows the electrino.

The "natural" "rest frame" is just the electrino relative rest frame cited in the previous sections. There are not many natural frames for single electrinos. There will be more natural frames of reference when we study echons.

G. Why Unitons Do Not Implode or Explode

The imploding force of the uniton's spinless electric charge cannot be counteracted by inertia. There must be some other basis for the stability of the electrinos. What is it? The non-relativistic acceleration of the electric shell of the uniton is finite and large. What stops the implosion?

The aether particles traveling through the uniton or electrino in both directions, however, travel at $\pm$ c. There is a natural relativistic barrier at c. Particles going slower than c do not naturally go faster than c, and vice versa.

The shell of charge is at the speed of light barrier in both directions, both in and out. More than the charge in the particles, which may cause elastic Coulomb scattering, the speed of light barriers in the electrinos may be the hard core in the particles, such as electrons, which cause scattering upon collisions. At any rate, the speed of light barriers in electrinos are the firm foundation upon which their electric charge and mass rest.

H. Electrino Spins

Electrinos (unitons, semions, quartons, and octons) are spherically symmetric particles. According to Postulate 5 they therefore cannot have detectable spin. Thus the radii of electrinos are indeterminate from the angular momentum, and must be determined independently by general relativistic calculations.

III. Deriving Electrons

A. Spin, Mass, and Radius

We have the smallest particles of matter, electrinos. We now wish to put them in orbit around each other to study their properties. We start with orbiting semions. We have already concluded that electrons are composed of two orbiting semions. $spin_{observable} = \hbar / 2$. The semions orbit at the speed of light, and have additional motion relative to the aether in the radial direction. Semions are $f = \frac{1}{2}$ particles. We can use Postulate 8 to put semions in orbit in electrons in the echon rest frame. We can use Eq. (6-1a) as a close approximation of Eq. (6-1b) and substitute in the equation the data for the electron. The orbital velocity for the electron is c_t (for other particles see [13, 14]). For the electron and its semions, the postulate 8 is as follows:

$$s_e \approx \left[r_e \times \left(m_{(1/2)(electron)} c_t \right) \right] \qquad (6\text{-}23)$$

$$\approx 2 \left[\frac{1}{2} \hbar / 2 \right] 1_J = \hbar / 2 \, 1_J.$$

The observable spin of the electron is $\hbar/2$ and the mass calculates from the observable spin.

$$m_e = 2 m_{(1/2)(electron)} \approx \frac{\hbar}{2 r_e c}. \qquad (6\text{-}24)$$

Let us do the parallel thing for the photon. The photon is made up of an orbiting uniton and anti-uniton. $f = 1$.

$$s \approx \left[R_0 \times \left(M_0 c \right) \right] \qquad (6\text{-}25)$$

$$\approx (2)(1) \hbar / 2 1_J = \hbar 1_J; \quad M_\gamma \approx \frac{\hbar}{R_0 c}.$$

It is much easier to measure the mass of the electron than the radius of the electron. Thus we can solve for the radius of the electron.

$$r_e \approx \frac{\hbar}{2 m_e c}. \qquad (6\text{-}26)$$

Published[7] values of c, $\hbar$, G, and m_e are

$$c = 299 \ 792 \ 458 \ m \ s^{-1} \ (exact), \qquad (6\text{-}27)$$

$$\hbar = 1.054\ 571\ 628(53)\ x\ 10^{-34}\ J\ s, \qquad (6\text{-}28)$$

$$G = 6.674\ 28(67)\ x\ 10^{-11}\ m^3\ kg^{-1}\ s^{-2}, \qquad (6\text{-}29)$$

$$m_e = 9.109\ 382\ 15(45)\ x\ 10^{-31}\ kg. \qquad (6\text{-}30)$$

From equations (6-26) – (6-30), we obtain

$$r_e \approx 1.931\ x\ 10^{-13}\ m. \qquad (6\text{-}31)$$

This value is $1/(2\alpha)$ or 137.035 999 679(94)/2 times the Classical value[8] which was incorrectly calculated. Our calculation would be accurate only if the electron g-factor were taken into account. The magnitude of the magnetic moment of the electron, m = iS = ecr/2 = -928.476 377(23) x 10^{-26} J T^{-1}, shows that the actual radius of the electron electrino orbits is 1.001 159 652 1811(15) times the radius we would expect from $r_e = \hbar/(m_e c)$. This discrepancy factor is half the electron g-factor[9] of $2(1 + \alpha/2\pi + ...)$, where α is the fine structure[10] constant equal to 1/137.035 999 679(94). This discrepancy factor, however, will not make any difference in the relativistic calculations in the next sections, because we calculate the values only to four place accuracy in this introductory paper.

The non-relativistic radius of a semion orbit may be on the order of 10^{22} times the radius of a uniton, and real. Nature treats the sphere of radius r_e as a shell of charge, rather than reckoning the radius of the semion as the size of the charge distribution. This greatly cuts down on the mass of the semions.

The radius of the electron is not relativistically contracted and the mass is not relativistically increased, because in the electron the calculations are not relativistic. This is true in the radial direction as well as in the tangential orbital direction. The electron circumference is

$$C_e = 2\pi r_e \approx 1.213 x \, 10^{-12} \, m. \qquad (6\text{-}32)$$

B. Characteristic Equations for an Electron Echon at Rest

The characteristic equation for an electron echon at rest is

$$V_e^2 = c^2 + v_e^2, \qquad (6\text{-}33)$$

where V_e is the total velocity of the electron's orbiting electrino, c is the ordinary speed of light, and v_e is the velocity of the aether in and out at the surface of the particle (at the radius of the electrino orbit).

An electron is an echon composed of two semions and is in state

$$f = 1/2 \qquad (6\text{-}34)$$

The total velocity squared for an electron V_e^2 is slightly greater than the speed of light squared.

C. Echon Velocity Vectors

Semions travel orbitally (or tangentially) at the speed of light c relative to the aether particles. It is essential at this point, however, to remember that elementary particles have masses and therefore have gravitational fields, however weak. Actually, the particles are so small that their mass to volume ratios are high and therefore general quasi-relativity is important in the dimensions of the particles. In Chapter 4 we learned that every gravitational body accelerates the non-zero-mass aether particles minus radially to the escape velocity $v_{escape} = v$ at every point in the gravitational field. The escape velocity for $v \ll c$ is

$$v = \left(\frac{2Gm}{r} \right)^{1/2}. \qquad (6\text{-}35)$$

For electrons, v at the radius of the electrino orbits is a very small fraction of the speed of light c, but it is not negligible. The total electrino velocity V is the vectorial sum of c and v in an echon at rest with the observer. V is close to c, slightly greater than c, and is much much greater than v. s is the electron semion. r_e is as follows:

$$r_e \approx \frac{\hbar}{2m_s c} \approx \frac{\hbar}{m_e c}, \text{ where } \hbar = \frac{h}{2\pi}, \qquad (6\text{-}36)$$

for an electron, V in Eq. (6-37) is the total velocity of the semions in an electron. v is the radial aether velocity from gravity. It is directed inward for the aether particles falling in to the echon and outward for the aether particles that have fallen through the echon and are on their way out. The vectorial difference of V and v is c, the speed of light. V is close to c, slightly greater than c, and is much much greater than v.

$$V^2 = c^2 + v^2. \qquad (6\text{-}37)$$

D. Relativistic Transformations in the Electrino-Echon System

Relativistic transformations in the electrino-echon system are complex, since we find out that all the rest frames in a particle system do not all have the same rest masses in them, as in Einstein's theories of relativity. These relativistic peculiarities require extreme care in calculations.

While the tangential velocity of the relative rest frames of electrinos (semions) is c with respect to the electron echon rest frame, the total velocity of the electrinos in the echon frame must be greater than c because the electrino system experiences a radial aether velocity (inward and outward) of $\pm$ v_e, the square of which is v_e^2, and the tangential orbital velocity squared, c^2, plus v_e^2 equals V_e^2. The true total velocity for the electron V_e must be measured with respect to the aether particles. (See Chapters 3 and 4.) Each electron is a small gravitational body which accelerates the aether inward toward the center of the particle. The net velocity V_e of the electrinos in the electron is

$$V_e = c\left(1 + \frac{v_e^2}{c^2}\right)^{1/2} = c\left(1 + \frac{Gm_e}{r_e c^2}\right)^{1/2}. \qquad (6\text{-}38)$$

(The term in the radical is Gm/rc^2 instead of $2Gm/rc^2$ whenever $v \approx c$ or the transformation calculated thereby is to relativistic velocities like c.)

The electron electrino system, governed by the equation $V_e^2 = v_e^2 + c^2$, where V_e is the net velocity of the electrinos, v_e is the radial gravitational velocity of the aether at the electrino orbital radius, and c is the speed of light (the tangential orbital velocity of the electrinos), is relativity transformation possibility rich.

In Chapter 3 we learned the importance of transforming into and out of the aether frame for correct transformations. Ordinarily that is true. In the electrino system, however, that presents us with a problem. The electrino, as seen from the aether frame, travels at the speed of light and has infinite relative mass increase. Nature hates infinities. It attempts to renormalize quantities so that they are finite.

That can be done, for instance, by adding v_e^2, the gravitational aether velocity squared, to the electrino system. Then the electrino velocity squared V_e^2 is more than c^2, and the relative mass increase of the electrinos is finite, though minus imaginary.

Echons are whole particles. They obey Postulate 2 and experience full quasi-relativity. Quartons and semions are not whole particles and do not obey Postulate 2. If these were whole particles, Postulate 2 of Special Quasi-Relativity would react such that these rest masses were conserved in every frame. It is a good thing that it doesn't apply here. If it did, the masses of everything would be $-i$ 10^{22} times heavier than they are, and existence, as we know it, would not be possible.

Fractional particles (quartons and semions) do not react the way whole particles (echons and yachons) do. The reason for this is whole particles travel slower than the speed of light, whereas fractional particles travel faster than the speed of light. The deeper reason is, in less than speed of light transformations, the system can remember the rest mass in the smooth continuous differential steps in Eq. (3-26), but can not remember the rest mass in the discontinuous differential steps crossing the infinities of the speed of light barrier. Therefore the particle assumes as its "rest mass", when moving in the aether, the reduced mass, which is the first stage of the quasi-relativity process (see Chapter 3).

As explained earlier, there are two possible rest frames—the rest frames of the electrino (in this case the semion). One of the rest frames is the absolute electrino rest frame. In it the aether (bosons) are at rest relative to the electrino. In this frame the rest mass of the electrino equals 0. The other rest frame of the electrino is the relative rest frame of the electrino—the frame in which there is no x-, y-, or z- motion of the electrino, but there are gravitational aether motions into and out of the electrino. In this frame, the electrino does have strong mass. The uniton has a strong mass -i 2.176 44 x 10^{-08} kg. in this frame. The semion has -i 1.088 22 x 10^{-08} kg strong mass, and the quarton has -i 5.441 1 x 10^{-09} kg strong mass in this frame. (Strong mass is imaginary mass.)

E. Length Contraction of Electrino Orbits in Echon Frames

Let us calculate the relativistic length contracted circumference C^*_e (where t = 0) of the electrino orbit in electron intrinsic spin (we will use the superscript * on a number of variables to designate the relativistically contracted quantities for those variables):

$$C^*_e = C_e \left(1 - \frac{V_e^2}{c^2} \right)^{1/2} = C_e \left(1 - \left(1 + \frac{v_e^2}{c^2} \right) \right)^{1/2}, \qquad (6\text{-}39)$$

$$= C_e \left(-\frac{Gm_e}{r_e c^2} \right)^{1/2}, \qquad (6\text{-}40)$$

where

C^*_e = Relativistically contracted circumference for the electron;

C_e = Non-relativistic circumference of the electron (simple semion echon type);

V_e = Net orbital velocity of electrinos in the electron;

v_e = Gravitational aether radial velocity at radius of the electron;

m_e = Mass of the electron at rest in the echon;

r_e = Radius of electrino orbit in the electron;

G = Gravitational constant;

c = Speed of light.

From the mass formula of the electron (6-24), we obtain the relationship

$$m_e \approx \frac{h}{C_e c}, \text{ where } C_e = 2\pi r_e.$$
(6-41)

Now continuing from above, substituting for m_e and changing C_e to $2\pi r_e$ in equation (6-40), we obtain

$$C^*_e = C_e \left(-\frac{Gm_e}{r_e c^2} \right)^{1/2} \approx 2\pi r_e \left(-\frac{Gh}{r_e^2 c^3} \right)^{1/2}.$$
(6-42)

We give the zeroth order approximation of the relativistic circumference $2\pi R_0$ the symbol λ_0.

$$R_0 = \frac{\lambda_0}{2\pi} = i \left(\frac{Gh}{c^3} \right)^{1/2}$$
(6-43)

$$\approx i \, 1.616 \, 252 \, x \, 10^{-35} \, m.$$
(6-44)

$$\lambda_0 \approx i \, 1.015 \, 521 \, x \, 10^{-34} \, m.$$
(6-45)

The radius R_0 is here presented for comparison purposes only and as part of the length contracted circumference λ_0. In actuality, while the circumference of the electron is relativistically contracted, the radius of the electron, mass, and time in the electron are not relativistically transformed.

F.   Gravitational Constant

   The gravitational constant can now be expressed in terms of other fundamental constants:

$$\lambda_0^2 \approx -2\pi Gh / c^3 \approx -4\pi^2 Gh / c^3. \qquad (6\text{-}46)$$

$$G \approx -\lambda_0^2 c^3 / (2\pi h) \approx \frac{-R_0^2 c^3}{h}. \qquad (6\text{-}47)$$

G.   Relationship Between Mass and Radial Aether Velocity in the Electron

Notice by the following derivation how mass in the electron rest frame is related to the gravitational aether velocity in the electron rest frame at the radius of the electron:

$$v^2 = \frac{2Gm_s}{r_e}, \qquad (6\text{-}48)$$

$$r_e \approx \frac{h}{2m_s c}, \qquad (6\text{-}49)$$

from Eq. (6-36).

$$v^2 \approx \frac{Gm_e^2 c}{h}. \qquad (6\text{-}50)$$

$$m_e^2 \approx \frac{v^2}{c^2} \frac{hc}{G}, \qquad (8\text{-}51)$$

$$m_e \approx i M_0 \frac{v}{c}, \; or \; |M_0| \frac{v}{c}. \qquad (6\text{-}52)$$

H. String Theory

Notice when the semions in orbit in an electron are relativistically inverse transformed to the echon rest frame, the dimensions of the semions are transformed to the echon rest frame. The x dimension of the semion diameter is transformed to the radius of the electron, 3.861×10^{-13} m. The y and z dimensions of the semion diameter are unchanged at i 1.616×10^{-35} m. The imaginary widths in the real frame are probably interpreted as zero widths. But even the imaginary widths are extremely small relative to the x dimension of the particle. Notice the x dimension of the particle is curved about the circumference of the electron. Thus the deeper structure in electrons appears to be curved, long, thin strings. There is a Theory of Everything (TOE), a Grand Unification Theory (GUT) that is based on string theory. Our model bears some resemblance to the string theory at this level. But notice strings require several parameters to define them: the radius of curvature of the string, the length of the string, and the widths of the string. This is not very parsimonious for the absolutely most fundamental particles. By contrast, all that has to be known about electrinos is their radius (even their charge can be inferred from that quantity). That is far more parsimonious for the absolutely most fundamental particles. Then through particle structure and relativity, deduced from our postulates, the underlying electrinos can take the string form and structure. The Electrino Fusion Model of Elementary Particles is one step deeper than The String Theory.

I. Quantization of Spin

Ever wonder why particle spins come in integral values times $\frac{1}{2}\hbar$? Why not in 0.3712 $\hbar$? or in 1.0172 $\hbar$? You say because it is a quantum mechanical rule that spins come in integral values times $\frac{1}{2}\hbar$. Yes, but why is it a rule? and how does mother nature know to obey the rule?

The reason all observed particle spins come in integral values times $\frac{1}{2}\hbar$ is that every particle in the Universe was created from copies of a single particle–namely an octon-anti-octon pair spinning with a combined $\frac{1}{8}\hbar$ spin. (That is in absolute value terms only. Actually, since the anti-octon has negative mass, it has negative spin when orbiting the same direction as the octon with positive spin. Thus in those terms, the octon-anti-octon pair has 0 net spin.) From copies of that single octon-anti-octon pair were fused various whole particles. (It takes eight octons or anti-octons in some fusion state to make a whole particle. Two octons fused make a quarton. It takes two pairs of quartons orbiting the opposite way of the quartons to make a pion. It has 0 net spin, though the quartons each orbit with $\pm\frac{1}{8}\hbar$ spin. But the quartons are in a super deep binding energy well with no current way of blasting them appart. Thus only whole particles can be detected, and the $\frac{1}{8}\hbar$ spins of the quartons cannot be detected. The net 0 spins of the pions are detected.

Similarly with semions. Four fused octons make a semion. Two semions, orbiting about each other with a combined $\frac{1}{2}\hbar$ spin, form an electron. Again, the $\frac{1}{4}\hbar$ spins of the semions are not detectable, but the $\frac{1}{2}\hbar$ spins of the whole particle electrons are.

Similarly with unitons. If eight octons are fused into one particle, a uniton is formed. A uniton forms the core particle of baryons. Also uniton-anti-uniton pairs form photons with combined $1\hbar$ spin. All matter particles, gravitons, and light can be formed of states of, or combinations of the above particles. Thus all observed spins are either 0 or integer combinations of $\frac{1}{2}\hbar$.

The equations in Postulate 8 shed some light on these questions.

$$s_p = \left[r_p \, x \left(m_{f_s} c_t \right) \right] = \sum_i \left[f_p \hbar / 2 \right]_i 1_J .$$ (6-53)

For charged leptons,

$$s_p = 2 f_p \hbar / 2 = \hbar / 2.$$ (6-53a)

The property that particles come in integer values times $\hbar/2$ is guaranteed by the right hand side of the equation because only charged leptons yield non-zero spins in particles and composite particles other than whole orbital spins.

IV. Deriving the Photon

A. Photon, Echon, Yachon, and Charge Structure

The photon is not a single echon like the electron. While the echon is an orbiting system of particles, it is a whole particle (made up of parts). It carries a whole charge. The photon is actually made up of two orbiting whole particles (yachons—unitons). They are oppositely charged. They are matter and antimatter. They are composed of a whole positive electric charge and a whole negative electric charge orbiting each other at various angles relative to the light path axis, and traveling together at the nominal speed of light c linearly along the light path axis.

Though there are differences, there are significant calculational similarities between photons and electrons. And also significantly, in natural electrino fusion in the laboratory, photons are formed when the semions in electrons fuse and anti-semions in positrons fuse simultaneously. It is natural to study their charge, radii, mass, and spin relationships in the fusing process. The photon derivations also present another data point in the calculations necessary to the work of generalizing the formulas, and finding the system for calculating the masses of the elementary particles. Thus the calculations in this section are also valuable.

The calculations in this section follow the outline in the last section in a condensed way.

B. Spin, Mass, and Radius

The particles we wish to put in orbit in photons are unitons. We know the photon has spin $\pm 1\hbar$. Mass, energy, and radius of the photon are related by the following formulas:

$$m_\gamma = \frac{E_\gamma}{c^2} = \frac{\hbar}{r_\gamma V} \approx \frac{\hbar}{r_\gamma c}; \quad m_\gamma = 2m_{uniton}. \qquad (6\text{-}54)$$

C. Characteristic Equation for Photon in Motion

Photon unitons have two velocity components. u is the linear velocity of light c. For circularly polarized photons, v_t is the transverse tangential orbital velocity of the unitons. The total velocity of the photon charges is V, faster than the nominal speed of light. (See Chapter 14, Sections E through H.) v transforms away in the equations, therefore v will not occur in the following characteristic equation for the photon.

The velocity components for a photon in motion are

$$V^2 = u^2 + v_t^2 = c^2 + v_t^2. \tag{6-55}$$

v_t^2 equals c^2 in the photon. (See Chapter 14, Sections E through H.) The net result is:

$$V^2 = c^2 + v_t^2 = c^2 + c^2 = 2c^2. \tag{6-56}$$

We can see by this latter equation that V is faster than the nominal c.

D. Escape Velocity

For transformations near the speed of light,

$$v = \left(\frac{Gm_\gamma}{r_\gamma} \right)^{1/2}. \tag{6-57}$$

E. Relativistic Transformations in the Photon System

While the total velocity of the electrinos in photons (unitons and anti-unitons) is greater than c, the longitudinal velocity of the photons u must be c exactly. The true total velocity V must be measured with respect to the aether particles. (See Chapters 3 and 4.) Each photon is a small gravitational body which accelerates the aether inward to the center of the photon. The net velocity V of the photon charges is

$$V = c\left(1 + \frac{v^2}{2c^2}\right)^{1/2} = c\left(1 + \frac{Gm_\gamma}{2r_\gamma c^2}\right)^{1/2}.$$

(6-58)

F. Length Contraction of Photon

Let us calculate the relativistic length contracted radius in the x longitudinal direction r^*_γ (where $t = 0$) of a photon traveling at velocity V just slightly faster than the nominal c.

$$r^*_{\gamma x} = r_{\gamma x}\left(1 - \frac{V^2}{c^2}\right)^{1/2} = r_{\gamma x}\left(1 - \left[1 + \frac{v^2}{2c^2}\right]\right)^{1/2}$$

(6-59)

$$= r_{\gamma x}\left(-\frac{Gm_\gamma}{2r_\gamma c^2}\right)^{1/2}.$$

(6-60)

From the photon's angular momentum Eq. (6-54) we obtain the approximation of m_γ to solve the above equation:

$$m_\gamma = 2m_{unitons} \approx \frac{\hbar}{r_\gamma c}.$$

(6-61)

Now continuing from above, substituting for m_γ in Eq. (6-60), we obtain

$$r^*_{\gamma x} \approx r_{\gamma x}\left(-\frac{G\hbar}{r_\gamma^2 c^3}\right)^{1/2} \approx R_0.$$

(6-62)

$r*_{\gamma x}$ does not equal $r*_{\gamma y}$, but $r_{\gamma x}$ does equal $r_{\gamma y}$. That is why the above approximation can be made.

G. Time in Relativistic Photon Frame

We do a similar thing for time as we did above for the radius of the photon.

$$t*_{\gamma} = \gamma\, t_{\gamma}, \ where\ x = 0 \tag{6-63}$$

$$= \frac{t_{\gamma}}{\left(1 - \dfrac{V^2}{c^2}\right)^{1/2}} = \frac{t_{\gamma}}{\left(1 - 1 - \dfrac{v^2}{2c^2}\right)^{1/2}} \tag{6-64}$$

$$= \frac{t_{\gamma}}{\left(-\dfrac{Gm_{\gamma}}{2r_{\gamma}c^2}\right)^{1/2}}. \tag{6-65}$$

H. Relativistic Mass

Just as there is a length contraction of photons in the relativistic frame, there is a relative mass increase in the photon when transformed to the relativistic frame.

$$m* \approx m_{\gamma}\left(1 - \frac{V^2}{c^2}\right)^{-1/2} \approx m_{\gamma}\left(1 - \left(1 + \frac{v^2}{2c^2}\right)\right)^{-1/2} \tag{6-66}$$

$$m^* \approx m_\gamma \left(-\frac{Gm_\gamma}{2r_\gamma c^2} \right)^{-1/2}. \tag{6-67}$$

From Eq. (6-54),

$$m^* \approx m_\gamma \left(\frac{-Gm_\gamma^2 c}{4\hbar c^2} \right)^{-1/2} \tag{6-68}$$

$$m^* \approx \left(-\frac{4\hbar c}{G} \right)^{1/2} \approx 2M_0, \text{ where} \tag{6-69}$$

$$M_0 \approx -i \, 2.176\,44(11) \times 10^{-08} \, kg. \ [7] \tag{6-70}$$

This is a very large number on particle scales, but it is minus imaginary. We transformed m_γ, the whole effective mass of the photon. Thus we obtained m* in the transformation, which is approximately equal to $2M_0$. We are rewarded that the total strong mass of the photon in the relativistic frame is approximately $2M_0$, an indication that indeed there are two unitons in the photon—a positive, whole matter charge, and a negative, whole antimatter charge—revolving about each other at some angle relative to their mutual flight at the speed of light in one direction.

Here we close our treatment of the photon in this volume. For additional calculations on the photon, relating v_t and v to V and r, as well as red shift models, see Chapter 14.

V. Synthesizing and Generalizing the Model—Fields

A. Why Electrons Don't Explode

There has long been a question why electrons don't explode. Charles Kittel, Walter D. Knight, and Malvin A.

Ruderman wrote, "We do not know the structure of the electron in detail. The model we have outlined cannot be entirely satisfactory, for what keeps the charge in the electron together? Why doesn't it fly apart under the Coulomb repulsion of like-charge elements? At present we have no theory of why there is an electron."[11] *Electrino Physics* presents a model of the structure of the electron. We are now in a position to answer that mystery—why electrons don't fly apart.

Gravity operating on the observed electron mass m_e is nearly 45 orders of magnitude too weak to hold together a spherical shell of electric charge. The exponential strong force (commonly called the strong force) does not reach out to the dimensions of an electron radius. What can hold the electron together?

In the first place the electron cannot have a spherical charge distribution, because it has spin. In our model the electron is composed of two semions orbiting about each other, each of approximate mass $M_{1/2} \approx$ -i 1.088×10^{-8} kg, for a combined mass of -i 2.176×10^{-8} kg. When the mass calculates to be minus imaginary, the radius calculates to be imaginary. An imaginary radius would effect not only gravitational force equations, but also the strong electric force equations. With R_0^2 in the denominator of the strong electric force equation, Eq. (6-71) is a central force.

$$\frac{q_1 q_2}{4\pi\varepsilon_0 \alpha \left(i\left(|R_0|\right)\right)^2} = -\frac{q_1 q_2}{4\pi\varepsilon_0 \alpha |R_0|^2}. \qquad (6\text{-}71)$$

The imaginary radius has the effect of reversing the strong electric repulsion into an electric attraction—converting an outward force into an inward force. The strong gravitational force remains attractive also with imaginary radius, because the relativistic inner frame mass is minus imaginary also.

$$-\frac{G\left(-i|M_1|\right)\left(-i|M_2|\right)}{\left(i|R_0|\right)^2} = -\frac{G|M_1||M_2|}{|R_0|^2}. \tag{6-72}$$

The strong electric force and the strong gravitational force are not only of the same sign in an electron echon, they are also of the same magnitude.

$$-\frac{1/4\,G\left(-\dfrac{\hbar c}{G}\right)}{-\left(\dfrac{G\hbar}{c^3}\right)} = \frac{1/4\,e^2}{4\pi\varepsilon_0\alpha\left(-\dfrac{G\hbar}{c^3}\right)} = -1/4\,\frac{c^4}{G}. \tag{6-73}$$

The questions naturally arise: Are the strong electric and the strong gravitational forces equal but separate and distinct forces that must be added together? Or are they equivalent expressions for the same force? We realize the strong gravitational numbers come from General Relativity and warped space. The strong electric numbers are calculated in flat space. It should be improper to add a number from warped space to a number from flat space. We should first transpose the flat space number into a warped space number, or vice versa. The strong gravitational number in the above equation may be the warped space version of the strong electric force. The electric force in flat space orbits the electrinos. The orbiting electrinos may warp the space along their line of flight, giving rise to the General Relativistic warped space. Einstein struggled for 30 years trying to unite gravity and the electric force, and failed. It has been thought that those forces may only be united at the high Planck mass region ($\hbar c/G$), which we have been working with all this chapter. If this is not the expected equivalence of forces at this mass, then what is it? The author will unite all the forces in the next chapter. The Planck mass energies, the aether system, and energy states are sufficient to do that.

Let us now finish balancing the forces in an electron. Since we are dealing with a two body problem, we must use the reduced mass to accurately calculate the two body orbits as a one body problem. The one body form of the equation using the reduced mass μ is

$$\frac{q_1 q_2}{4\pi\varepsilon_0 \alpha(ir)^2} 1_r = \mu \frac{d^2 r}{dt^2}. \qquad (6\text{-}74)$$

We have already solved for the left hand expression. Since the semions are equal in mass, the μ is 1/2 (1/2M_0). The acceleration of the semions is c^2/R_0. However the inertial force is the negative of the force applied to the particle. Do the central force and the inertial force add to 0?

$$\left(-1/4 \frac{c^4}{G}\right) + \left(-1/4 (-i)\left(\frac{\hbar c}{G}\right)^{1/2} \frac{c^2}{i\left(\frac{\hbar G}{c^3}\right)^{1/2}}\right) = 0. \qquad (6\text{-}75)$$

The attractive strong electric force and the inertial force are equal and opposite. They add to zero. The two semions are in an orbit of radius approximately R_0. The electron is bound in the relativistic frame.

The attractive strong forces in the inner region translate to the attractive forces in the outer radius region. To find the attractive equivalent forces in the outer frame, use imaginary radii and masses and see what forces those are equivalent to in real radii and masses. Again the strong gravitational force and the strong electric force are equal and equivalent for orbiting electrinos. Only one strong force needs to be considered in this case. The orbital forces perfectly balance in the outer frame as well as the relativistic inner frame.

$$\left(1/4\frac{e^2}{4\pi\varepsilon_0\alpha(ir_e)^2}\right)+\left(-1/4(-im_e)\frac{c^2}{ir_e}\right)=0. \qquad (6\text{-}76)$$

$$\left(-1/4\frac{\hbar c}{r_e^2}\right)+\left(1/4\frac{\hbar c}{r_e^2}\right)=0. \qquad (6\text{-}77)$$

The attractive central force in the electron frame is equal and opposite to the inertial force. They add to zero. Thus the attractive strong electrical force or the equivalent strong gravitational force orbits the electron semions at the relativistic electron radius R_0, which transforms to $r_e = 3.861 \times 10^{-13}$ m (Eq. (6-31)). Thus the electron is held together and does not explode.

B. The Derivation of the Fine Structure Constant

To derive the fine structure constant, all one has to do is equate the strong gravitational force and the strong electric force in fine structures as was done in the last section and Eq. (6-73), and solve for α.

$$-\frac{1/4\,G\left(-\dfrac{\hbar c}{G}\right)}{\left(-\dfrac{G\hbar}{c^3}\right)}=\frac{1/4\,e^2}{4\pi\varepsilon_0\alpha\left(-\dfrac{G\hbar}{c^3}\right)}. \qquad (6\text{-}78)$$

$$\alpha=\frac{e^2}{4\pi\varepsilon_0\hbar c}. \qquad (6\text{-}79)$$

The value of α was known before, but not calculable from any theory or model. But recently a physicist in England, James G. Gilson, devised a theory for α as a coupling constant that can be calculated to any degree of precision.[12] (He supplied his theoretical α to the author to 63 places.) But his calculation may not be correct. In the author's model, the value of α obtains from the unification of the strong electric force with the strong gravitational force in fine structures.

C. The Significance of α and Imaginary Numbers in Forces

The last section derived α by equating a strong gravitational force with a strong electric force in fine structures. Neither of these forces may have been known previously, but both fell out naturally from the previous calculations in this chapter. Three decades ago it may not have been known that the strong mass of all elementary particles may be calculated from the charges and fields in them when the Fine Structure Constant α is included in the electric force equation. But now we see it. It is because the masses of the particles primarily are due to the charges in the hyper-optic velocity spaces where α is part of the electric force and field equation. (The Fine Structure Constant is not a part of the force and field equation in the sub-optic velocity spaces in the Coulomb force.) The neat thing about the Fine Structure Constant is that it unifies the forces—making the strong electric force equal to the strong gravitational force. The next most important aspect of α is that it is much different than 1, allowing different states and structures. The electron is a fine structure governed by alpha, but atoms are not. The electrons in atoms obey the Coulomb force and make sub-optic orbits around the nuclei. Not so the semions in electrons. The Fine Structure

Constant allows for important differences of scale in matter.

Imaginary and real numbers in forces are another important factor in the scale of systems and the sense of the equations. Generally imaginary numbers arise in the electric and gravitational forces due to relativity, where hyper-optic velocities give rise to imaginary radii and masses in the transformations. But here is where the aether model of relativity comes into such important play. For instance, the current Standard Model of the photon has the photons of each color traveling precisely at c. This model has the photon linear velocity at c, but the total photon charge velocity made up vectorially of v_t, a transverse orbital velocity, and c, the linear velocity of the photon. The total net V is faster than the speed of light c, pushing the relativity into the hyper-optic region. This is all not possible without an aether system.

The hyper-optic velocities and aether relativity give rise to imaginary radii and masses, but not charges, in the force equations. But there is also extreme length contraction and relative mass increase from the relativistic equations on the back-side of the speed of light barrier. The imaginary radii squared reverse the sense of the forces from repulsive to attractive; and the extreme length contraction of the radii make the attractive forces extremely strong. Thus in the hyper-optic region the forces are strong for two reasons: 1) the introduction of α in the denominator of the electric force equation, and 2) the extreme length contraction in the relativistic frame.

The force line-up and unification in this new Electrino Fusion Model of Elementary Particles is different than in the current Standard Model. For a listing, definition, and unification of the forces, please see the next chapter.

[1]M. Gell-Mann, "A Schematic Model of Baryons and Mesons," *Physics Letters*, Volume 8, number 3, 1 February 1964, pp. 214, 215.

[2]Z.J. Xiao, "Preon Model with Complementarity," *Communications in Theoretical Physics*, 16(1):115-118 (1993).

[3]Author unknown.

[4]Francis Halzen and Alan D. Martin, *Quarks & Leptons: An Introductory Course in Modern Particle Physics* (New York: John Wiley & Sons, 1984), p. 13.

[5]http://physics.nist.gov/cuu/Constants/

[6]William Taussig Scott, *The Physics of Electricity and Magnetism*, Second Edition (New York: John Wiley & Sons, Inc., 1966), p. 76.

[7]*http://physics.nist.gov/cuu/Constants/*

[8]John R. Reitz, Frederick J. Milford, and Robert W. Christy, *Foundations of Electromagnetic Theory* (Reading, Massachusetts: Addison-Wesley Publishing Company, 1980), pp. 429, 465.

[9]H. Haken and H. C. Wolf, *Atomic and Quantum Physics: An Introduction to the Fundamentals of Experiment and Theory*, Translated by W. D. Brewer (Berlin: Springer-Verlag, 1984), p. 65.

[10]*http://physics.nist.gov/cuu/Constants/*

[11]Charles Kittel, Walter D. Knight, and Malvin A. Ruderman, *Mechanics*, Berkeley Physics Course--Volume

1 (New York: McGraw-Hill Book Company, 1965), p. 270.

[12]J. G. Gilson, "Calculating The Fine Structure Constant," *Physics Essays*, Volume 9, number 2, 1996, pp. 342-353.

[13]G. L. Ziegler and I. I. Koch, "Prediction of the Masses of Charged Leptons," *Galilean Electrodynamics*, November/December 2009, Vol. 20, No. 6, pp.114-118.

[14]G. L. Ziegler and I. I. Koch, "Prediction of the Masses of Every Particle, Step 1," *Galilean Electrodynamics*, Summer 2010, Vol. 21, Special Issues No. 3, pp. 43-49.

Problem Set 6

1. Why is it virtually impossible to experimentally probe one step deeper into the structure of matter?

2. What alternative approach does the author take?

3. Why cannot Einstein's aetherless model of relativity unlock the riddle of particle structure?

4. Why should a spherically symmetric charge distribution not have detectable spin?

5. According to the application of Postulates 4 and 5, how many particles must be in a ½ spin electron? ...in a 0 spin pion meson? How many particles are required in the simplest 0 spin system?

6. What can account for heavier electrons (muons, tauons, ...)? What can account for heavier neutrons (Λ, Ξ^0, ...)?

7. What can account for different varieties of neutrinos (electron-, muon-, tauon-, ...)?

8. Which particles are more massive--the physically larger ones or the smaller ones?

9. What is speed c in the relativistic frame relative to the non-relativistic frame? How long would this relativistic velocity of light take to go 15 billion light years of non-relativistic light? How long would it take to go the radius of an electron? In the dimensions of a particle, this velocity appears to be instantaneous.

10. Are electrinos absolute point charges?

11. What causes the smaller size limit of electrinos?

12. Which has a greater charge to volume ratio--a semion or a quarton? In terms of a uniton, what would be the radius, mass, and charge of an octon?

13. Why is the aether frame not the most convenient for calculating many things about particles?

14. How can the relativistically contracted circumference, radius, and increased mass of an echon be expressed in terms of fundamental constants? ...in terms of meters and kilograms? About how different are they in absolute value from the non-relativistic, non-contracted circumference, radius, and mass of the electron?

15. What is the relationship between mass and the radial aether velocity at the radius of an electron?

16. Explain why one does not observe the mass M_0 in an electrino orbiting system, but only the much smaller m.

17. Why don't electrinos implode or explode?

18. Why don't electrons explode?

19. How can the fine structure constant be derived?

20. Why do electrinos appear to be strings in the next to the last level of particle structure?

21. Is a string model the most parsimonious model of particle physics?

22. What is the most parsimonious model of particle physics?

23. What quantitizes intrinsic spin?

24. How can particles of more than one energy state all have ½ spin?

25. What is the quantization of intrinsic spin similar to?

26. What is the mass of the photon in the relativistic frame?

27. Do photon unitons orbit close to the speed of light transverse to the light path axis?

28. Account for circular, elliptical, and line polarizations of photons in the electrino model.

29. Do all colors of light travel precisely at c?

30. Do you add the strong electric and strong gravitational forces in particle calculations? Why or why not?

31. How can quartons, semions, and unitons be formed from octons?

32 All the copies of all the particles in the Universe could be formed from how many copies of what one particle pair?

33. What alternative to the Big Bang Theory does this suggest for the origin of the Universe?

34. What is the spin of a single semion in an electron? What is the spin of a single quarton in a pion? Why are these spins not observed? Only what spins are observed in particle physics?

Chapter 7

UNITING THE FORCES

A. Introduction

Recently science listed three forces it hoped to unite in a unified field theory: gravity, electro-weak force, and the strong force. This list is too few, and it is too many. That is, there are more identifiable forces, but these should all be united in one.

The author has identified eight forces plus an infinite series of interactions. In this chapter these shall all be united. In the next chapter, in Section M, we will see that the forces and interactions are on the same footing in the particle g/2-factors. So forgive if we sometimes call interactions forces.

The strong nuclear force is not, just as observed, an elementary force. The strong nuclear force is a particle (pion) mediated force of underlying elementary forces (the strong electric or strong gravitational forces). The weak interactions are also particle (magneton) mediated. In the author's model, there are an infinite number of weak interactions. This agrees with recent experiments that show that weak forces are not constant, but vary with distance—a property called running. [9, 10] Even without these interactions, the following list of forces appears obscenely too many to the physicist. But the forces holding the electron together have never been studied before. And the forces necessary to account for the electron and muon g-factors, have never been fully explored before. This chapter and the next shall, in part, explore these forces. The forces and interactions in their types and relative strengths are:

Core Forces

Strong electric force
Electric force
Magnetic force
$Weak_1$ force
$Weak_2$ force
etc.
Gravity

Miscellaneous Forces

Strong gravity
Strong nuclear force
Meso-electric force
Inertia

The core forces are involved in the derivation of the terms in the electron g/2-factor (see Chapter 8, Section M). The miscellaneous forces are not involved in those terms. The 'weak force' is apparently the sum of all subscripted weak interactions—an infinite series. Most mediated subscripted weak interactions are approximately 0.233 515 281 2032 . . . times as strong as the previous subscripted weak interaction. (See end of C. and the end of Chapter 8.) All the weak interactions are necessary to account for the electron g/2-factor.

B. Definition of Forces

Strong electric force

In fine structures and hyperoptic particle velocities, a slightly different force equation applies than the Coulomb force

$$F = \frac{q_1 q_2}{4\pi\varepsilon_0 \alpha R^2} 1_R \quad (in\ SI\ units), \quad (7\text{-}1)$$

where R is the hyperoptic super length contracted imaginary radius and

$$\alpha = \frac{e^2}{4\pi\varepsilon_0 \hbar c} \quad (7\text{-}2)$$

The α makes the force stronger, and the length contracted imaginary radii from hyper-optic total velocities make the force super strong.

Electric force

The Coulomb force law is

$$F = \frac{q_1 q_2}{4\pi\varepsilon_0 r^2} 1_r \quad (in\ SI\ units). \quad (7\text{-}3)$$

This force law is active outside of echons like electrons. It is the electric force at macroscopic distances, and is operative in electron orbits in atoms and molecules. This law applies to the Coulomb force with real radii.

Magnetic force

There are many equations regarding magnetism. The conceptually simplest may be for uniformly moving charges with velocities v_1, v_2, respectively,

$$F_m = \frac{\mu_0}{4\pi} \frac{q_1 q_2}{r^2} v_1 \times \left(v_2 \times 1_r\right). \quad (7\text{-}4)$$

"Also as in the case of the electrostatic force, it is convenient to abstract from the properties of the 'test charge' by defining a magnetic *field*; in this case not only the test charge q_1 but also its velocity v_1 must be factored out:

$$F_m = q_1 v_1 \times B, \qquad (7\text{-}5)$$

where the *magnetic induction* **B** is

$$B = \frac{\mu_0}{4\pi} \frac{q_2}{r^2} v_2 \times 1_r. \qquad (7\text{-}6)$$

If a number of moving source charges is present, the magnetic forces and fields are additive.... If both an electric field and a magnetic field are present, the total force on a moving charge is $F_e + F_m$,

$$F = q_1 \left(E + v_1 \times B \right), \qquad (7\text{-}7)$$

known as the *Lorentz force.*"[1]   (Formulas from the reference were converted to the style of this book.)

Weak forces

All the above magnetic formulae apply to the root aspect of the weak forces, except everywhere there is a magnetic induction **B** in the above formulae there must be substituted an $\alpha^k B$ (and also when B is broken into its component parts, there must be multiplied an α^k), and everywhere there is an r in the formulae there must be substituted an R (super length contracted imaginary radius). Also α = 0.007 297 352 5376. . . .—the Fine Structure Constant.[2]   Its k = (p + q), where k is the order of mass singularity, where 1 represents a singularity, 2 represents a singularity within a singularity, etc.; p is the number of a particle in a sequential list, or order of singularity, or number of subscripted weak interaction. q is the number of singularity shell the particles are in (starting with 0). (The singularities in particles are in quantum shells similar to electron shells in elements. Not all mass singularities fit in a single shell.) All "elementary" particles are miniature mass singularities. And higher order particles are deeper level singularities. It has been said that no signal can escape a black hole. But as it turns out, weak interaction signals can escape a singularity. The Fine Structure Constant α is the coupling constant of relative signal strength of a weak interaction from a singularity to a receptor outside the singularity. α^2 is the coupling constant for a singularity within a singularity and a receptor outside the singularity. α^5 is the coupling constant for a 5 level singularity with the outside, etc. (This is not considering some other effects, which we will consider next.)

But with the weak interactions some other interesting things happen. The magnetons act as mediator particles for the weak forces. Just as we shall see later in this chapter that the pion mediated strong nuclear force is 32 times weaker at its peak than the strong force, the mediated weak force is 32 times *stronger* than would be expected without the magnetons. Two generations of mediation make it 1024 times as powerful as expected. Five generations of mediation make it 32^5 or 2^{25} times as strong as expected. But the Fine Structure Constant coupling is so weak, that even with the mediation, the coupling is only 0.233 515 281 2032 . . . for each order of mass singularity. The signal still gets weaker and weaker the higher order of singularity the signal is emitted from.

Interestingly enough, the magnetons do not enhance the $weak_1$ and $weak_2$ interactions. That is because the first and second particles (electrons and muons) have radial magneton velocities at their surfaces much much less than the speed of light c. Enhancement of the weak interactions cannot take place unless the magnetons travel radially near c–as is the case for the weak$_3$ and higher weak forces. Since the enhancement is cumulative, this means the strength of enhancement is $(32)^{\{k_p-2\}}$, where k_p is the k value for the particle under consideration.

Another phenomenon that occurs with weak interactions is that all but one weak interaction are multiplied by an integer variable n, where

p | 0 1 2 3 4 5 ...
n | 0 1 3 6 10 15 ...

in g/2-factors. The n is a mass distinguishing constant in the g/2-factors. But taking into account n, B_{wk} looks like Eqn. (8-20) in the next chapter for weak$_2$ and higher weak forces.

These are lessons learned from a careful study of the electron and muon g/2 factors. (See the next chapter.) Other lessons learned: exponents of α (k, p, q, etc.) do not add like vectors. Terms in the electron g/2 factor show that such exponents add like simple numbers. Terms also indicate that there are more as yet undiscovered charged leptons.

Just as the electric force in the hyper-optic super length contracted region is the strong electric force, the magnetic force in the hyper-optic super length contracted region is the weak interactions. There is that tie between the magnetic force and weak interactions, and the weak interactions and the strong force. The weak interactions all have imaginary radii for fine structures. Equations for the **root aspect** of the weak interactions are as follows:

$$F_{wk} = \frac{\mu_0 \alpha^k}{4\pi} \frac{q_1 q_2}{R^2} v_1 \ x \left(v_2 \ x \ 1_R\right). \qquad (7\text{-}8)$$

$$F_{wk} = q_1 v_1 \ x \ B_{wk}, \qquad (7\text{-}9)$$

where the *weak induction* $\mathbf{B_{wk}}$ is

$$B_{wk} = \frac{\mu_0 \alpha^k}{4\pi} \frac{q_2}{R^2} v_2 \ x \ 1_R.$$ (7-10)

If a number of moving source charges is present, the weak forces and fields are additive. If both a strong electric field and a weak field are present, the total force on a moving charge is $\mathbf{F_{se}} + \mathbf{F_{wk}}$,

$$F = q_1 \left(E_{se} + v_1 \ x \ B_{wk} \right) = q_1 \left(\frac{E}{\alpha^k} + v_1 \ x \ \alpha^k B \right).$$ (7-11)

When $k > 0$, we have the root aspect of the weak forces—weaker for higher level particles because of α^k. (Remember, this is only for the **root aspect** of the weak forces. The net values of the weak forces include terms for mediator particle enhancement and factor of n engrossment of the weak terms, as described above.) When we have α^k and $k = 0$, the magnetic force is obtained, thus uniting the weak forces with the magnetic force.

The introduction of the factor α^k in the denominator of Eqs. (7-1) and (6-4) requires the introduction of the opposite factor α^k in the numerator of the weak forces, Eq. (7-8), and the weak induction, Eqn. (7-10), because

$$\varepsilon_0 \mu_0 = \frac{1}{c^2};$$ (7-12)

and if we introduce an α^k in the equation with the ε_0 term, we must introduce its inverse with the μ_0 term to keep the speed of light c constant:

$$\left(\varepsilon_0 \alpha^k \right) \left(\mu_0 \frac{1}{\alpha^k} \right) = \frac{1}{c^2}.$$ (7-13)

The relativistic constancy of c to all observers balances our higher state forces for us.

Gravity

Gravity was the first force to be codified. Sir Isaac Newton derived the Classical expression for gravity:

$$F = -\frac{Gm_1 m_2}{r^2} 1_r. \qquad (7\text{-}14)$$

This form of the equation is derived for flat space. General relativity gives an equivalent expression based on warped space. The above equation applies to ordinary gravity when m_1 and radii are real.

Strong gravity

$$F = -\frac{GM_1 M_2}{R^2} 1_R. \qquad (7\text{-}15)$$

The above formula applies to strong gravity when the masses and radii are imaginary.

Strong nuclear force

Not all the forces currently studied are elementary in this model. The strong nuclear force is not elementary in this model. It is a pion mediated strong gravitational force—at best 1/32 as strong as the underlying strong gravitational force, but also exponentially decayed with distance.

The strong nuclear force (exponential strong force) between nucleons is roughly of the form

$$F_{nuc}(r) = \frac{\partial}{\partial r} U_{nuc}(r), \text{ where} \qquad (7\text{-}16)$$

$$U_{nuc}(r) = -D\frac{e^{r/r_0}}{r} = \frac{1}{32}\frac{GM_0^2 e^{r/r_0}}{r},\qquad(7\text{-}17)$$

$r > 0.7$ fm, where the constant D is of the order of magnitude 10^{-27} joule-m $\approx$ $-(1/32)$ GM_0^2; r_0 is a constant having the dimensions of a length ($r_0 = \hbar/m_p c \sim 2 \times 10^{-15}$ m).

Mesoelectric force

In fine structures without hyper-optic speed of the particles (ordinary sub-optic speeds of particles), the Fine Structure Constant is in the denominator of the electric force equation, and there are real radii. This gives rise to a force stronger than the Coulomb force, but much weaker than the strong electric force with length contracted imaginary radii. The unitons and anti-unitons in photons are held together by this force. The neutrinos and neutrons are held together by this force. This is too important of a force to omit.

Inertia

Inertia takes many forms. For linear acceleration

$$F = -ma.\qquad(7\text{-}18)$$

For circular orbits

$$F = m\frac{v^2}{r}1_r.\qquad(7\text{-}19)$$

The above equations apply to ordinary inertia when the masses and radii are real, and also to strong inertia when the masses and radii are imaginary, and the radii are very small.

C. Uniting the Forces

Uniting gravity and inertia

Inertia has been previously considered a fictitious force, and not counted in the list. But in the aether model it is as real as gravity and obeys the same equations as gravity:

$$F = m\left[\frac{\partial}{\partial} P_3 + \nabla P_t\right], \qquad (7\text{-}20)$$

where P_3 is the three vector momentum, and P_t is the time component of the momentum—the energy. Where $v \ll c$,

$$F = m\left[\frac{dv}{dt} + 1/2 \nabla\left(v^2\right)\right], \qquad (7\text{-}21)$$

where v is the aether velocity relative to the observer. The equations are derived and explained in Chapter 5. The equations unite gravity and inertia—a feat possible with an aether system, but not without one.

Uniting electricity and magnetism

This feat was done in the nineteenth century by James Clerk Maxwell through what are now called the Maxwell Equations. These equations, by the way, were formulated with a luminiferous aether in mind, and are certainly compatible with an aether system. "Maxwell's equations have the following form in rationalized MKSC units:

$$\nabla \times E = -\frac{\partial B}{\partial t} \qquad (7\text{-}22)$$

$$\nabla \cdot D = \rho \qquad (7\text{-}23)$$

$$\nabla \times H = J + \frac{\partial D}{\partial t} \qquad (7\text{-}24)$$

$$\nabla \cdot B = 0 \qquad (7\text{-}25)$$

"In these equations, **E** and **B** are the electric-field intensity and magnetic induction, respectively, and are defined in a given reference frame in terms of the ordinary force **f** acting upon a charge q moving with ordinary velocity **u:**

$$f = q\left(E + u \times B\right) \qquad (7\text{-}26)$$

"The quantities **D** and **H** are defined in terms of **E** and **B** and the polarization vectors **P** and **M:**

$$D = \varepsilon_0 E + P \qquad (7\text{-}27)$$

$$H = \frac{B}{\mu_0} - M \qquad (7\text{-}28)$$

". . . The quantities ρ and **J** are the electric-*charge and -current densities*, respectively. These are defined by the equations

$$\rho = \lim_{\Delta V \to 0} \frac{\sum_i q_i}{\Delta V} \qquad (7\text{-}29)$$

$$J = \lim_{\Delta V \to 0} \frac{\sum_i q_i u_i}{\Delta V} \qquad (7\text{-}30)$$

where the summations are to be carried out over all charges lying within ΔV, and $\mathbf{u}_i$ is the ordinary velocity of q_i. The limit $\Delta V \to 0$ is intended to apply in a physical sense only— the volume element must be kept large enough that a sufficiently large number of elementary charges q_i will be contained inside it to assure that ρ and $\mathbf{J}$ will be, macroscopically, smooth functions of position. ρ and $\mathbf{J}$ are said to be the *sources* of the fields $\mathbf{E}$ and $\mathbf{B}$."[3]

Uniting the electric (Coulomb) force and the strong electric force

There are two differences between the Coulomb electric force and the strong electric force. One is an additional factor α (the Fine Structure Constant) in the denominator of the strong electric force making it a factor of about 137 times as strong as the Coulomb electric force. The factor α was derived in the last chapter simply by equating the strong electric force and the strong gravitational force. The second difference is the radii are imaginary and extremely length contracted in the strong electric force and real in the Coulomb electric force.

There are significant similarities, however, between the strong electric force and the Coulomb electric force. The form of the force equations is basically the same. For the Coulomb electric case, α is raised to the k power, where k is just equal to 0. In the strong electric force case, $k = 1$.

Uniting gravity with the strong electric force

Gravitational forces are caused by masses. But all masses are caused by electric fields and their motions in elementary particles. Intrinsic charge mass $|m_q| = |m_{kin}| =$

½m$_p$, where m$_p$ is the mass of the particle. This equation works also for strong masses M$_p$, M$_q$, and M$_{kin}$ and imaginary radius R$_p$.

$$1/2\, m_p = |m_q| = |m_{kin}| = \frac{e^2}{8\pi\varepsilon_0\, ar_p c^2} \qquad (7\text{-}31)$$

Uniting the strong electric force with the strong gravitational force

This was done in the last chapter simply by equating them in fine structures. Space warp is not just the result of mass, it is the result of orbiting electric charges traveling at the speed of light. Space warp, and all General Quasi-Relativity that goes with it, is rightly an electric phenomenon, related to the strong electric force. Also it is an aether phenomenon, for the aether travels at ± c at the small radius R$_0$ at the surface of a uniton, mediating the gravitational strong force as well as gravity.

Uniting gravity and the strong gravitational force

$$M_0 = m_p\left(1 - \frac{V_p^2}{c^2}\right)^{-1/2} ; \ R_0 = r_p\left(1 - \frac{V_p^2}{c^2}\right)^{1/2} . \qquad (7\text{-}32)$$

The difference between gravity and the strong gravitational force is only a matter of which relativistic frame you are measuring in.

Uniting gravity and the strong nuclear force

Nature measures gravity at a distance from an object and the strong gravitational force up close to the object. How does nature know how to tell whether to measure by the far away formula or by the close up formula? In echons the cutoff is abrupt at the surface of the echon. But in the strong nuclear force mediated by pions, nature fades out the strong gravitational force exponentially and ingrows the gravitational force exponentially. We can write the combined net force as follows:

$$F = -G\left(\frac{-iM_1}{ir}e^{\frac{-r}{2r_0}} + \frac{m_1}{r}\left(1 - e^{\frac{-r}{2r_0}}\right)\right) x \qquad (7\text{-}33)$$

$$\left(\frac{-iM_2}{ir}e^{\frac{-r}{2r_0}} + \frac{m_2}{r}\left(1 - e^{\frac{-r}{2r_0}}\right)\right).$$

As $r \to 0$, the second terms drop out. The force becomes

$$F = -\frac{GM_1M_2}{r^2}e^{\frac{-r}{r_0}}, \qquad (7\text{-}34)$$

which is strong nuclear force like. If $r \to \infty$, the first terms become 0 and the second terms lose their exponentials. Then

$$F = -\frac{Gm_1m_2}{r^2}, \qquad (7\text{-}35)$$

which is the familiar gravitational force.

By simply exponentially decaying the strong gravitational force and ingrowing the gravitational force, we have gravity at a distance and an exponential strong-like force at close range. We say exponential strong-like force because the measured exponential strong force is only 1/32 as strong as Eq. (7-34). That is because the pion mediates this force. The pion is composed of four orbiting quartons: two orbiting one way and two orbiting the other way. A quarton in the pion attracts mass M_1 with a force. The quarton on the opposite side of the pion attracts M_2 with the like force. The center link of force is within the pion itself. The resultant force between M_1 and M_2 will be the weakest link of force, which will undoubtedly be within the pion itself, which has the smallest masses and greatest particle separation. The quartons in the pion are aligned only about half the time with particles M_1 and M_2. The other half of the time they are in a line perpendicular to M_1 and M_2. Thus the quartons mediate the force effectively only half the time, which is like having their mass equal to $\frac{1}{8}M_0$ and mediating all the time. (Remember M_0^2 is negative.) The force in the pion is

$$F = -\frac{G\left(1/8\ M_0\right)^2}{(ir)^2}e^{\frac{-r}{r_0}} \tag{7-36}$$

$$= \frac{1}{64}\frac{GM_0^2}{r^2}e^{\frac{-r}{r_0}}, \tag{7-37}$$

for one pair of orbiting quartons in the pion, where ir is not the radius of the pion, but the total separation of quartons on opposite sides of the pion. r is about the separation of M_1 and M_2. The pion has two such pairs of orbiting quartons, therefore the total force in the pion between opposite sides is

$$F = \frac{1}{32}\frac{GM_0^2}{r^2}e^{\frac{-r}{r_0}}, \tag{7-38}$$

where $M_0^2 = -\hbar c/G$. Eq. (7-38) is the correct value of the exponential strong force. This being the weakest link in a chain of three stretching forces, this will be the force between M_1 and M_2.

*Uniting   magnetism   and   the   subscripted   weak
interactions*

All of the forces in this subsection act at right angles to
the radius. The principle difference between the ordinary
magnetic force and the subscripted weak forces is the
magnetic force is not length contracted, but the weak interactions are extremely length contracted with imaginary
radii. These differences arise due to relativity and different
frames. One other difference concerns the multiplication of
α^k in the numerators of the forces, where k = 0 for the
magnetic force, but where k equals positive integers for all
considered weak interactions. This model holds that there
are an infinite number of weak interactions.
As mentioned earlier in this chapter, weak interactions
are mediated by magnetons in going from one level mass
singularity to another. The magnetic force and $weak_1$ and
$weak_2$ interactions are not enhanced, because their radial
magneton velocities are much less than the near speed c for
enhancement.
Different parameters associated with the different levels
of singularities are all that separate and differentiate one
weak interaction from another.   The following master
formula for weak interaction terms in g/2 factors is proof of
the unity of the weak interactions. It unites a compound
infinity of interactions.

$$t_{wk} = \frac{1}{4\pi}(-1)(-1)^p \, n(32)^{k_p-2} \, \alpha^{k_p+1}, \text{ where } k_p \geq 2, \qquad (7\text{-}39)$$

where k_p is the exponent value for the given particle and
interaction, n is the mass distinguishing constant of the g/2-
factor, and p is an integer number of force or mass
singularity order.

D. The Unifying Aspect of a Single Common Aether

The single common aether is not a sea of a single particle, but a single sea of a mixture of every type of boson—such as spin 2 gravitons and spin 1 bosons, doped slightly with spin 0 bosons. Even spin 2 gravitons come in more than one type. There are electrons in tight orbits with positrons for one type of graviton. There are also neutrinos in tight orbits with anti-neutrinos for another type of graviton. The spin 1 bosons also come in more than one type. There are pions in tight orbits with anti-pions for one spin 1 boson. There are also magnetons (electrons orbiting anti-electrons, where one of the particles is inverted, canceling out the intrinsic spin, and having only the spin 1 of the particle orbit). These magneton particles are very important, however, in magnetism and the weak forces, as will be discussed later.

All free particles in the aether sea, or in matter, are combined particles with particles in orbit—which brings us to the first type of force mediation: The aether sea cannot be electrically polarized if the aether particle orbits are only always circular. But if a static electric stress upon the aether cloud can induce an elliptical eccentricity to the aether particles' orbits, then all particles—in the aether and in all matter—can be electrically polarized. The orbiting charged particles in the aether particle will spend a higher percentage of time toward the opposite charge causing the polarization. With random phases in orbits for aether particles, there is a net polarization of the aether cloud.

In the cases of the electric force, the meso-electric force, and the strong electric force, a static electric action-at-a-distance does not have to occur directly. The intimate aether can be polarized by the given static electric charge. Then the rest of the aether can be polarized appropriately everywhere through intimately adjacent polarization adjustments. In this way, the aether can mediate the three electric forces.

Most aether particles, however, cannot be magnetically polarized. For instance, take electron-positron gravitons. Electrons have intrinsic spin -½. Positrons also have spin -½. They orbit each other with an additional spin -1, for a total graviton spin of -2. The electrons spin with negative charge, and the positrons with positive charge. Thus north will be in opposite directions, and south in opposite directions. The magnetic field cancels out in the gravitons and most aether particles. These particles cannot mediate the magnetic field.

Fortunately, there are some particles in the aether that can be magnetically polarized. The author calls them magnetons. Particle physicists call them ω particles.[4] The ω(782) particle is the same as an electron-positron spin 2 graviton, except either the electron or positron has been knocked upside down, so the intrinsic spin cancels out, and there remains only the spin 1 of their mutual orbit. The magnetic fields do not cancel out in this particle. They reinforce constructively. The particle is like a permanent bar magnet. Magnetically polarizing the aether, then, is merely aligning all the magneton particles. Electron-positron spin 2 graviton particles are very abundant particles. Statistically, ω(782) particles should be some fraction of that—occurring naturally. Magnetons, however, would have a finite density in space, so there would be a finite maximum magnetic field strength.[5] Again, magnetic polarizations in space do not have to be direct action-at-a-distance effects from a distant magnetic source. Instead, the magnetically polarized aether can mediate the magnetic force.

The subscripted weak forces differ from the magnetic force chiefly in the exponent parameter k—differ in velocity of the aether and the degree of smallness (order of black hole) of the source emitting particles and the receptor or target particles. The subscripted weak forces are not altogether fundamentally different from the magnetic force.

The same magneton aether particles can mediate the subscripted weak forces as the magnetic force.

Slow and fast, the magnetons mediate the magnetic and weak forces. When the magneton velocity is much much less than c, the magnetons mediate the force singly. But when a magneton has a velocity near c, either way in the direction of the magnetic poles, the magnetic field in the perpendicular direction is increased,[6] and magetons want to attach to the initial magnetons. The constancy of the speed of light governs the increased magnetic field. To preserve a constant speed of light, when α^k is in the denominator of the electric forces, α^k must be in the numerator of the magnetic-weak forces. Likewise, when 32 is in the denominator of the mediated strong nuclear force, 32 must be in the numerators of the mediated weak forces. Actually, it is one power of 32 for each weak force after the $weak_2$ force. Then, considering the coupling constant α, when the weak forces go from one order of a black hole to one less order of black hole, the net coupling constant for low velocity magnetons is α, but for magnetons traveling near c the net coupling constant is 32α.

We have seen how the aether can mediate all forces except gravity and inertia and their related strong gravitational and strong nuclear forces. Both gravity and inertia obey Eqs. (7-20, 7-21). That is, the gravitational force on an object may be due solely to one half the gradient of the aether velocity square times the mass—not on an additional action-at-a-distance. If aether particles were themselves massive particles and obeyed Eqs. (7-20, 7-21), and the velocities of the aether particles intimately adjacent to the mass were controlled by the mass, the aether can cascade the appropriate gradients through reactions of intimately adjacent aether particles—all through the aether—to the second mass in consideration. Thus the aether, by another means—other formulae—can mediate gravity, inertia, and strong gravity and the strong nuclear force. Thus by different means the single common aether mediates every different force.

E. The Essence of the Unified Field Theory

This chapter unites the forces. This unified field theory is not limited to a series of Gauge theories, where a different elementary particle mediates each separate force. No force (this includes the strong nuclear force mediated by the pion) is mediated by just one boson. "The 'graviton', or quantum emitted in gravitational radiation, cannot be assigned a definite spin. It can be shown that this particle must involve a super-position of spin 0, spin 1, and spin 2 contributions.[7][8] There is not a single graviton— such as a spin 2 graviton. Gravitons are *all* integral spin bosons. Also, contrary to popular opinion, the zero mass photon does not mediate on the lowest level the electric and magnetic forces. The electric force is mediated by all non-zero, imaginary mass, integral spin bosons. (The zero mass photons travel linearly only at c. But the non-zero, imaginary mass, integral spin bosons can travel anywhere from low velocities to much faster than the speed of light. It is true that the photon carries momentum and stirs up an electromagnetic wave-train, but the wave is of the luminiferous aether, and the non-zero, imaginary mass, integral spin bosons mediate the electric forces.) Even the magnetic field is not mediated by just one magneton—the $\omega(782)$ particle. All the omega particles are magnetons— such as $\omega(1420)$ and $\omega(1600)$ particles. This model holds that there is an infinite number of higher mass omega particles to be discovered. All omega particles are magnetons. The magnetic forces are not mediated by a single Gauge particle. While on a high level, the weak forces may be seen to be mediated by the W and Z particles, the weak forces are mediated on the low fundamental level by the electric and magnetic bosons. Even the strong nuclear force, mediated on the higher level by the pion, is mediated by all the non-zero mass bosons on the lower fundamental level. Thus this unified field theory is not limited to a series of Gauge theories.

This unified field theory is in the old spirit of one parent force being related to every other force by one aether. As will be seen in the next chapter, the parent force is gravity. The aether is a sea of bosons. It is not possible to unify all the forces without solving the structure of an "elementary" particle such as the electron. The forces are bound up in the elementary charge structures. In summary it is seen that all matter is electrical in character. As will be seen in the next chapter, all forces are gravitational in underlying basis. And the unifying aether is electrical in nature. The same electric aether mediates all the forces, including the weak forces and the strong gravitational force, the underlying force to the exponential strong force (strong nuclear force), secondarily mediated by the pion.

F. Summary and Forward Look

This volume started with the duality mystery of light, considered aetherless and aether relativity, united gravity and inertia in an aether system, solved for the structure of an electron and its sub-particles, thereby uniting the forces. The next chapter in this book will further unify the physical universe, deriving the forces and principal constants from a single defining formula. This volume will continue to go on from here, building on this beginning to formulate a unified Universe and derive supernatural powers. This volume will continue with particle structure inferred from chonomics. Electrino fusion will be introduced with its practical applications, and its implications in cosmology and the origin and evolution of the Universe. This volume will continue in a derivation of how to reverse aging, disease, and decay processes. Then it will formulate the structure of known matter and the masses of charged leptons, predicting the masses of un-discovered particles.

This volume will barely introduce the science of constants calculation from first principles, but will point the way to greater deduction and induction of physical truths.

---

[1]John R. Reitz, Frederick J. Milford, and Robert W. Christy, *Foundations of Electromagnetic Theory*, Third Edition (Reading, Massachusetts: Addison-Wesley Publishing Company, 1980), pp. 160, 161.

[2]http://physics.nist.gov/cuu/Constants/.

[3]Robert B. Leighton, *Principles of Modern Physics* (New York: McGraw-Hill Book Company, Inc., 1959), pp. 40-41.

[4]*CRC Handbook of Chemistry and Physics*, 80th Edition, Editor-in-Chief David R. Lide, Ph.D. (Boca Raton: CRC Press, 1999-2000), p. **11-8**.

[5]William Taussig Scott, *The Physics of Electricity and Magnetism*, Second Edition (New York: John Wiley & Sons, Inc., 1966), p. 410.

[6]*Foundations of Electromagnetic Theory, op. cit.,* p. 497.

[7]V. Ogievetsky and I. Polubarinov, *Ann. Phys.* (USA) **35,** 167 (1965).

[8]H. A. Atwater, *Introduction to General Relativity* (Oxford: Pergamon Press, 1974).

[9]Variation in weak force found by E158, CERN COURIER, Jul 18, 2005; http://cerncourier.com/cws/article/cern/29370.

[10] SLAC E158:   Measuring the Electron's WEAK Charge, Running Strengths of Interaction, http://www-project.slac.stanford.edu/ e158/running-unification.html

Problem Set 7

1. All the forces can be united to what parent force?

2. Would this be possible without an aether?

3. What is the fundamental difference between most forces?

4. What is the greatest unifying factor in this unified field theory?

5. What type of numbers are used for the radii and masses of fine structures? What type of numbers are used for the radii and masses of macroscopic structures?

6. What particle mediates the weak forces? What evidence is there of this?

7. Why are the $weak_1$, $weak_2$, and magnetic forces not enhanced by 32?

8. What integer variable multiplies all weak interaction terms above the first one?

9. What relationship do particles sustain to mass singularities?

10. What, if any, forces can escape a mass singularity?

11. What, if any, signal coupling constant is there between adjacent levels of mass singularities before enhancement? After enhancement?

12. What formulae unite gravity and inertia?

13. What unites gravity and the strong electric force?

14. What unites strong gravity and the strong electric force?

15. What unites gravity and the strong nuclear force?

16. How much weaker is the strong nuclear force than the strong force? Why is that?

17. How much stronger is an enhanced weak force than the non-enhanced weak force?

18. How many weak interactions are there? Other than those, how many forces has the author identified?

19. What are the p, q, and k values for the magnetic force?

20. What unites the electric and strong electric forces?

21. What unites gravity and strong gravity?

22. What is the meso-electric force? How is it united with the electric and strong electric forces? What particles are held together by this force?

23. What unites electricity and magnetism? Is this system compatible with an aether?

24. Why should the unified field theory not be a series of Gauge theories?

Chapter 8

UNIFYING THE UNIVERSE

I. Introduction

A. Defining Formula

The author sought a unified universe in this volume (and papers that antedated this volume) through the application of eight postulates. These led to a unified particle theory and a unified field theory. But because of this work the author found a way to further unify science. The author learned how to unite and predict many of the constants, forces, and particles, given but one simple formula:

$$\frac{0 \ arbitrary \ mass \ unit}{0} \equiv M_0 (arb. \ ma. \ u.), \qquad (8\text{-}1)$$

M_0 is the strong mass of a whole particle in the relativistic frame as seen by an observer at rest, 0 in the numerator on the left side of the equation is the mass of a whole particle (uniton) at rest, and 1/0 is the gamma factor transforming the non-relativistic frame to the relativistic frame when the uniton travels at exactly the speed of light. Many fundamental constants, forces, and particles in the Universe can be derived from this simple definition. This chapter shall begin that work.

B. Significance of the Uniting Definition

Since M_0 is in terms of $\hbar$, c, and G, the application of the definition shows that the speed of light c, the gravitational constant G, and Planck's constant over 2π $\hbar$

are not dependant on matter density in the Universe, temperature, or pressure, or any other thing. They only appear complex due to the units we measure them in—kg, m, s. In natural units they are all 1 and changeless. The Universe has an underlying simplicity to it.

The uniting definition has shockingly obvious clues in it as to the origin of the Universe and the Universe sustaining principle. Many books have been written regarding the origin of the Universe, with diametrically opposed origin theories. These will have to be re-written after the revelation of the uniting definition. To avoid distraction from the overall objectives of this book, however, these clues and interpretations of the uniting definition will be deferred to later books and papers by the author on this subject.

The singular system solution of the forces and particles in the Universe brings to light forces and powers never known by man before. Mankind has been living in a low, natural level of abilities due to his scientific ignorance. But now are revealed other supernatural abilities through now understood laws of physics that have heretofore been in the realm of religion, the psychic, and paranormal. These powers can now be de-mystified, tamed for mankind, and employed for individual and collective interests.

A tremendous benefit of this discovery, also, is the great harmony this can bring to some of science. Much is hard derived from one uniting master definition. Enjoy the beauty and elegance of these derivations.

II. Deriving the Universe

A. The force of gravity

We start this derivation of the Universe with the selection of the form of the force of gravity with our one given, Eq. (8-1). While 0/0 is indeterminate and must be defined, 0 x 0 = 0 without question. We can make use of this truism to deduce the form of the force of gravity. The 0 in the numerator in the left side of Eq. (8-1) refers to the mass of the uniton in the non-relativistic frame. Gravity is a force in the non-relativistic frame. Therefore we wish to select a form of the force of gravity that will satisfy the condition that 0 x 0 = 0 for the zero mass unitons. We want a force form between any two arbitrary masses separated by a radius r. We can guess a form

$$F_g = -\frac{G}{r^2} m_1 m_2.$$ (8-2)

As long as both masses are zero, the force will be zero in this form irrespective of r and the constant G. That satisfies the truism that 0 x 0 = 0. There are other forms we could guess that would not satisfy the truism, such as $F = G + m_1 + m_2 - r$ or $F = G^{m1\ m2\ /r}$. Our reason for selecting an inverse r^2 comes from the very geometry of our problem, and expecting lines of force to become more widely separated with distance. The exact constants of that geometry we lump into our constant G, which as yet is undefined. (Don't worry about that. We will define this constant later all consequent to Eq. (8-1).)

The form of the force of gravity in Eq. (8-2) works for unitons at rest with the aether. But what about other non-zero mass particles? We have no additional condition to derive the form of the force equation for non-zero mass particles. Thus, in an effort to keep the Universe from getting less parsimonious, we make Eq. (8-2) work for all masses. Just from Eq. (8-1) and the truism that 0 x 0 = 0, we have selected the form of the force of gravity. From gravity all other forces can be derived.

B. Strong gravity

We need also the form of the force of strong gravity. We have no other definitions upon which to base any other form or any other constant than

$$F_G = -\frac{G}{R^2} M_1 M_2. \qquad (8\text{-}3)$$

If we select any other form for strong gravity, the foundation of the Universe can not be derived from just one definition. So we adopt the form of Eq. (8-3) for the force of strong gravity.

Notice that M_1 and M_2 are not generally zero. We cannot start with the 0 x 0 = 0 truism in this case. We cannot start with strong gravity as the primary force from which all other forces are based. We need to start with gravity, and establish strong gravity by analogy.

C. Covariance of physical laws

The above relationship between gravity and strong gravity demonstrates complete covariance of the law of gravity with different velocities. This is the most parsimonious case. To have otherwise would require more a priori postulates or definitions than just our one. There is enough information in that one definition to derive all the forces. But not unless there is the covariance of physical laws for all forces and laws.

D. Constancy of c

We shall derive the exact formula for c in a little bit. There is enough information in the single definition to derive c. But not a smear of different c's. Therefore the only way the speed of light can be derived for all observers from the single definition, is to require that the speed of light c be constant to all observers. As you know, this puts amazing constraints and conditions on things at high velocities.

E. Special Quasi-Relativity in an Aether

All of Special Relativity can be derived from the above two conditions, except it must be known whether there is or is not to be an aether. Relativity without an aether leads to contradictions, but relativity in an aether does not. Relativity without an aether would require too many remedial definitions to define. The only way the foundation of the Universe can be deduced from one definition is if there is an aether. But there must be only one aether—electrical and magnetic in character as well as gravitational and inertial.

F. Force against inertia

From the force of gravity law we find that force is mass proportional. It's units are (length/time squared) times mass. We find the units of the multiplying term are the same as acceleration. We would like some formula of the form $F = k_1 ma$. But our single definition does not provide for the derivation of any k_1. In order to preserve the single definition foundation for the Universe, k_1 must be taken to be 1. We then have the valuable result

$$F = ma. \tag{8-4}$$

G.   Escape velocity

In calculating the escape velocity, we find that we have here a two body problem, and not a single body problem. Therefore we must introduce a reduced mass μ into the calculations.   For simplicity sake, let us consider two particles of equal mass at rest ($m_1 = m_2 = m$) at a separation of infinity.   Let them mutually attract and accelerate each other gravitationally.   Let their final relative velocity be 2c. Measure both particles from the frame of particle 1. Particle 1 will always be at rest, but particle 2 will have twice the escape velocity of the particles in the negative direction.   With respect to frame 1, particle 2 will go from rest to twice the speed of light.   It will experience relative mass increase with respect to particle 1 [γm, where $\gamma = 1/(1 - ((1/2)v_r)^2/c^2) = 1/(1 - v_e^2/c^2)$].   The two body problem will become more like a one body problem as the relative velocity v_r approaches 2c, the escape velocity v_e approaches c, and the relative mass increase approaches ∞. For this transition we need to consider the reduced mass μ. For low relative velocities, $1/\mu \equiv 1/m_1 + 1/m_2$.[1]   For relativistic relative velocities for equal masses at rest, we have $1/\mu \equiv 1/m + 1/(\gamma m)$.   To begin with, at low relative velocities, the masses are almost equal, and (m/μ) is almost equal to 2.   As the acceleration of gravity proceeds, and the relative velocity of the two masses is approximately 2c (the escape velocity is approximately c), the one particle is observed as having an extreme mass increase relative to the other particle.   Then the problem becomes essentially a one body problem with one mass approximately an infinity times the other mass.   In that instance, where $v \approx c$, (m/μ) becomes approximately 1, because $m/\mu \equiv m/m + m/(\gamma m)$. Near c, the last term is approximately 0, and $(m/\mu) \approx 1$. Now the above approximations are true and convenient for many physics calculations.   But we are making an exact derivation of the foundation of the Universe.   We can make no approximations.   We must take into account every term in the equation. Therefore we must include the term m/μ in the calculations.

Let us evaluate m/μ and solve for v_e^2/c^2

$$\frac{m}{\mu} \equiv \frac{m}{m} + \frac{m}{\gamma m} = 1 + \left(1 - \frac{v_e^2}{c^2}\right)^{1/2}.$$ (8-5)

$$\left(\frac{m}{\mu} - 1\right)^2 = 1 - \frac{v_e^2}{c^2}$$ (8-6)

$$\frac{v_e^2}{c^2} = 1 - \left(\frac{m}{\mu} - 1\right)^2$$ (8-7)

$$\frac{v_e^2}{c^2} = 2\frac{m}{\mu} - \frac{m^2}{\mu^2} = \frac{m}{\mu}\left(2 - \frac{m}{\mu}\right).$$ (8-8)

In the final equation in Eqs. (8-8), m/μ varies from 2 to 1, and (2 - m/μ) varies from 0 to 1. Both terms must accompany v_e^2/c^2 where it occurs in equations. For instance, the exact form of the gravitational formula for the escape velocity must be

$$\frac{v_e^2}{c^2} = \frac{k'Gm_2}{rc^2}; \quad v_e^2 = \frac{k'Gm_2}{r},$$ (8-9)

where m_2 is allowed to vary in excess of m_1, and k' is a constant to be determined. $Gm_2/(rc^2)$ varies from 0 to 1. That takes care of the final term in Eqs. (8-8). But where is the next to last term, which must vary from 2 to 1? That term must be added in k'. We make k' = m_1/μ. Then the exact gravity formula for the escape velocity is

$$\frac{v_e^2}{c^2} = \frac{m_1}{\mu} \frac{Gm_2}{rc^2}.$$

(8-10)

The velocity calculated in this section is the escape velocity, which is a function of r but not of t about any gravitational body. Notice the γ in the above equations implies there is an aether traveling with respect to m_2 at the escape velocity.

H. General Quasi-Relativity in an aether

Given an aether, the escape velocity, and Special Quasi-Relativity, all of General Quasi-Relativity in an aether can be computed to any desired degree of precision. [See Chapter 4.]

I.   Derivation of the value c and units of measure

Velocity is in units of length divided by time. c is a velocity. We want the master definition, Eq. (8-1) to predict for us the speed of light. Where c = K length/time, we want the model to predict K. There is nothing in the definition that will predict K. Therefore, if we do not want to increase the number of necessary definitions in the Universe, we will have to take K = 1 in the natural system of units. And physicists have long done this.

We can, however, define K to be any arbitrary value and thereby define a system of units of length and time. Recently scientists defined c to be 299792458 meters/second. That was because that value was more precise than could be measured and calculated from our standard units of meters and seconds, yet was consistent with the standard units of meters and seconds. When the value of c was defined exactly, then ε_0 could also could be calculated exactly, because $\varepsilon_0\mu_0 = 1/c^2$ and μ_0 was also defined arbitrarily in the MKSC and SI systems of measurement ($= 4\pi \times 10^{-7}$ H m^{-1}). The exact definition of ε_0 will be useful for us in calculating exactly the other constants. In the natural system of measurement, ε_0, μ_0, and c are all equal to 1. But in the MKSC system of units (the author's favorite), they all have different values. To 31 significant figures, those constants are as follows:

c   299 792 458.0 m s^{-1} (exact)

μ_0 1.256 637 061 435 917 295 385 057 353 311... x 10^{-06} H m^{-1}

ε_0 8.854 187 817 620 389 850 536 563 031 712... x 10^{-12} F m^{-1}.

J. Definition of charge, electric and strong electric forces

The master definition of the foundation of the Universe defines mass at two different velocities. But what is mass made of? What kind of substance? We can say there is some substance that matter is made of. We can call that substance charge. We can say that all matter will be electrical in character. Matter attracts or repels matter in both gravity and strong gravity. And matter is composed of charge. Therefore we can say that charge attracts or repels charge in both the electric and strong electric (in relativistic frame) forces. What should be the form of the forces? Both gravity and strong gravity are inverse square forces. Therefore we require that the electric and strong electric forces both be inverse square forces. What should be the proportionality constant in the electric force? The master definition says nothing about that. Therefore, to avoid the need for further definitions, we are forced to take the electric proportionality factor in natural units to be 1. Other defined units, such as MKSC (SI) units, can have an electric force scale factor $1/(4\pi\varepsilon_0)$ (ie. $f = q_1q_2/(4\pi\varepsilon_0r^2)$).

What should be the proportionality factor for the strong electric force? We allow that the scale factor could be different for the strong electric force than for the electric force. The strong electric force is in a different order black hole than the electric force. The scale factor for different black holes is α^k. Therefore we will assign the proportionality factor $1/\alpha^k$ ($F = q_1q_2/(4\pi\varepsilon_0\alpha^k r^2)$) to the generic electric force, where k is the order of black holes, and α is the Fine Structure Constant in the measuring system. Where k = 0, the ordinary electric (Coulomb) force is obtained. Where k = 1, the strong electric force is obtained.

There must be some link between the gravity forces and the electric forces. If we would define a standard situation, we must use standard uniton particles. However the non-relativistic mass of unitons is zero. Therefore gravity with unitons is undefined except 0 x 0 = 0. Thus if we equate the forces, we are forced to equate strong gravity and the strong electric forces. In the MKSC (SI) units, we have

$$-\frac{GM_1M_2}{R^2} = \frac{q_1q_2}{4\pi\varepsilon_0\alpha R^2}. \qquad (8\text{-}11)$$

$$-\alpha = \frac{q_1q_2}{4\pi\varepsilon_0GM_1M_2}. \qquad (8\text{-}12)$$

For the defining case of two unitons, this is

$$-\alpha = \frac{1^2\,e^2}{4\pi\varepsilon_0G(1^2)M_0^2} = \frac{e^2}{4\pi\varepsilon_0GM_0^2}. \qquad (8\text{-}13)$$

We will solve for α in a grand solution of fundamental constants.

K. Spin and angular momentum

If there are only two unitons in the Universe, and no other particles, there cannot be spin and angular momentum. But if there are two unitons and an aether, then there can be orbital rotations and angular momentum. Angular momentum and orbits are essential in keeping particles from imploding, of holding planets and moons in orbit around the sun, and rotating the entire Universe, thereby keeping it from imploding or exploding. Therefore we desire to have angular momentum in the Universe. We desire natural units of spin to include a product of mass, radius, and velocity. The master definition does not say anything about a scale factor for angular momentum. Therefore we are forced to make the scale factor 1 in natural units. In MKSC (SI) units we can define a scale factor $\hbar \equiv M_0 R_0 c$. We will solve for $\hbar$ in the grand solution of fundamental constants. But first notice that if we take the most parsimonious structure for unitons—an extremely thin spherical shell of charge of radius R_0—we can solve for $M_{1/8}$, M_0, $R_{1/8}$, and R_0 in terms of fundamental constants $\hbar$, G, and c. We know from special relativity in an aether, which we have already derived, that $E = Mc^2$. Let us calculate the mass energy of relativistic unitons:

$$M_0 c^2 = \frac{1^2 e^2}{4\pi\varepsilon_0 \alpha R_0} = M_0\left(\frac{-GM_0}{R_0}\right). \qquad (8\text{-}14)$$

But α is also defined as

$$\alpha \equiv \frac{e^2}{4\pi\varepsilon_0 \hbar c}. \qquad (8\text{-}15)$$

Therefore

$$\hbar c = -GM_0^2; \quad M_0^2 = -\frac{\hbar c}{G}. \qquad (8\text{-}16)$$

$$R_0 = \frac{\hbar}{M_0 c}; \quad R_0^2 = -\frac{\hbar G}{c^3}. \qquad (8\text{-}17)$$

Fracton masses are simple fractions of the uniton mass. We see that $M_{1/8} = M_0/8$ and $R_{1/8} = R_0/8$. By the same token, we can take $M_{1/4} = M_0/4$, $R_{1/4} = R_0/4$, $M_{1/2} = M_0/2$, and $R_{1/2} = R_0/2$.

We see there are four distinct elementary particles (or electrinos): octons, quartons, semions, and unitons and their anti-particles. These can all be derived from the uniton and the anti-uniton, however, if two octons can fuse to one quarton, two quartons can fuse to one semion, and two semions can fuse to one uniton. This provision will have profound effects upon the Universe.

L. Magnetism and the weak forces

We desire that charge in motion, and especially orbiting charge, shall have a force. We define a force $f \equiv k''qv \times B$. The master definition doesn't say anything about the scale factor k''. Therefore we are forced to take it as one in natural units. We have then $f \equiv qv \times B$. B depends on the type of particle structure system we select for the Universe. Do we want only four types of particles in the Universe, with strong masses $M_{1/8}$, $M_{1/4}$, $M_{1/2}$, and M_1 (where the author has called M_1 M_0)? Or do we want an infinite number of possible masses? Four types of particles can be constructed with a single order of mass singularity. But an infinite number of particles requires an infinity of orders of mass singularities.

Various ways of having the aether speed at the speed of light are various ways of having mass singularities.

Orbiting charges form current loops.   There are two different ways to increase the current in the loops: 1. make the charges travel faster, and 2. make the loops smaller. Nature appears to employ the latter in the hierarchy of particles. Each higher order particle is smaller—a higher order singularity—while orbital particle velocity remains constant at c on the event horizon.

The exponent k and orbital velocity v_2 affect the definition of B. The scaler equivalent of the magnetic field is

$$B = \frac{K' \sum_i q_i \left(- v_2 \right) \alpha^k}{r_i^2},$$ (8-18)

where $k \approx 0$ and $v = v_2 \ll c$ for the magnetic field in scaler form. For all the subscripted weak forces, $v_2 = c$. The master definition says nothing about the scale factor K'– which therefore must be 1 in natural units. In MKSC (SI) units,

$$K' = \frac{\mu_0}{4\pi}, \quad B = \frac{\mu_0}{4\pi} \frac{\sum_i q_i \left(- v_2 \right) \alpha^k}{r_i^2}.$$ (8-19)

The scale factor μ_0 is defined in the following manner: $\varepsilon_0 \mu_0 = 1/c^2$. Also $(4\pi\varepsilon_0)(\mu_0/4\pi) = 1/c^2$.

We wish also to define B_{wk} for the weak fields. That is a complex job. There are an infinite number of subscripted weak interactions. Each has a different scale factor, α^k. All but two have force enhancement by magnetons, multiplying them by $(32)^{\{k_p - 2\}}$, where k_p is the k value of the given particle.   Finally,   all but one weak interaction are susceptible to engrossment by an integer variable n, where

p | 0 1 2 3 4 5 ...
n | 0 1 3 6 10 15 ...

The parameter p is the number of the force (as subscripted weak force) in a sequential list. As a net result, B_{wk} looks like the following for weak$_2$ and higher weak forces:

$$B_{wk} = \frac{\mu_0}{4\pi} \frac{\sum_i q_i c (-1)(-1)^p n (32)^{k_p - 2} \alpha^{k_p}}{R_i^2}. \qquad (8\text{-}20)$$

M. g/2-factors

1. g/2-factor Theory

Our model of half the g-factor is a sum of terms—each term for a force—each term an integration of a force with respect to r or R (in other words, an energy), divided by the standard energy of that particle system in the strong relativistic frame or the non relativistic frame, as appropriate (in other words, $E = Mc^2$ or $E = mc^2$, respectively).

The forces in the strong relativistic frames we integrate to the standard radius in the inner relativistic frames, namely R.   The forces in the non-relativistic frames—electricity and magnetism--we integrate to the standard in non-relativistic frames, namely $2\pi r$. For the weak forces, we integrate to the standard $2\pi R$, because there are no magnetic monopoles.   In the following general term derivations, the integration limit will be expressed as lim, to be substituted from Table 8-1. lim will either be 1 for the strong force and gravity, or 2π for all other forces. For the magnetic force and weak forces, the $2\pi r$ or $2\pi R$ is for only one r or R in r^2 or R^2 (in other words r($2\pi r$), etc.

The inertial force is not included in this calculation. Strong gravity and the strong electric force are equated in the model. They are not additive. They are equivalent calculation of forces. Therefore we need to consider only one of the two—strong electric or strong gravity forces—in half the g-factor. The following is the pattern of terms we will encounter in the calculated g/2 factors.

$$g_p / 2 \equiv \left(\sum t_{se} + t_e + t_{mg} + t_{wk1} + t_{wk2} + \ldots + t_g \right)_p, \quad (8\text{-}21)$$

where t here is defined to be the term, se stands for strong electric, e electric, mg magnetic, wk1 weak$_1$, etc., and g gravity. We will calculate each term separately.

In all the magnetic (and weak) type forces and interactions, the integration is done already. We only have to adjust the resultant energy for the R or r limits and weak terms, calculate, and simplify.

The magnetic and weak g/2 terms have in the calculations a velocity v_2, which in these calculations is orbital. For all subscripted weak interactions, $v_2 = c$. For the magnetic force, $(v_2 \ll c) \approx 0$.

The exponent of α k is the sum of two parameters p and q: p is the sequential number of a list of forces; q is the number of shell the mass singularity is in. As with electron shells, singularities in particles do not all come in a single shell. For electrons and muons, q = 0. For tauons and a few higher charged leptons, q = 1. Be aware that q will switch to 2 and higher numbers as p increases beyond the first few numbers.

Like $v^2 = 2GM/r$ for speeds much, much less than c, and GM/r for speeds about c, the denominators of some g/2 factor magnetic and weak terms have an extra factor of two in them. In the calculations, a parameter Δ substituted from Tables 8-1 or 8-3 accounts for these differences. For the magnetic term and weak$_1$ term for the electron g/2-factor, Δ = 2. For all other weak terms for any particles, Δ = 1. The appropriate values of Δ will be recorded in Tables 8-1 and 8-3.

Please note that r^2 is positive, R^2 is negative, m^2 is positive, M^2 is negative, rm is positive, and RM is positive. In the expression two charges are multiplied together—ee. The test charge is negative—the electron—-e. If the main charge is also –e, the negative signs cancel, and the Sign is +. If the main charge is positive, the Sign is -. In the electric force, r and m are used (the outer non-relativistic values). In the strong electric force, R and M are used (the inner relativistic values). In the equations, the radius is abbreviated as rad.

2. Calculation of g/2-factor terms

First we calculate the general terms, employing the above data for the g/2-factor. Then we will calculate the individual force terms for the electron g/2-factor and for the muon g/2-factor.

a. General electric term.

$$t_{ge} \equiv \frac{-\int_{\infty}^{\lim rad} \dfrac{sign\ e^2}{4\pi\varepsilon_0 \alpha^k rad^2}}{mass\ c^2}, \qquad (8\text{-}22)$$

$$= \frac{-\hbar c}{\lim rad\ \alpha^{(k-1)}\ mass\ c^2}, \qquad (8\text{-}23)$$

$$= \frac{-\hbar c\ c}{\lim c^2 \alpha^{(k-1)}\hbar}, \qquad (8\text{-}24)$$

$$= \frac{-1}{\lim \alpha^{(k-1)}}. \qquad (8\text{-}25)$$

For the electric force, lim = 2π, k = 0. Therefore, for the electric force, $t_e = -\alpha/(2\pi)$.

For the strong electric force (equivalent to the strong gravitational force), lim = 1, k = 1. Therefore $t_{se} = -1$. This is the same as the strong gravitational term t_{sg} calculated below:

b. Strong gravitational term.

$$t_{sg} \equiv \frac{-\int_{r=\infty}^{R_0} \dfrac{GM_0^2}{r^2} dr}{M_0 c^2}, \qquad (8\text{-}26)$$

$$= \frac{GM_0^2}{R_0 M_0 c^2} = \frac{-M_0^2}{M_0^2} = -1. \qquad (8\text{-}27)$$

Only one strong term is employed in the g/2-factor.

The explanation of the minus signs in Eq. (8-26) is as follows: The minus sign is because of the backward integration. Another minus sign comes from the integration of r^{-2} to r^{-1}.

c. Magnetic term.

The magnetic force makes a split of the energy state of $e\hbar B/m$ separation when in a magnetic field for electrons in atoms, adding $e\hbar B/2m$ to the maximum energy state in atom orbits. We are interested in how much energy a magnetic field can add to the maximum energy state of a particle. The charge of electrinos is a fraction of e, but the charge sums to e in common particles. Also the relativity reduced mass of the electrinos is a fraction of the mass of the particle m, but the mass sums to m in the particle. Taking the energy $e\hbar B/2m$ without integration, but putting in the limits of the 'would have been' integration, and remembering the Δ term for magnetic type terms, we obtain the magnetic term. In the general electric term, α is always positive. But in the magnetic and weak terms, α is alternately negative or positive, depending on the power p. For the magnetic term, we will take $p \approx 0$ and $k_p \approx 0$. So we also make note of that in the following terms. (The parameters are defined in Chapter 7, pp. 217-221.)

$$t_{mg} = \frac{e\hbar(-1)^P B}{2\Delta \ mass \ mass \ c^2},$$ (8-28)

$$= \frac{e\hbar(-1)^P \alpha^{k_p}}{2\Delta \ mass^2 \ c^2} \frac{\mu_0 e(-v_2)}{4\pi \ rad \ lim \ rad},$$ (8-29)

$$= \frac{e^2}{4\pi\varepsilon_0\alpha} \frac{\hbar(-1)^P \alpha^{(k_p+1)}(-v_2)}{2\Delta c^4 \ lim} \frac{1}{r^2 m^2}.$$ (8-30)

But $$r^2 m^2 = \frac{\hbar^2}{c^2}.$$ (8-31)

Therefore $$t_{mg} = \frac{\hbar^2 c^2 (-v_2/c)(-1)^P \alpha^{(k_p+1)} c^2}{lim \ 2\Delta c^4 \hbar^2},$$ (8-32)

$$= \frac{(-v_2/c)(-1)^P \alpha^{(k_p+1)}}{lim \ 2\Delta}.$$ (8-33)

But $p = k_p = 0$, lim $= 2\pi$, and $\Delta = 2$, (8-34)

so $$t_{mg} = \frac{(-v_2/c)\alpha}{8\pi}.$$ (8-35)

Where v_2 is $\ll$ c (as in magnetism), it is from one order less mass singularity hole than c is. We already have an example of what one order less singularity does to the g/2 factor term. The electric is one order less a singularity than the strong. The electric term is $-\alpha/(2\pi)$. The strong term is -1 The ratio is $\alpha/(2\pi)$. For the magnetic term, we take $v_2/c = \alpha(2\pi)$. Then $t_{mg} = -\alpha^2/(16\pi^2)$.

d. General weak term.

Weak forces are similar to magnetic forces, except they have more terms to account for magneton mediation and mass distinguishing constant engrossment. Also $v_2/c = 1$ in all cases, so there is an extra (-1) for ($-v_2/c$) in the terms. In the following calculations, Eq. (8-20) will be substituted for Eq. (8-19). The equation for the general weak term is

$$t_{wk} = \frac{e\hbar(-1)^P B_{wk}}{2\Delta \ mass \ mass \ c^2}, \tag{8-36}$$

$$= \frac{e\hbar(-1)^P \alpha^{k_p}}{2\Delta mass^2 c^2} \frac{\mu_0 e(-c) n(32)^{k_p-2}}{4\pi \ rad \ lim \ rad}, \tag{8-37}$$

$$= \frac{e^2}{4\pi\varepsilon_0 \alpha} \frac{\hbar(-1)^P \alpha^{(k_p+1)}(-c) \ n(32)^{k_p-2}}{2\Delta c^4 \ lim} \frac{R^2 M^2}. \tag{8-38}$$

But $e^2/4\pi\varepsilon_0\alpha = \hbar c$, $RM = \hbar/c$, lim = 2π, and $\Delta = 1$

$$\tag{8-39}$$

for weak$_2$ and higher weak forces. Thus the master weak term for weak$_2$ and higher weak forces, where $k_p \geq 2$, is

$$t_{wk} = \frac{1}{4\pi}(-1)(-1)^p \, T(32)^{k_p-2} \, \alpha^{k_p+1}. \qquad (8\text{-}40)$$

For lower order weak terms it is best to determine the terms through experimental calculations and best fit to the g/2 factor.

e.   Gravity term.

$$t_g \equiv \frac{-\displaystyle\int_{\infty}^{r} -\frac{Gm^2}{r^2}\,dr}{mc^2} = -\frac{Gm^2}{rmc^2}, \qquad (8\text{-}41)$$

$$= -\frac{G}{\hbar c}m^2 = \frac{m^2}{M_0^2}. \qquad (8\text{-}42)$$

The author's guess is that the gravitational term of the g/2-factor also has r limits of integration. It will require 45 place accuracy to disprove this. For exact g/2-factors, the gravity term is required. However, if less than 45 place accuracy is required, the gravity term can be neglected.

3.   Evaluating the Electron g/2-factor

We begin to evaluate the electron g/2-factor by completing Table 8-1.

Table 8-1
Parameters Used in Calculating the Electron g/2-factor Terms

force	sign	rad	mass	Δ	lim	v_2	p	q	k_p	n	electron g/2-factor term
strong	+	R	M	1					1		-1
electric	+	r	m		2π				0		$-\alpha/(2\pi)$
magnetic		r	m	2	2π	≈ 0	0	≈ 0	≈ 0	1	$-\alpha^2/(16\pi^2)$
weak$_1$		R	M	2	2π	c	1	≈ 0	1	1	$\alpha^2/(8\pi)$
weak$_2$		R	M	1	2π	c	2	≈ 0	2	1	$-\alpha^3/(4\pi)$
weak$_3$		R	M	1	2π	c	3	≈ 0	4	1	$(32\alpha)^1\alpha^3/(4\pi)$
weak$_4$		R	M	1	2π	c	4	1	5	1	$-(32\alpha)^3\alpha^3/(4\pi)$
weak$_5$		R	M	1	2π	c	5	1	6	1	$(32\alpha)^4\alpha^3/(4\pi)$
weak$_6$		R	M	1	2π	c	6	1	7	1	$-(32\alpha)^5\alpha^3/(4\pi)$
weak$_7$		R	M	1	2π	c	7	1	8	1	$(32\alpha)^6\alpha^3/(4\pi)$
weak$_8$		R	M	1	2π	c	8	1	9	1	$-(32\alpha)^7\alpha^3/(4\pi)$
etc.											
gravity		r	m	1							$m_e^2/(M_0)^2$

The right hand column of Table 8-1 can be evaluated if we obtain a value for α, the Fine Structure Constant. The most recent (2006) CODATA value[2] is 0.007 297 352 5376(50). The author used the CODATA value to evaluate the g/2-factors. For the evaluation of the electron g/2-factor, see Table 8-2 below.

Table 8-2
Electron g/2-factor Evaluation

force	electron g/2-factor term	numerical value
strong	-1	$-1.000\ 000\ 000\ 000\ \ldots$
electric	$-\alpha/(2\pi)$	$-0.001\ 161\ 409\ 727\ \ldots$
magnetic	$-\alpha^2/(16\pi^2)$	$-0.000\ 000\ 337\ 218\ \ldots$
weak$_1$	$\alpha^2/(8\pi)$	$+0.000\ 002\ 118\ 804\ \ldots$
weak$_2$	$-\alpha^3/(4\pi)$	$-0.000\ 000\ 030\ 923\ \ldots$
weak$_3$	$(32\alpha)^1\alpha^3/(4\pi)$	$+0.000\ 000\ 007\ 221\ \ldots$
weak$_4$	$-(32\alpha)^3\alpha^3/(4\pi)$	$-0.000\ 000\ 000\ 393\ \ldots$
weak$_5$	$(32\alpha)^4\alpha^3/(4\pi)$	$+0.000\ 000\ 000\ 091\ \ldots$
weak$_6$	$-(32\alpha)^5\alpha^3/(4\pi)$	$-0.000\ 000\ 000\ 021\ \ldots$
weak$_7$	$(32\alpha)^6\alpha^3/(4\pi)$	$+0.000\ 000\ 000\ 005\ \ldots$
weak$_8$	$-(32\alpha)^7\alpha^3/(4\pi)$	$-0.000\ 000\ 000\ 001\ \ldots$
etc.		
gravity	$m_e^2/(M_0)^2$	$-0.000\ 000\ 000\ 000\ \ldots$
Total with eight weak forces		$-1.001\ 159\ 652\ 163\ \ldots$

The calculated theoretical electron g/2-factor is -1.001 159 652 163 . . . . The 2002 CODATA electron g/2-factor is -1.001 159 652 1811(08). The theoretical value differs from the measured value by 1.81×10^{-11}. (The difference with 1998 CODATA data was 2.38×10^{-11}, and with 1986 CODATA data was 5.70×10^{-11}.) The fit of the theoretical values with the measured values continues to get better. Still, the measured value of the electron g/2-factor is more accurate than the measured α used to try to calculate it. We need a more accurate α.

4. Evaluating the Muon g/2-factor

The muon g/2-factor differs from the electron g/2-factor in one way. All but one weak interaction terms are multiplied by n ≈-3 for muons. We evaluate the muon g/2-factor by completing Tables 8-3 and 8-4.

Table 8-3
Parameters Used in Calculating the Muon g/2-factor Terms

force term	sign	rad	mass	Δ	lim	v_2	p	q	k	n	muon g/2-factor
strong	+	R	M	1					1		-1
electric	+	r	m		2π				0		$-\alpha/(2\pi)$
magnetic		r	m	2	2π	≈0	0	≈0	≈0	-1	$-\alpha^2/(16\pi^2)$
weak$_1$		R	M	1	2π	c	1	≈0	1	-1	$-\alpha^2/(4\pi)$
weak$_2$		R	M	1	2π	c	2	≈0	2	-3	$+3\alpha^3/(4\pi)$
weak$_3$		R	M	1	2π	c	3	1	4	-3	$-(32\alpha)^1\alpha^3/(4\pi)$
weak$_4$		R	M	1	2π	c	4	1	5	-3	$+3(32\alpha)^3\alpha^3/(4\pi)$
weak$_5$		R	M	1	2π	c	5	1	6	-3	$-3(32\alpha)^4\alpha^3/(4\pi)$
weak$_6$		R	M	1	2π	c	6	1	7	-3	$+3(32\alpha)^5\alpha^3/(4\pi)$
weak$_7$		R	M	1	2π	c	7	1	8	-3	$-3(32\alpha)^6\alpha^3/(4\pi)$
weak$_8$		R	M	1	2π	c	8	1	9	-3	$+3(32\alpha)^7\alpha^3/(4\pi)$
etc.											
gravity		r	m	1							$m_\mu^2/(M_0)^2$

Table 8-4
Muon g/2-factor Evaluation

force	muon g/2-factor term	numerical value
strong	-1	$-1.000\ 000\ 000\ 000\ \ldots$
electric	$-\alpha/(2\pi)$	$-0.001\ 161\ 409\ 727\ \ldots$
magnetic	$-\alpha^2/(16\pi^2)$	$-0.000\ 000\ 337\ 218\ \ldots$
weak$_1$	$-\alpha^2/(4\pi)$	$-0.000\ 004\ 237\ 608\ \ldots$
weak$_2$	$+3\alpha^3/(4\pi)$	$+0.000\ 000\ 092\ 769\ \ldots$
weak$_3$	$-3(32\alpha)^1\alpha^3/(4\pi)$	$-0.000\ 000\ 021\ 663\ \ldots$
weak$_4$	$+3(32\alpha)^3\alpha^3/(4\pi)$	$+0.000\ 000\ 001\ 181\ \ldots$
weak$_5$	$-3(32\alpha)^4\alpha^3/(4\pi)$	$-0.000\ 000\ 000\ 275\ \ldots$
weak$_6$	$+3(32\alpha)^5\alpha^3/(4\pi)$	$+0.000\ 000\ 000\ 064\ \ldots$
weak$_7$	$-3(32\alpha)^6\alpha^3/(4\pi)$	$-0.000\ 000\ 000\ 015\ \ldots$
weak$_8$	$+3(32\alpha)^7\alpha^3/(4\pi)$	$+0.000\ 000\ 000\ 003\ \ldots$
etc.		
gravity	m_μ^2/M_0^2	$-0.000\ 000\ 000\ 000\ \ldots$
Total with eight weak forces		$-1.001\ 165\ 912\ 489\ \ldots$

The calculated theoretical muon g/2-factor is -1.001 165 912 489 . . . The 2006 CODATA muon g/2-factor is -1.001 165 920 7(06). The theoretical value differs from the measured value by 8.21×10^{-9}.

The electron and muon theoretical g/2-factors are close but not exact fits with the measured g/2-factors. When α and the g/2 factors are known more precisely, it will be interesting to see if the measured values are closer to the theoretical values.

The number of identifiable precise terms in the g/2-factors is greatly increased with this scheme. We have good confirmation of the fundamental forces of the Universe by the g/2-factors. These very accurate calculations are without recourse to renormalization theory of Quantum Electrodynamics. They are without the quark hypothesis. These accurate calculations are one test for a new boson-aether model of physics in which charges are divided into e, e/2, e/4, e/8, 0, -e/8, -e/4, -e/2, and –e.

Acknowledgment. I am very grateful to Dr. James G. Gilson for discussions on the electron and muon g/2 factors.

N. Grand Solution of Fundamental Constants

All we need now to tie together the concepts of this chapter is to solve for the many scale factors to obtain the primary fundamental constants of the Universe. We have introduced many constant scale factors in this chapter: M_0, G, α, $R_{1/8}$, $R_{1/4}$, $R_{1/2}$, R_0, $M_{1/8}$, $M_{1/4}$, $M_{1/2}$, $\hbar$, c, ε_0, μ_0, e, g_e, g_μ. If we know M_0 and R_0, we know the other M's and R's. Also, if we know α, we know g_e and g_μ. Also c, ε_0, and μ_0 are already known exactly. All we have to solve for are M_0, R_0, G, α, $\hbar$, and e.

$\alpha = e^2/4\pi\varepsilon_0\hbar c$. Thus, all we have to solve for is M_0, R_0, G, $\hbar$, and e. They are 1, 1, 1, 1, 1 in natural units. All we have to do is solve them in SI units. Attempts to solve them, however, go in circles because the SI system defines only macroscopically units of mass (kilograms) as well as length (meters) and time (seconds), and charge (Coulombs), whereas c in SI units employs only meters and seconds, and does not define either. Mass, length, and charge have been defined macroscopically. We want them defined microscopically. Then we will be able to calculate all the above constants.

Eq. (8-1) is a good place to start defining units of mass. M_0 can be anything we like. The definition will affect all units of mass in our system of measurement. M_0 is a calculated value based on measured values of G and $\hbar$ and a defined value of c. Let us calculate M_0 from measured quantities of G and $\hbar$ and a defined value of c, define the result as the exact value, and then calculate G and $\hbar$.

$$- M_0 = -\left(-\frac{\hbar c}{G}\right)^{1/2} \tag{8-43}$$

$$\approx i\left(\frac{1.054571628 \ x \ 10^{-34} \ 299792458}{6.67428 \ x \ 10^{-11}}\right)^{1/2}$$

in units of kg. M_0 is approximately equal to -i 2.176437462 x 10^{-08} kg. Only the first five numbers are significant numbers, due to only five significant figures in the measured G. We could add one digit and define it exactly, but that would seriously disturb the final digits in what is thought to be the known $\hbar$. Therefore, to be in close harmony with the existing fundamental constant recommendations as much as possible, let us take the first nine digits and define M_0 as $M_0 \equiv$ -i 2.17643746 x 10^{-08} kg exactly. Now we can back calculate G, $\hbar$, and R_0 in terms of the exact M_0 and c.

Whereas c is defined exactly, both the units of length (meters) and units of time (seconds) are undefined microscopically. They are not fundamental constants, and are subject to slight variation. We need a microscopic definition of either meters or seconds. If we define one, we will have the other because c is defined. From strong gravity it would be easier to define R_0 than T_0 (the time it takes to go R_0). So let us make a definition of $R_0 = 8 \ R_{1/8}$ by the method used for M_0 above:

$$R_0 = \left(-\frac{\hbar G}{c^3}\right)^{1/2} \tag{8-44}$$

$$\approx \left(\frac{1.054571628 \ x \ 10^{-34} \ 6.67428 \ x \ 10^{-11}}{299792458^3}\right)^{1/2}$$

in units of positive imaginary meters. The calculated result is $R_0 \approx$ i 1.616252457 x 10^{-35} meters. The calculation is significant only to the first five digits because the measured G we used was significant only to the first five digits. We could add one digit and define it exactly, but that would seriously disturb the final digits in what is thought to be the known $\hbar$. Therefore, to be in close harmony with the existing fundamental constant recommendations as much as possible, let us take the first nine digits and define R_0 exactly. $R_0 \equiv i\ 1.61625246 \times 10^{-35}$ kg exactly. We can now solve exactly for G.

$$G = \frac{R_0}{M_0} c^2 = 6.67427984...x\ 10^{-11} \qquad (8\text{-}45)$$

exactly, in units of $m^3/kg\ s^2$.

We can now solve exactly for $\square$.

$$\hbar = \frac{-R_0^2 c^3}{G} = 1.054571653...x\ 10^{-34} \qquad (8\text{-}46)$$

exactly, in units of J s.

All we have left to solve for are α and e. If we know α, we know also e, and vice versa, because

$$\alpha = \frac{e^2}{4\pi\varepsilon_0 \hbar c}. \qquad (8\text{-}47)$$

M_0, R_0, G, $\hbar$, c, and e are all 1 in natural units. To obtain MKSC values for these constants, M_0, R_0, and c had to be defined. e also must be defined to obtain MKSC electro-magnetic units. Actually an accurate value of e may be already back calculated from our already very accurate α (known to twelve figures). Our value of α, and thus e, will only get better with time.

If the reader is unsatisfied with the defined units in this chapter, he/she is welcome to define the units differently (but in harmony with MKSC measurements). At this point it is analogous between deciding on O^{16} based atomic mass units or deciding on C^{12} based atomic mass units. It is not what any one individual will decide, but upon the convention of the majority of scientists.

No more quantities need to be defined to calculate all the constants listed on page 184, Section N, paragraph 1. The author has shown how in this chapter to calculate them all accurately. That is far better than current science, where G is known only to five significant figures. This is a triumph of a Unified Field Theory, Unified Constant Theory based on a single definition.

---

[1]Charles Kittel, Walter D. Knight, and Malvin A. Ruderman, Mechanics: Berkeley Physics Course–Volume 1 (New York: McGraw-Hill Book Company, 1965), p. 276.

[2]http://physics.nist.gov/cuu/Constants/

Problem Set 8

1. The Master Definition defines very few things. How can important scale factors for other entities be defined without increasing the number of definitions necessary to derive the Universe?

2. How many times was this method utilized in this chapter? List the relations thus defined.

3. According to the Master Definition, should $\hbar$, c, or G vary with location or time in the Universe?

4. Why do they seem complex to us?

5. What are they in natural units?

6. How can a mass that is 0 at rest be a large quantity at the speed of light?

7. What is the only force that can be the parent force of all other forces? Why?

8. The relationship between gravity and strong gravity demonstrates what law of physics?

9. What puts constraints on things at high velocities?

10. How many conditions are necessary to derive Special Quasi-Relativity in an Aether? (The author used different conditions to do this in Chapters 3 and 8.)

11. Why is the escape velocity tricky?

12. What elements are necessary to derive General Quasi-Relativity in an aether?

13. Can there be angular momentum if there are only two unitons in the Universe and no other particles? What if there is also an aether?

14. Why are B_{wk} more complex than B?

15. What procedure and model is common to all g/2-factor terms?

16. Why for the weak interactions do we integrate to the standard $2\pi R$?

17. How many subscripted weak interactions are there?

18. What has heretofore been measured as the weak force?

19. Are there separate terms for strong gravity and the strong electric force in the g/2-factors? Why or why not?

20. How many digits in the evaluated g/2-factor would be required before the gravity term would be non-zero?

21. What three term values are common to all g/2-factors?

22. Meters and kilograms are defined macroscopically. How do they need to be defined?

23. If they were thus defined, what else would we be able to calculate exactly?

Chapter 9

UNITED THEORY

A. Significances of the Unified Particle Theory and the Unified Field Theory

1. Further Unification of Physics and Chemistry. The author of this Volume believes Chapter 7 contains a Unified Field Theory suitable to all mass singularity order p particles we consider. In addition, Chapter 6 of this Volume, as amplified in the remainder of this book, contains a Unified Particle Theory. It is more parsimonious than String Theory and Many Dimensional Theories (this theory requires only three space dimensions and time and mass and their imaginary complements). It is more parsimonious also than the Standard Model of Quarks and Leptons. The Standard Model currently requires 61 quarks, anti-quarks, leptons, and anti-leptons and other particles to make up all known matter. By contrast, it takes only one copy of an octon-anti-octon pair, in the Electrino Unified Theory, to create all copies of all possible particles in the Universe through different ionization states, fusion states, and combination states. One cannot get more parsimonious than that.

One beautiful feature of the Electrino Unified Particle Theory and Associated Unified Field Theory is that they potentially can greatly further unify physics and chemistry. As the method of predicting particle masses is further refined, the process of calculating particle masses might be made into a simple computer program. If the masses of particles can be calculated accurately, so too can their radii, their magnetic moments, and other particle parameters. As the constituents of the boson aether are better understood, the half-lives and lifetimes, also, of particles can be calculated. Thousands of measured physical constants

268

could be reduced to just a few fundamental defined constants. There could come a Unified Constant Theory.

2. Revolution in Origin Theories. One thing the Electrino Unified Particle Theory does, it revolutionizes our theories of the origin of the Universe. Instead of needing to start with a Big Bang, the Universe could be created a step at a time from controlled octons. Instead of the interstellar red shift calculating a time for the Big Bang through the Doppler effect, the Unified Theories present a new mechanism for interstellar red shifting—making the Universe much older than it is now thought to be. (See Chapter 14.) A Big Bang should make equal quantities of matter and anti-matter, which is not the case in the Universe. The Unified Particle Theory shows how to make more matter than antimatter in the Universe. (See Chapter 12.)

3. New Powers. So many have sought the Unified Field Theory for so long, that someone questioned if the theory would yield any new powers to man, or only be some quaint useless mathematical curiosity. The author is here to tell that this theory indeed does yield new powers and abilities to man.

This Theory shows how to reverse the order to disorder arrow in the second law of thermodynamics, thereby giving interesting properties to space. (See Chapter 16.)

This Theory shows how to convert matter into antimatter, or vice versa, making anti-matter rockets possible, making new kinds of power reactors possible that can annihilate matter for power, or to clean up and safely dispose of high level radioactive wastes and toxic chemicals. This theory points in the direction of how to make inertia-less craft that would not have a speed of light barrier, but could hurtle through space at almost infinite speeds. (See Chapter 5.)

B. Tests for the New Model

1. Derivation of Constants. Ultimately it is hoped the United Theory (Electrino Fusion Model of Elementary Particles) can predict many constants like the masses, radii, charges, magnetic moments, and half lives of particles. Already the New Model can predict those quantities for all four electrinos and their anti-particles. Also the model predicts the g/2-factors of charged leptons. The electron g/2-factor and the muon g/2-factor are two tests for the model. Other tests we expect soon.

2. Creation of Antimatter. New antimatter (without pair production) can be created by fusing the constituents of electrons. It is actually the semions in electrons that fuse to anti-unitons—the core particles of anti-protons and anti-neutrons. When the anti-unitons are produced, the rest of the constituents of anti-protons and anti-neutrons are scavenged from the boson sea, so that whole anti-protons and anti-neutrons are formed. These products can be detected experimentally.

Electron semions can be fused by accelerating the electrons in head to head linear accelerators—or synchrotrons—to over 938 MeV each (such as 1 GeV each), and colliding them in an electron-electron collider. For the semions to fuse, one electron must have up axial spin with respect to its accelerator, whereas the other electron must have the opposite down axial spin with respect to its accelerator, so both electrons have the same axial spin in the center of mass frame of the collider.

Presently no accelerator in the world has this configuration and capability. Therefore the work will have to be done in a new facility or an old facility with a major overhaul.

3. Creation of Matter. New matter (without pair production) can be created by fusing the constituents of

positrons obtained from pair production. This is just the reverse case of the above. Anti-semions in positrons are fused to the core particles of protons and neutrons. After scavenging from the graviton sea, whole protons and neutrons are formed, which can be experimentally detected.

The positron beams, once formed, are accelerated in head to head linear accelerators—or synchrotrons—to about 1 GeV each positron. They are collided with the apparatus and spin orientation as above.

Presently no accelerator in the world has this configuration and capability. Therefore the work will have to be done in a new facility or an old facility with a major overhaul.

4. Reversing the Second Law of Thermodynamics. The third test (the fusion of the constituents of positrons) also violates the second law of thermodynamics by a strong force reaction. The reaction is theorized to reverse the order to disorder arrow, but not the entropy arrow, of the second law of thermodynamics. In order to show that in a limited test area, the researcher would want to fuse a net 1.0 to 10 picoamps per beam at an expected 1.0 beam collision efficiency to obtain the effect in a radius of 100 meters or less. (The less the efficiency and the less the beam current the greater an area where the effect will be felt. It would be difficult not to affect a wide area, at least in transients, when turning the machine on and focusing the beams.) Any old positron-positron collider would suffice to test the reversal of the order to disorder arrow in the second law of thermodynamics. No coherent beam heroics are necessary.

Problem Set 9

1. The much celebrated Unified Field Theory may be found in what chapter of this book?

2. Does this book have also a Unified Particle Theory?

3. What theories are the Unified Particle Theory more parsimonius than?

4. How many particles does the Standard Model currently require to construct known matter? How many particles does the Unified Particle Theory require to construct all particles?

5. How does the Electrino Unified Particle Theory revolutionize our theories of the origin of the Universe?

6. What new powers are brought to light by the Unified Field Theory?

7. What four tests are there for the new model?

8. Roughly what energy of particles is required to fuse semions?

9. The second law of thermodynamics is made up of how many arrows of time? What are they?

10. Would coherent beam technology be required in a positron-positron collider? Why or why not?

Chapter 10

INTRODUCTION TO CHONOMICS

I. Introduction

For centuries man has been attempting to identify the most elementary particles of matter. At first atoms were thought to be the smallest subdivision of matter. Then subatomic particles such as electrons, protons, and neutrons were discovered. For several decades these have been called "elementary" particles. However, particles in this class have proliferated into the hundreds. Many scientists have reasoned there are too many "elementary" particles to be truly elementary. In our model we have identified two levels of structure below "elementary" particles. We have named them echons and electrinos. Echons are systems of orbiting electrinos. We would believe electrinos are the truly elementary particles. So then what do we call what are now named "elementary" particles? Unfortunately there is no name for them as a class other than "elementary" particles. There needs to be such a name. The one offered here may not stick, but it is descriptive of their structure. All "elementary" particles are constructed of yachons (unitons) and echons (electrinos in orbit). Both yachons and echons end with "chons". So why not call "elementary" particles "chonstructs"?

In Chapter 6, we derived electrinos and semion echons--their masses, charges, spins, radii, and orbital velocities. We wish to do the same thing for the broad class of "elementary" particles we call chonstructs. In later chapters and Appendix B, a number of less technical things we can do are done to set the stage for such calculations, as inducing the yachon-echon structure of all particles discovered. We did some of that already in Chapter 6. But we can do a much more thorough job of it in the study of "chonomics" in the appendices. In the Appendix A, this study is done for leptons.

II. Chonomics' Definitions

A. Yachons and Echons

A large part of chonomics is the balancing of yachons and echons in particle decay schemes. We will need new descriptive tools to do this job.

Let us introduce new pictorial symbols which shall be used in a new kind of pictorial equation. We need first of all symbols for the common three kinds of yachons and echons: minus or plus half spin semion echons, zero spin orbital quarton echons, and zero spin nearly point charge uniton yachons. Let us differentiate them by their spins. We could call a plus ½ spin particle a +½, and a minus ½ spin particle a -½. But this is too cumbersome in the equations. Let us imply the ½$\hbar$ spin and label an up spin semion echon simply a +, and a down spin semion echon a -. Zero spin orbital quarton echons we can call o particles. (The symbol o stands both for their 0 spin and their orbital characteristic. We use the letter o instead of the numeral 0 because it is more round, representative of the quarton orbits, and to facilitate an adaptation of the convention in dealing with multiple pions in particle resonances.) What should we call zero spin nearly point charges? To differentiate them from o particles, let us call them dots (•). All "elementary" particles can be formed by combinations of minuses, pluses, zeros, and/or dots (-, +, o, •).

B. Positional Grid

The minuses, pluses, zeros, or dots can be of either electrical polarity. Also the minuses, pluses, and zeros can have three or more orbital energy states. To differentiate the charge or energy state of the minuses, pluses, zeros, or dots, we introduce a positional grid:

The positional grid has a vertical line in the middle. All echons and yachons (-, +, o, •) to the right of the vertical line are electrically positive. All echons and yachons (-, +, o, •) to the left of the vertical line are electrically negative.

The positional grid has horizontal lines. These separate spaces at different levels in the grid. These represent the different possible orbital energy states a non-dot echon can have. The bottom level is the ground state or 0 state. The middle level is the 1 state, and the top level is the 2 state. As higher energy particles are discovered, we will need more energy state levels to explain particle energies. We would simply make more levels in the grid. But three levels explains all common particles.

For example, an electron can have either a minus ½ spin or a plus ½ spin (can be either a - or +) and is in the ground state for a plus or minus semion echon. It is electrically negative. We could write it as either

e⁻ $\qquad$ or $\qquad$ e⁻

depending on its spin. Let us just focus on the - spin right now. The same echon could have the higher energy state

, as a muon,

or the still higher energy state

$$\tau^-$$
$$-|-$$
$$\underline{\perp}, \text{ as a tauon.}$$
$$|$$

This model regards the muon and tauon the same as electrons, except for differences of energy states of the electrinos.

Similarly π-, K-, and D-mesons are related to each other. They are simply composed of negative or positive zero echons in various states.

$$\pi^+ \qquad \pi^- \qquad K^+ \qquad K^- \qquad D^+ \qquad D^-$$

The bar over a particle symbol indicates it's an antiparticle. But we do not give both the bar and the charge sign. It is like a double negative.

Zero spin nearly point charge uniton dot yachons can only have one energy state. They are always only nearly point charges. Therefore we write them as dots in the bottom level of the grid, or ground state level or 0 state:

$$U^- \qquad\qquad U^+$$

We can call these U particles for unitons. As we calculated in Chapter 6, their masses isolated are $M_0 = -i\ 2.177 \times 10^{-8}$ kg—too massive to be isolated by current man-made machines.

C. Matter, Antimatter

Notice the difference between matter and antimatter—particles and antiparticles: Negatively charged semion echons (minuses or pluses) are matter, whereas positively charged semion echons are antimatter. With quarton echons (zeros) it is just the opposite. Positively charged quarton echons (zeros) are matter, whereas negatively charged quarton echons are antimatter. Positively charged dots (uniton yachons) are matter, whereas negatively charged dots are antimatter.

D. Multi-echon Particles

We must now begin to figure the composition of multi-echon elementary particles. To begin with, neutrinos seem to be two-component particles.[1] One component is a zero and the other a minus or plus:

$$\nu_e \qquad \nu_e \qquad \nu_\mu \qquad \nu_\mu \qquad \nu_\tau \qquad \nu_\tau$$

```
 νe        νe        νμ        νμ         ντ        ντ
  |         |         |         |        -|        |-
 ─┴─       ─┴─       ─┴─       ─┴─       ─┴─       ─┴─
 _|_  ,   _|_  ,    -|_  ,    _|_  ,    _|_  ,    _|_  .
 -|o       o|-        |o       o|-        |o       o|
```

E. Spin Hatches

Whereas the positional grid can keep track of the spins of one component particles, it cannot correctly keep track of multi-component particles if they orbit about one another. Some component particles orbit around each other with some spin (whole units of spin—generally spin ± 1). To make our chonomic pictorial equations complete with all the pertinent spin data, we need to add something to the scheme. Let us add an additional hatch below the positional grid to be used for spin numbers:

$$\nu_e$$

$$
\begin{array}{c|c}
\hline
- & \circ \\
\hline
1 & \frac{1}{2} \\
\end{array}
$$

．

The top left number in the hatch is the orbital spin (if any) of the component echons. In the case of the electron neutrino above, it is one since the two echons orbit about each other. The top right number in the spin hatch is the total particle spin. It is the sum of the orbital spin (in the top left corner of the hatch) and the total spin shown in the positional grid (intrinsic echon spin). In the above case one echon is a o (has 0 spin) and one echon is a - (has -½ spin). The total intrinsic echon spin is therefore -½. Add that result to the orbital spin (top left hand number in the hatch), and the result is +½—the total expressed in the top right number in the hatch.

Besides the total particle spin, particles in a reaction can contribute or take away angular momentum due to the off-centeredness of their collisions or lines of retreat. This angular momentum is part of the total angular momentum in the system. Total angular momentum is conserved, but particle spin is not always. Therefore we need to keep track of off-centered collision momentum (positional-kinetic angular momentum) in particle equations, and this we will do in the bottom left position of the hatch. The bottom right number in the hatch will be the total spin or angular momentum a particle will contribute to a reaction or carry off from it. It is the total of the particle spin in the top right position of the hatch and the positional-kinetic angular momentum in the bottom left position of the hatch.

F. Mass, Mean Life, Percent Pathway, and Maximum Momentum Per Particle

The symbolic pictorial scheme lacks one more thing. It lacks a relative indication of the masses of the particles. Let us put a mass value (in MeV) under the particle symbol and above the positional grid in each particle description in our chonomic equations. This will give us a feel for conservation of energy in the reactions. Mean life, percent pathway, and maximum momentum per particle are important particle parameters also. But for simplicity we will not list them in the chonomic pictorial equations. These constants will be listed in the headers above the chonomic equations in the appendices. We will use referenced values in this volume.[2] [It is difficult to keep up with the data in this science. This chapter was first based on 1990 data; adjustments had to be made with 1992, 1996, and now 1998 data. Physical constants like masses continue to change. However this chapter and book are valuable in demonstrating a basic concept.]

G. Balancing a Simple Reaction

We are almost ready to do particle decay schemes. The plan of balancing the echons and yachons in reactions or decay schemes is that echons and yachons should be conserved in reactions. There should be the same number of positive echons and yachons before and after a reaction. And there should be the same number of negative echons and yachons before and after a reaction. As it turns out, however, the energy state and type of echon or yachon may not be conserved in a reaction. Echons can be bumped up or down in energy states in reactions. Also echons may be fused, changed from - or + to dots, etc. Yet the total number of negatively charged echons and yachons should be conserved in a reaction, and the total number of positively charged echons and yachons should be conserved in a reaction.

Let us try balancing a simple reaction.

$$\pi^+ \rightarrow e^+ + \nu_e$$

π^+	e^+	ν_e
139.56995	0.51099907	<0.000010

$$\pi^+ \qquad \rightarrow \qquad e^+ \qquad + \qquad \nu_e$$

$0 \mid 0$	$0 \mid -\frac{1}{2}$	$1 \mid \frac{1}{2}$

The equation as it is written does not balance. Something is missing on the left hand. There are one negatively charged - echon and one positively charged - echon on the right side of the equation that do not appear on the left side of the equation. It is just as though a

particle is missing on the left hand side of the equation. H. C. Dudley in "Is There an Ether?"[3] theorized there is no such thing as spontaneous radioactive decay, but all radioactive decay results from unseen neutrinos combining with the parent particles according to particle-neutrino cross sections. We also need an unseen particle to combine with the parent particle in our equation. But we don't need a neutrino. A neutrino is a combination of a o echon and a - or + echon. We need the minus-minus particle above. It has spin -1 on the face of the grid. A minus-minus particle may be obtained by having an orbital spin of -1. The total particle spin, then would be -2. Our minus-minus particle that we are missing then would be a graviton with low or imaginary mass. Gravitons are plentiful. Apparently our missing particle is a graviton.

Unseen particles of one kind or another are in decay schemes of all unstable elementary particles. Often more than one particle combines with the parent particle in the decay process. Other particles that we need are

```
   |                      |                        |
   |                      |                        |
 --+--                  --+--                    --+--
 - | -  ,                • | •  ,                 o | o  .

 1 | 0                   1 | 1                     1 | 1
   |                      |                         |
```

The respective particles are a π^0 particle, a photon, and a spin 1 graviton. (There are also spin 2 gravitons.) Pions and photons have been previously known. Spin 1 gravitons are previously unknown. They are plentiful in decay schemes, however.

We have a confusing array of gravitons. Gravitons may be composed of -, +, or o echons, and may be in different energy states. We need compact symbols to differentiate these particles. Let us call gravitons g^{xy} where x is the type of echon (-, +, or o), and y is the energy level of the graviton. As an abbreviation we may omit the energy level for ground state. The following are examples of that terminology:

```
   g⁺              g⁻¹              g°              g°¹

    |               |                |                |
    |               |                |                |
  --+--           --+--            --+--            --+--
  + | +            - | -  ,         | |              o | o  .
                     |             o | o
  1 | 2           -1 | -2           1 | 1           -1 | -1
    |                |               |                |
```

Let us now try to go back and balance our decay equation.

$\pi^+ \rightarrow e^+ + \nu_e$.

π^+	g^-	q.p.	e^+	ν_e
139.569	>0i	?	.5109	<.000010

We insert an intermediate step in the reaction, a combined particle called q.p. for quasi-particle. The graviton combines with the parent particle to form a jumbo particle which redivides into the daughter particles. This intermediate step will appear more necessary in later decay schemes.

Now let us try to balance the spins in the equation. The echons on the face of the positional grids are balanced. The total particle spins, however, are not balanced. That means there must be some positional-kinetic angular momentum in the particles coming in to or going out of the reaction, or both, to make the total spins balance. We wish to write our decay schemes in the frame of reference of the parent particles. Therefore, we never assign positional-kinetic angular momentum to a parent particle. We can put a 0 in the bottom left hand corner of the spin hatch for the π^+. Adding the spins for the π^+, that means there will be a total 0 spin in the bottom right hand corner of the spin hatch for the π^+ as well.

The graviton probably has +1 positional-kinetic angular momentum, because, in this problem, there is no other incoming particle that can have positional-kinetic angular momentum. And positional-kinetic angular momentum can be at most +1, or at least, -1 per particle. So we can write 1 for the bottom left hatch position for the g^-. The bottom right positions in the hatches are just particle totals, the sum of the upper right numbers and the lower left numbers. Thus there will be a 1 and -1 for the bottom left and bottom right hand positions in the spin hatch for the graviton. Our equation now looks like the following:

$\pi^+ \rightarrow e^+ + \nu_e$.

```
    π⁺           g⁻                 q.p.          e⁺            νₑ
 139.569        >0i                  ?           .5109      <.000010
    |            |                    |            |             |
    __          __                   __           __            __
    |     +      |         →          |    →       |    +        |
    __          __                   __           __            __
   |o          -|-                  -|o-          |-           -|o

   0|0         -1|-2                  |           0|-½          1|½
   0|0          1|-1                  |           0|            1|
               |_____|
               unobserved
                particle
```

Now we can find the spin hatch numbers for the quasi particle. Total spin is conserved. So the sum of the bottom right hatch numbers on the left hand side of the equation is the total spin (bottom right hand position of the hatch) of the quasi particle. That number is -1. There is never any positional-kinetic angular momentum for a quasi-particle. So we can immediately fill the bottom left hatch position with a zero. The top right hatch number is the difference between the bottom right and bottom left numbers. So in this instance the top right number is -1. The top right number minus the spin on the face of the positional grid equals the top left number in the spin hatch. In this case we have two - echons and a o echon in the positional grid. The o echon contributes no spin. Each - echon contributes -½ spin, so the total on the face of the positional grid is -1. The top right hand spin hatch number is -1. Therefore the top left hand spin hatch number is 0 net (total particle spin minus intrinsic spin on the face of the positional grid), which we denote as 1-1 (to keep track of oppositely orbiting particles within the system). This is the net orbital spin in the quasi particle. The - echons are orbiting in the - direction with -1 spin. Because of the positional-kinetic angular momentum in the graviton, the minus echons orbit with +1 spin relative to the o echon. Now we can write the quasi particle as

$$
\begin{array}{c}
\texttt{q.p} \\[-2pt]
\underline{\ |\ \underline{}} \\[-2pt]
\underline{\ |\ } \quad \cdot \\[-2pt]
-\ |\ \texttt{o}-
\end{array}
$$

$$
\begin{array}{c}
\underline{\texttt{1-1}\ |\ \texttt{-1}} \\
\texttt{0}\ |\ \texttt{-1}
\end{array}
$$

Now we can try to balance the right hand side of the equation. The total spin of the quasi particle is -1. But the total of the particle spins on the right side of the equation is 0. So we need positional-kinetic angular momentum to balance the equation. The v_e probably should not have positional-kinetic angular momentum because it is a balanced two-component system. Thus we should add a positional-kinetic angular momentum of -1 to the e^+ particle. The equation now balances perfectly.

$$\pi^+ \rightarrow e^+ + \nu_e \ .$$

π^+	g^-	q.p.	e^+	ν_e
139.56	>0i	?	.5109	<.000010

```
    |            |            |            |            |
 ___|___      ___|___      ___|___      ___|___      ___|___
    |    +       |      →     |      →     |      +     |
 ___|___      ___|___      ___|___      ___|___      ___|___
   |o           -|-          -|o-          |-          -|o


  0|0          -1|-2        1-1|-1         0|-½         1|½
 ─────        ──────       ──────        ─────        ────
  0|0           1|-1         0|-1        -1|-1½         0|½

              |_____|
               unobserved
                particle
```

These pictorial equations are powerful, for they show not only what echons are involved in the reactions, but whether they are upside down or right side up, what energy state they are in, what are their intrinsic spins and orbital spins, and whether or not and in what direction the particles are off-centered in the collisions and recoil paths. The mass numbers above the positional grids show whether particle reactions are possible. In all systems where the second law of thermodynamics is in force, the combined mass of the particles on the left hand side of the equation must be greater than the combined mass of the particles on the right side of the equation, so that energy may be released in kinetic energy and heat to increase entropy. The mass relationships can be seen at a glance with these equations.

Notice in the above equation, echons are conserved. What comes in goes out. That quality does not exist in current quark and lepton models of physics. Echons can construct all leptons. Yachons and echons can construct all hadrons. Each lepton does not have to be a separate elementary particle. In yachons and echons (•, o, +, -), we have found something truly elementary.

III. Electrino Process Types

There are hundreds or thousands of possible particle decay schemes. This section will illustrate a few basic types.

A. Wrenching Echons From Orbits to Straight-Forward, Simple Echon Recombinations.

The sample equation in Section II.G., Balancing a Simple Reaction, fits in this category. The following are two more examples.

1. $n \rightarrow p + e^- + \nu_e$

Let us consider the weak nuclear force and beta decay. One echon cannot mediate the weak nuclear force between echons. For a mediating particle, such as the graviton, there must be positive and negative echons with whole orbital spins. When a graviton gravitationally combines with a semion echon, the magnetic moment of the semion echon tries to realign the magnetic moments of the positive and negative orbital echons. At a distance from the semion echon, the gravitons have zero magnetic moments, since the orbital magnetic moments of the positive and negative echons cancel. But as the graviton approaches the semion echon, its magnetic moment tries to align the orbital moments of the positive and negative echons. This is impossible without wrenching the echons out of orbit, which is done by the weak magnetic force.

Let us review two particles introduced in Chapter 6, the proton and the neutron. Experimental tests seem to indicate that the proton is a three component particle. It is a baryon, and baryon number is conserved in reactions. That is, it is a heavy particle. In all reactions where there is a heavy particle, there will always result a heavy particle after the reaction. The proton must therefore have a different stable constituent than have leptons and mesons which we have studied already (composed of -, +, and o echons in their various states). A proton, then, must contain a dot (•) yachon. The proton has three components and it has spin ± ½. How about the following for the formula for a proton?

$$
\begin{array}{c}
p \\
938.27231 \\
\hline
\mid \\
\hline
\mid \\
\hline
+ \mid \bullet \circ \\[4pt]
\underline{1{-}1 \mid \tfrac{1}{2}} \quad . \\
\mid
\end{array}
$$

The + echon orbits about the dot yachon with -1 orbital spin. The o echon orbits about the dot yachon with +1 orbital spin. Thus we have a total of 1-1 for the orbital spin number.

The neutron is also a baryon, but has one less positive charge compared to the proton. It has ∓ ½ spin. A proton can trade a π^{+} (o echon) with a neutron, converting it into another proton. Therefore it is logical that the formula of the neutron should be

$$
\begin{array}{c}
n \\
939.56563 \\
\hline
\mid \\
\hline
\mid \\
\hline
+ \mid \bullet \\[4pt]
\underline{-1 \mid -\tfrac{1}{2}} \quad . \\
\mid
\end{array}
$$

With these formulae we can balance the decay scheme for an upside down neutron.

$$n \rightarrow p + e^- + \nu_e$$

```
    n        g°       g-     q.p.      p        e⁻       νₑ
 939.566    >0i      >0i      ?     938.272   .510   <.000010
   ⊥         |        |       ___|___    |        |        |
   |    +    |   +    |    →    |    →    |   +    |   +    |
   |         |        |         |        |        |        |
  -|•       o|o      -|-     --o|•o-    -|•o      -|       o|-


  1|½       1|1     -1|-2    2-1|-½   1-1|-½    0|-½     1|½
  0|½       0|1      0|-2     0|-½     0|-½     0|-½     0|½
        |_____|
            unobserved
            particles
```

A g° graviton and a g⁻ gravitationally combine with an upside down neutron. The dot in the neutron and the - in the neutron go to the upside down proton. The negative - echon in a graviton goes to the electron. The positive - echon and negative o echon from the gravitons go to the electron anti-neutrino with +1 orbital spin. The positive o echon goes to the proton. With two different gravitons, this reaction takes place without any extra positional-kinetic angular momentum. This is the first instance we have seen more than one unobserved particle in a reaction. This will be prevalent in the future.

Straight-forward, simple echon recombinations is the simplest of a variety of electrino processes possible. We will give another example of this category of electrino process.

Before we balance the next decay scheme, let us consider a prevalent graviton—a neutrino orbiting about an anti-neutrino. The graviton can be composed of a - echon orbiting about a o echon in the neutrino, with +1 orbital spin; with a - echon orbiting about a o echon in the anti-neutrino, with +1 orbital spin; and the neutrino orbiting about the anti-neutrino with another +1 orbital spin. The total orbital spin of this graviton is +3, and the total intrinsic spin is -1. Thus the total spin of the particle is +2. The particle is a graviton. We will assign it the symbol g^{o-}. A graviton with opposite intrinsic spins and orbital spins would be g^{o+} with -2 net spin.

$$g^{o-} \qquad\qquad (\nu_e \qquad \nu_e)$$

$$\frac{3\,|\,2}{0\,|\,2} \qquad\qquad\qquad \frac{3\,|\,2}{0\,|\,2} \qquad .$$

2. $\mu^- \rightarrow e^- + \nu_e + \nu_\mu$

A negative muon gravitationally attracts a g^{o-} graviton, which combines with it, making a quasi particle. The echons redivide into an electron, electron anti-neutrino, and a muon neutrino. The orbital angular momentum of the graviton is carried away by the electron anti-neutrino, the muon neutrino, and positional-kinetic angular momentum of the electron. This is a straight forward recombination of echons in their original states as we would expect from the quasi particle.   There are no Pauli Exclusion Principle violations in this reaction. That is, no two identical fermion echons are in the same positional grid space in the quasi particle.   Pauli Exclusion violations tend to make the reactions more violent—such as blowing the particles apart, knocking an echon to a different energy state, or even fusing echon electrinos.

B. Echons Knocked To Other Energy States

1. $\tau^- \rightarrow \mu^- + \nu_\mu + \nu_\tau$

```
    τ⁻       g⁰⁺       q.p.      μ⁻        ν_μ          ν_τ
 1777.05    >0i         ?      105.65    <.17         <18.2
  +|          |         +|        |         |          +|
 ___         ___       ___       ___       ___         ___
   |    +      |    →   +|+   →   +|    +    |+    +      |
 ___         ___       ___       ___       ___         ___
   |       +o|o+       o|o        |        o|           |o

 0|½      -3|-2      -3|-1½   0|½    -1|-½         -1|-½
 ___      ___        ___     ___    ___           ___
 0|½       0|-2       0|-1½  -1|-½   0|-½          0|-½
            |_____|
          unobserved
          particle
```

We here balance one decay scheme with + echons (upside down particles) instead of - echons to show it can be done. In this decay scheme, a graviton composed of a neutrino and an anti-neutrino orbiting about each other gravitationally combine with a tauon, forming a quasi-particle. The + echons in the graviton are knocked to the 1 state by the tauon, while the + echon in the tauon retains its state.   The echons redivide, with the muon (from the negative + echon in the graviton) carrying away -1 orbital spin in positional kinetic angular momentum.

2. $\Lambda \rightarrow p + \pi^-$

Let us introduce a new particle here, the Lambda particle. It is heavier than a neutron and has the same spin as a neutron. In yachon-echon balancing particle decay schemes it seems to have the same yachons and echons as a neutron. It must be a neutron in a more energetic state. It must be a

$$\Lambda$$
$$1115.683$$

$\Lambda \rightarrow p + \pi^-$

Λ	g^o	q.p.	p	π^-
1115.68	>0i	?	938.27231	139.56995

A g^o graviton knocks the + echon of a Λ particle to the ground state as it forms a quasi particle with it. The echons redivide into a p and a π^-.

C. Reversal of Orbital Spins

Some decay schemes demonstrate the reflection of orbiting yachons and echons in the collisions of different particles. The following are a few examples.

1. $K^+ \rightarrow \pi^+ + \pi^0 + \gamma$

Here we introduce a new particle—a π^0. The π^+ and π^- are composed of o echons. That is all they are. A π^+ is a positively charged o echon, and a π^- is a negatively charged o echon. A π^0 also has about the same mass and also has zero spin, but has 0 charge. To have 0 charge the π^0 must have two oppositely charged echons in it. But the echons cannot be o echons, for the echons must have ± 1 orbital spin or the particles will annihilate. If the echons were o echons, the total particle spin would be ± 1. That is the formula for a g^0 graviton. We need 0 spin for the π^0. The only way we can get that is to make the echon intrinsic spin cancel the echon orbital spin. This can be done if the echons are - or + echons and if the orbital echon spins are +1 or -1, respectively. Besides, the balancing of echon equations indicates that this is the formula for the π^0.

$$\pi^0$$
$$134.9764$$
$$\frac{|}{\frac{|}{-|-}}$$

$$\underline{1|0}$$
$$| \quad .$$

Let us now use this information to balance decay scheme 1 above.

```
K⁺        g⁻       Y        q.p.      π⁺       π⁰        Y
493.67   >0i      0         ?       139.569  134.976   0

 ⊥        |        |        |         |        |         |
 |o   +   |   +    |    →   |     →   |   +    |    +    |
 |       -|-      •|•      -•|•o-     |o      -|-       •|•

0|0     -1|-2    1|1    1-1|-1      0|0      1|0      -1|-1
0|0      0|-2    0|1     0|-1       0|0      0|0       0|-1
         |_____|
              unobserved
              particles
```

A graviton or a photon knocks a K⁺ echon to the ground state (π^+) as they combine with it to form a quasi-particle. The echons of the graviton and the yachons of the photon collide, exchanging their orbital spins. Thus a graviton is converted into a π^0 particle. The yachons and echons separate into π^+, π^0, and γ.

2. $K^+ \rightarrow e^+ + \nu + \pi^0 + \pi^0$

```
     K⁺        g⁻       g⁻       g⁺            q.p.
   493.677    >0i      >0i      >0i             ?

    |         |        |        |              |
    |o    +   |    +   |    +   |      →       |       →
    |        -|-      -|-      +|+           --+|o+--

   0|0     -1|-2    -1|-2    1|2           2-2|-1
   0|0      1|-1     0|-2    0|2            0|-1
           |_____|
           unobserved particles
```

e^+	ν_e	π^0	π^0
0.51099907	<.000010	134.9764	134.9764

$$\rightarrow \quad \boxed{\begin{array}{c} \mid \\ \mid \\ \mid - \end{array}} \quad + \quad \boxed{\begin{array}{c} \mid \\ \mid \\ - \mid \circ \end{array}} \quad + \quad \boxed{\begin{array}{c} \mid \\ \mid \\ - \mid - \end{array}} \quad + \quad \boxed{\begin{array}{c} \mid \\ \mid \\ + \mid + \end{array}}$$

$\dfrac{0 \mid -\frac{1}{2}}{-1 \mid -1\frac{1}{2}}$	$\dfrac{1 \mid \frac{1}{2}}{0 \mid \frac{1}{2}}$	$\dfrac{1 \mid 0}{0 \mid 0}$	$\dfrac{-1 \mid 0}{0 \mid 0}$

A graviton knocks the K^+ echon to the ground state as two g^- and a g^+ graviton combine with it to form a quasi particle. The echons of a g^+ and a g^- graviton collide in the quasi particle, and reflect, exchanging their orbital spins. They go out as a π^0 and its anti-particle.

D. Flipping Intrinsic Spins

Echons do not always come as $--$, $++$, or $\circ\circ$ in positional grids. Every conceivable combination is employed in chonomics, such as $-+$ or $+-$, as in the $\omega(782)$ employed in the decay schemes below.

1. $\omega(782) \rightarrow \pi^- + \pi^+ + \pi^0$

$\omega(782)$	g^0	q.p.	π^-	π^+	π^0
781.94	>0i	?	139.56	139.56	134.97

$$\boxed{\begin{array}{c} \mid \\ \mid \\ - \mid + \end{array}} + \boxed{\begin{array}{c} \mid \\ \mid \\ \circ \mid \circ \end{array}} \rightarrow \boxed{\begin{array}{c} \mid \\ \mid \\ - \circ \mid \circ - \end{array}} \rightarrow \boxed{\begin{array}{c} \mid \\ \mid \\ \circ \mid \end{array}} + \boxed{\begin{array}{c} \mid \\ \mid \\ \mid \end{array}} + \boxed{\begin{array}{c} \mid \\ \mid \\ - \mid - \end{array}}$$

$\dfrac{-1 \mid -1}{0 \mid -1}$	$\dfrac{1 \mid 1}{0 \mid 1}$	$\dfrac{1 \mid 0}{0 \mid 0}$	$\dfrac{0 \mid 0}{0 \mid 0}$	$\dfrac{0 \mid 0}{0 \mid 0}$	$\dfrac{1 \mid 0}{0 \mid 0}$

```
         |_____|
unobserved particle
```

A g^0 graviton combines with an $\omega(782)$ to form a quasi-particle. The $+1$ orbital spin of the graviton flips the $+$

echon of the $\omega(782)$ right side up to a - echon. After that, straight-forward recombinations yield π^-, π^+, and π^0. Balancing the spins in this decay scheme is tricky, however. Since the intrinsic spin of an echon is flipped, the intrinsic spins on the face of the grids in this scheme do not balance, as is usually the case. Therefore the orbital spins in the quasi-particle do not balance with the left hand side of the equation either. But the total spins still balance.

2. $\omega(782) \rightarrow \pi^0 + \gamma$

$\omega(782)$	γ		q.p.		π^0		γ
781.94	0				134.97		0

1|1	-1|-1		1-1|-1		1|0		-1|-1
0|1	-1|-2		0|-1		0|0		0|-1

|_____unobserved particle_____|

A photon with -1 positional-kinetic angular momentum strikes an $\omega(782)$, flipping over its + echon to be a - echon. Straight-forward recombination yields a π^0 and an outgoing photon.

3. $\tau^- \rightarrow \omega(782) + \pi^- + \nu_\tau$

τ^-	g^{0-}		q.p.	$\omega(782)$		π^-		ν_τ
1777.05	>0i		?	781.94		139.56		<35

0|-½	3|2		2|1½	1|1		0|0		1|½
0|-½	0|2		0|1½	0|1		0|0		0|½

|_____unobserved particle_____|

Because of the overall +1 spin of the g^{o-} graviton, the tauon knocks the positive - echon of the g^{o-} graviton up side down to a + echon as the graviton and tauon combine to form a quasi-particle. Straight-forward recombinations result in an $\omega(782)$, π^-, and ν_τ.

E. Fusion of Electrinos

A significant aspect of electrinos is that they may fuse. Elementary particle fusion is not theorized in any other model of physics. In the author's model, electrino fusion has several major impacts.

One is that electrino fusion unites the particles completely. All "elementary" particles can be constructed of quartons, semions, and unitons. But all those are only fusion states of positive and negative octon pairs. Orbiting octons have such high charge per mass ratios they can ionize octon pairs out of absolutely nothing.   This is significant in the creation and evolution of the Universe. All the Universe could be created from one octon pair. Quartons could be fused octons, semions fused quartons, and unitons fused semions.   The great variety of "elementary" particles can all be constructed from the quartons (in o echons), the semions (in - or + echons) and the unitons (• yachons).

A second major impact of the fusion of electrinos is that each time the electrinos are fused the particles switch from being matter to antimatter, or vice versa. Current popular models of physics do not predict that, but our model of chonomics shows this already occurs naturally with common particles of physics in laboratories without the physicists' knowledge. When this process is controlled, it is a way of making matter into antimatter (without pair production) for use in annihilating other matter as an energy source and other practical applications.

How do electrinos fuse?   There are two semion electrinos in a - or + echon, for example, in an electron. Electrons strongly repel each other because of the Coulomb electric force.   But if through high velocity collisions that repulsion is overcome, two electrons with four semions can come so close together that the strong gravitational force overcomes the electric forces, and the four particles all become attracted to each other.   When there are only two particles in an orbit, the attractive force keeps the particles in a stable orbit.   But where there are four particles all in one orbit, going the same way, that attract each other, they will inevitably not be equally spaced.   One semion from one electron and one semion from the other electron will be attracted to each other, and similarly with the other two semions.   The four semions will fuse into two unitons, which are whole particles (yachons), which will go their own ways.

Octons fuse to quartons, which fuse to semions, which fuse to unitons. Semion to uniton fusion is fairly easy and occurs naturally.   Quarton fusion is more difficult, however. Quartons are normally found in o echons. There are four quartons in a o echon, two orbiting one way and two orbiting the other way.   o echons have 0 net spin, therefore they act as bosons, not obeying the Pauli Exclusion Principle, which requires that there be only one particle in each energy state in a system.   Boson pions (quarton sources) can be piled one on top of another many times in particle resonances without fusing. It is difficult to fuse quartons in o echons. But quartons can be fused rarely through weak interactions.

The best way to describe natural semion fusion is to show some chonomic fusion decay schemes, and then to explain them.

1. $\eta'(958) \to \gamma + \gamma$

```
   η'(958)              g⁻              q.p.1
   957.78              >0i                ?
     |                  |                 |
    ___                ___              _____
     |           +      |         →      |          →
    ___                ___              _____
   -o|o-               -|-             --o|o--

   2-1|0              -1|-2             2-2|-2
   ─────              ─────             ──────
    0|0               0|-2              0|-2
                  |_____|
                  unobserved
                  particle
```

```
     q.p.2              g°            γ            γ
       ?               >0i           0            0
       |                |            |            |
     _____             ___          ___          ___
  →    |         →      |      +     |      +     |
     _____             ___          ___          ___
   o••|••o             o|o          •|•          •|•

   1-3|-2             -1|-1        -1|-1         1|1
   ─────              ─────        ─────         ───
   0|-2               -1|-2        0|-1          0|1
                  |_____|
                  unobserved
                  particle
```

A g⁻ graviton combines with a η'(958) to form a quasi particle. The four - particles fuse to four dots, forming a second quasi particle. The second quasi particle simply redivides into a g° graviton and two photons. The photons are observed, but the graviton is not.

2. $K^0_L \to \overline{\pi^0} + \gamma + \gamma$

The following classic example demonstrates echon recombination, knocking echons to other energy states, collision and reflection of echon spins, and electrino fusion all in one particle decay scheme.

We introduce a new particle here, the K^0-long. There seems to be two different K^0 particles, the K^0-short, and K^0-long, the K^0-short having short decay half lives, and the K^0-long having long half lives. Carefully balancing the echon equations for these two particles indicates that they are composed differently as follows:

```
        K⁰ₛ                    K⁰ₗ
      497.672                497.672
         |                      |
        ———                    ———
        -|-        ,           -|
         |                     |-

        1|0                    1|0
        ———                    ———
         |                      |
```

Now let us balance our classic decay scheme:

```
   K⁰ₗ         g⁺          g⁻            q.p.1
 497.672      >0i         >0i              ?
    |          |           |               |
   ———        ———         ———             ————————
   -|         |           |               |
   |-    +   +|+    +     -|-      →      --+|+--        →
             
   1|0       1|2        -1|-2            2-1|0
   ———       ———        ———              ———
   0|0       0|2         0|-2            0|0
             |_____|
                   unobserved
                   particles
```

q.p.2 $\overline{\pi}^0$ Y Y
? 134.9764 0 0

| | | |
| | | |
$\rightarrow$ ___ $\rightarrow$ ___ + ___ + ___ .
+ • • | • • + + | + • | • • | •

$$\frac{1-2\,|\,0}{0\,|\,0} \qquad \frac{-1\,|\,0}{0\,|\,0} \qquad \frac{-1\,|-1}{0\,|-1} \qquad \frac{1\,|\,1}{0\,|\,1}$$

First the 1 state + echon in the K^0_L is knocked to the 0 state while two oppositely spinning gravitons combine with the K^0_L to form the first quasi particle. The - echons fuse to dots (q.p.2). One set of the dot echons orbiting with -1 orbital spin collide with the + echons orbiting with +1 orbital spin. The echons reflect and exchange orbital spins. The second quasi particle redivides into an anti- π^0 with + echons and -1 orbital spin and two oppositely spinning photons.

F. Annihilation of Electrinos

Electrinos can also annihilate each other. The electrinos in the following particle systems will annihilate each other, provided they are in the same energy state:

| | | |
| | | |
___ , ___ , ___ , ___ .
o | o − | + + | − • | • •

$$\frac{0\,|\,0}{\,|\,} , \qquad \frac{0\,|\,0}{\,|\,} , \qquad \frac{0\,|\,0}{\,|\,} , \qquad \frac{0\,|\,0}{\,|\,}$$

Opposite charged like electrinos with equal and opposite intrinsic spin, no orbital spin, and like energy states annihilate each other. If the energy states are different, they have orbital spin, or their intrinsic spins are not equal and opposite, they will not annihilate so long as they retain their same states. The following are examples of particle systems that will not annihilate so long as the electrinos are in the following states:

$$
\begin{array}{c}
\dfrac{\mathrm{o}\,|}{|\,\mathrm{o}} \\[2pt]
| \quad ,
\end{array}
\qquad
\begin{array}{c}
\dfrac{|}{|} \\[2pt]
\mathrm{o}\,|\,\mathrm{o} \quad ,
\end{array}
\qquad
\begin{array}{c}
\dfrac{|}{|} \\[2pt]
+\,|\,+ \quad ,
\end{array}
\qquad
\begin{array}{c}
\dfrac{|}{|} \\[2pt]
-\,|\,+ .
\end{array}
$$

$$
\begin{array}{c}
\dfrac{\mathrm{0}\,|\,\mathrm{0}}{|}
\end{array}
\qquad
\begin{array}{c}
\dfrac{\mathrm{1}\,|\,\mathrm{1}}{|}
\end{array}
\qquad
\begin{array}{c}
\dfrac{\mathrm{0}\,|\,\mathrm{1}}{|}
\end{array}
\qquad
\begin{array}{c}
\dfrac{\mathrm{1}\,|\,\mathrm{1}}{|}
\end{array}
$$

Now that this book has introduced the science of chonomics, this book will put that science to work in the appendices. Appendix A will cover known lepton decays. Appendix B will list the theoretical chonomic structure of all known particles (as of 1998).

---

[1]Francis Halzen and Alan D. Martin, *QUARKS AND LEPTONS: An Introductory Course in Modern Particle Physics* (New York: John Wiley & Sons, 1984), p. 114.

[2]SUMMARY TABLES OF PARTICLE PROPERTIES, January 1 1998, Particle Data Group, *as quoted by CRC Handbook of Chemistry and Physics,* 80th Edition, David R. Lide, Ph.D, Editor in Chief   (Boca Raton: CRC Press, 1999), pp. **11**-3 to **11**-5.
[3]H. C. Dudley, *Ind Res.* 43, 44 (Nov. 15, 1974).

Problem Set 10

1. Balance decay scheme $\mu^- \rightarrow e^-\overline{\nu}_e\nu_\mu$. What electrino process does it demonstrate?

2. Balance decay scheme $\tau^- \rightarrow \pi^-2\pi^0\nu_\tau$. What electrino process does it demonstrate?

Note: Both of the above decay schemes are solved in full in Appendix A, Lepton Summary Table. If you have any problems with the above, refer to Appendix A. It would be good practice learning how to look up decay schemes, and learning how to read them.

Chapter 11

SUMMARY OF CHONOMICS

Never before in human history has the science of particle physics been as illuminating as now with the science of chonomics. Before we continue on in the development of particle processes and developments in particle physics, let us summarize what we have learned from chonomics.

First of all, we see that particles come in whole, half, fourth, and eighth charges, not in two thirds and one third charges as in the quark theory. All of the elementary particles and resonances can be composed of quarton, semion, and uniton echons and yachons better than quark structure. Furthermore each decay scheme can be balanced in chonomics--a feat not possible with quarks and leptons. Chonomic equations bring unprecedented power to the theorist: they tell not only what echons and yachons are in a particle, but in what energy states they are, what their intrinsic spins are, what their orbital spins are, and how much angular momentum is carried in or out of a particle reaction by off-centeredness in collision and lines of retreat.

Chonomics is much more parsimonious than the quark and lepton theory. It takes 24 quarks, anti-quarks, leptons, and anti-leptons plus 37 other particles to make up *known* particles. But this is a function of the present financial capabilities of scientists, not as a limit to possible matter particles. Could the Super Conducting Super Collider have been funded and built, it would have disclosed many more high energy particles, which would have required many more quarks and leptons to explain. On the other hand, one particle (an octon-anti-octon pair), in all its possible ionization states, fusion states, and warp states, can compose any elementary matter particle as well as light and any boson for the forces. Also there is but one postulate needed to derive the model, and only one force is required (gravity) to derive the infinite number of forces in the model. Including those listed in the last chapter, plus pair production and particle capture, there are eight possible electrino-echon-yachon processes. No wonder early particle physicists discovered the "eight-fold way." New particles would always come in eights.

You be the judge as to whether particle physics should continue the current quark-lepton model or switch to the electrino fusion model of elementary particles.

Problem Set 11

1. Have quarks ever been discovered isolated? Why has the quark hypothesis been primarily adopted?

2. Have electrinos ever been detected isolated? According to electrino theory, can they ever be detected isolated in the foreseeable future with man-made machinery? What is the reason for this?

3. Why is the electrino theory preferable to the quark theory? How many quarks, anti-quarks, leptons, and anti-leptons and other elementary particles does it currently take to compose all known matter? How many kinds of electrinos does it take to create all known light and matter and bosons for the forces?

4. What recurring number is there in particle physics in the number of electrino, echon, yachon processes and the number of new particles that occur together?

5. What unprecedented power does chonomic equations bring to the theorist?

Chapter 12

ARTIFICIAL ELECTRINO FUSION

A. Creation and Annihilation of Antimatter

Appendix C, Meson Mode Type Table, demonstrates, among other things, natural electrino fusion. Generally positive and negative semions fuse to positive and negative yachons, which go out as photons. As many positive yachons are created as negative yachons. That is why they go out as new photons. This chapter will discuss the creation of charged single yachons from the artificial fusion of semions from single polarity echons. Such yachons cannot go out as photons, but may become the core particles of protons, neutrons, or anti-protons or anti-neutrons.

Why couldn't we fuse the semion subparticles of two negatively charged + or - echons without the corresponding positively charged echons? Those negatively charged isolated + or - echons are just electrons. Why couldn't we fuse the semion electrinos of two electrons into two negatively charged dots? It may be possible, but there is an energy problem. Isolated electrons are light—only 9.109 3826 x 10^{-31} kg—whereas isolated unitons are extremely heavy— -i 2.177 x 10^{-8} kg—too heavy to be produced isolated in man-made machines. But with graviton assists, anti-protons or anti-neutrons could be created when the energies of the fusing semions exceed the energies of anti-protons or anti-neutrons. The constituents of electrons are fused to unitons in the presence of the other anti-proton sub-particles. This is without pair production. Thus if these anti-protons or anti-neutrons annihilated protons or neutrons in their environs, there would be a net loss of pre-existing mass in those environs. In the annihilation process the anti-protons and protons would be converted into

305

gravitons, photons, and neutrinos. Anti-neutrons and neutrons yield photons and gravitons.

Let us look at those reactions. First consider a two-dimensional echon diagram of the annihilation of hydrogen with graviton assist (see Figure 12-1). First of all, on the left hand side of the diagram the hydrogen is ionized. The electrons are accelerated separately in oppositely directed accelerators to high energies. Their relativistic energies must be above 938 MeV (each—about 2 GeV in the center of mass frame). The spins of the colliding electrons must be alike in the center of mass frame. One must be up axial and the other down axial with respect to their accelerators, so the semions orbit in the same direction as they collide.

In Figure 12-1 the collided + echons gravitationally attract two g^{0+} gravitons. In the resulting quasi-particle, two negatively charged + echons fuse to dots. The resulting quasi-particle then redivides into two electron neutrinos and two anti-protons. With photon assist, these two anti-protons annihilate two protons to yield four photons and two g^{0+} gravitons. The annihilation process is assisted by unobserved photons, so momentum may be conserved in the annihilation process. Two photons are emitted in opposite directions from the annihilation.

Let us now show how anti-neutrons are created and neutrons are annihilated. See Figure 12-2.

The protons in deuterium could be annihilated first, leaving the neutrons. But the neutrons would just decay with a twelve minute half-life to protons, electrons, and electron anti-neutrinos. The protons and electrons could combine to form hydrogen. The other way which we show in Figure 12-2 is to annihilate the neutrons, thus reducing deuterium to hydrogen. So either way the deuterium is reduced to hydrogen with one electrino fusion annihilation process.

The energy investment in accelerating the electrons is not lost. It can be recovered in the thermal energy caused by the annihilation photons. We see now also that *annihi-*

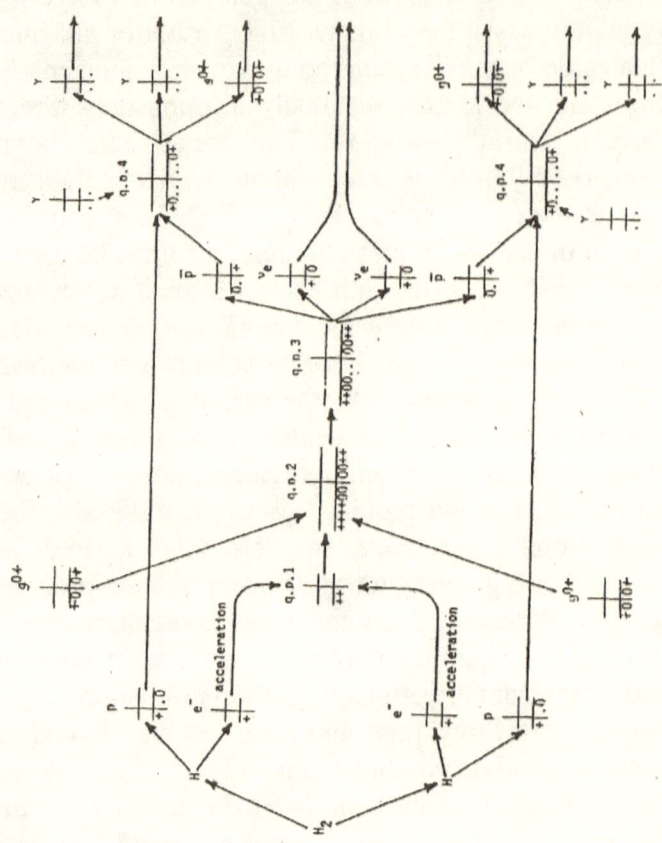

Figure 12-1. Hydrogen annihilation through fusion of electron semions.

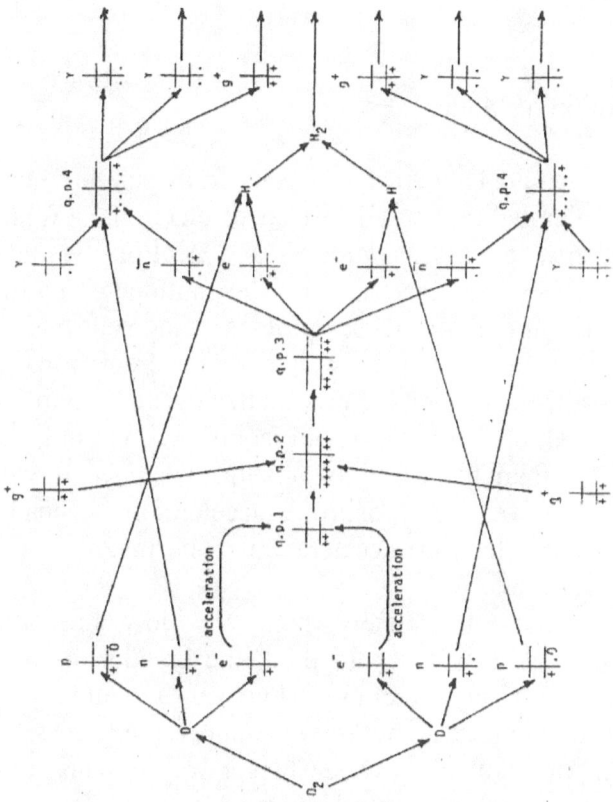

Figure 12-2. Deuterium reduction to hydrogen through anti-neutrons.

lation of particles is the wrong word for it. The echons are not destroyed or obliterated. They just go into tight combinations in photons, gravitons, and neutrinos. They can be redivided again in particle decay schemes and pair production.

B. Creation of Matter

What would happen if positrons from pair production were accelerated and fused instead of electrons? With g^+ and g^{0+} graviton assists, protons and neutrons would be formed. These are matter. With no antimatter particles around to annihilate these protons and neutrons and electrons (from neutron decay into protons, electrons, and electron anti-neutrinos), the new matter would be stable.

The creation of matter would require a different kind of accelerator than the creation of anti-nucleons. It would require positive ion (positron) accelerators instead of negative ion (electron) accelerators. The magnetic fields must be reversed.

Since we live in an energy hungry world where there is plenty of matter, one would think the creation of matter would not be of any direct use. However, it would not only be of academic interest but of value in reversing the second law of thermodynamics. (See Chapter 16.) Because of our experience with pair production, we would naturally assume there must be as much antimatter in the Universe as matter. We could assume there must be antimatter worlds balancing matter worlds. That would be a dangerous Universe. A hapless intergalactic traveler could be annihilated touching down on the wrong planet. Our electrino fusion model of elementary particles shows there does not have to be an equal amount of anti-matter in the Universe as matter. A Universe could be constructed entirely of matter, except for the anti-matter constituents in gravitons, neutrinos, and photons. All the matter we have monitored so far—this planet, the moon, the sun, meteors, other planets in this solar system—have been matter not anti-matter. Neutrinos detected from a distant super-nova have indicated the star was composed of matter. It could well be the whole Universe is matter.

C. Origin of the Universe

The electrino fusion model of elementary particles may revise our model of the origin of the Universe. While an all matter Universe could be constructed using electrino fusion techniques, an all matter Universe could not result from a Big Bang. There would be symmetry in the chaotic forces produced there. A Big Bang should produce as much anti-matter as matter. There is no electrical asymmetry which should dictate that all positrons should be fused to protons and neutrons and not that electrons should be fused to anti-protons and anti-neutrons.

Whether nucleons are matter or antimatter should be like a flip of a coin in a Big Bang. Heads for matter and tails for antimatter. If there are about 10^{80} nucleons in the Universe, what is the probability that they are all matter and that they came from a Big Bang?

$$P_{Big\ Bang} = \frac{1}{2^{10^{80}}}. \qquad (12\text{-}1)$$

This is a probability so small the author cannot evaluate it.

There could be an original asymmetric fusion of positrons and not electrons, however, resulting in the Universe as we observe it. This begs for an obvious interpretation affecting our models of the origin of the Universe. However, this volume will not interpret this evidence.

$$P_{asymmetric\ fusion} = 1 - \frac{1}{2^{10^{80}}}.$$

(12-2)

The Universe must have originated from an asymmetric fusion of positrons and not electrons.

Problem Set 12

1. How can unitons be created when their mass isolated is about 10^{22} times as heavy as an electron and minus imaginary?

2. About how much energy must there be in the center of mass frame to fuse electron semions into anti-proton and anti-neutron unitons? Why?

3. What particles are created when electron semions fuse? In our Universe would they be stable? What would result from fusing electron semions?

4. What particles are created when positron semions fuse? In our Universe would they be stable? What would result from fusing positron semions?

5. Could an all matter Universe be created through electrino fusion? Could it be created in a Big Bang?

Chapter 13

RELATIVITY WITH REAL MASSES

In Chapter 6, the general quasi-relativity of Chapter 4 was utilized in deriving the structure of electrons and their sub-particles (electrinos) at rest. We have already covered special quasi-relativity in Chapter 3. However, those calculations were for masses in the abstract. Considering relativity with real masses—with the particle structures derived in Chapters 6 and 10—is a fruitful exercise. It will give us a good perspective in the next chapter, Interstellar Red Shift, and later calculations. So we devote this chapter to the study of relativity in electrino systems.

A. Characteristic Equation

Electrinos orbit at speed c relative to the aether particles in particles. In Chapter 6, for an observer in the relative rest frame of particles, we found that for orbiting electrinos

$$V^2 = c^2 + v^2, \tag{13-1}$$

where V is the total velocity of the electrinos, v is the radial aether velocity at the radius of the electrino orbits, and c is the orbital velocity. c and v are at right angles. For particles not at rest in the aether, another velocity u of the system relative to the aether must be introduced. This must be done by velocity additions according to Quasi-Special Relativity, for bosons (which mediate relativistic effects) penetrate the electrino orbits, and even go through the super small electrinos. Also, the velocity u is not always perpendicular to c and v, though it is in the simplest case. That situation is when u is parallel to the normal of the c-v plane. The next step more complex problem is when u is parallel to the c-v plane. We will study also a third case: u

312

is at a general angle relative to the c-v plane. (Actually we will solve the general case, and the normal to the plane case and the parallel to the plane case will be special cases of that general case.) In each situation we must solve the equations such that the vectorial sum of c, v', and u is equal to V'.

In the general case, -u is at an angle φ to the normal to the c-v plane. When -u is perpendicular to the plane is the simpler case when $\varphi = 0$. When -u is parallel to the plane is the simpler case when $\varphi = \pi/2$. (See Fig. 13-1.)

The hard part is drawing a three dimensional representative intelligible figure. Once this is done, the quantities and trigonometric relations can be read directly from the figure. Since -u is at an angle φ to the normal to the plane, c and v cannot be determined directly from the relativity velocity addition formulae. One must first break c and v into c_x, c_y, c_z, v_x, v_y, and v_z, where the x direction is along the -u direction, and the y and z directions are perpendicular to it. Then these quantities must be substituted into the following relativistic velocity additions from Chapter 3, and then adapted for use here.

$$v_x' = \frac{v_x - V}{1 - v_x V / c^2}; \tag{13-2}$$

$$v_y' = \frac{v_y}{1 - v_x V / c^2}\left(1 - \frac{V^2}{c^2}\right)^{1/2}; \tag{13-3}$$

$$v_z' = \frac{v_z}{1 - v_x V / c^2}\left(1 - \frac{V^2}{c^2}\right)^{1/2}. \tag{13-4}$$

These formulae are written in the common terms of special relativity. We wish now to convert them to the terminology of this chapter and book, giving sense to the orbital parameters of electrino systems. V from the relativistic equations equals -u from our equations. V from our equations is the same symbol but has the different meaning of the total velocity of the electrino orbit—the vectorial sum of all the velocity components, including c for orbital velocity of the electrinos.

The rigorous calculations of c_x', c_y', c_z', v_x', v_y', v_z', $c_x'^2$, $c_y'^2$, $c_z'^2$, $v_x'^2$, $v_y'^2$, $v_z'^2$, and $c'^2 + v'^2$ from the above equations is ponderous, difficult, and expansive. The calculation will be done for the reader. The results will follow.

Using these guidelines, let us solve for $c'^2 + v'^2$. Reading from Fig. 13-1, let us first obtain c_x, c_y, c_z, v_x, v_y, and v_z for the General Case, Normal Case ($\varphi = 0$), and the Parallel Case ($\varphi = \pi/2$).

General Case	Normal ($\varphi = 0$)	Parallel ($\varphi = \pi/2$)
$c_x = c\sin\theta\sin\varphi$	$c_x = 0$	$c_x = c\sin\theta$
$c_y = c\cos\theta$	$c_y = c\cos\theta$	$c_y = c\cos\theta$
$c_z = c\sin\theta\cos\varphi$	$c_z = c\sin\theta$	$c_z = 0$
$v_x = v\cos\theta\sin\varphi$	$v_x = 0$	$v_x = v\cos\theta$
$v_y = v\sin\theta$	$v_y = v\sin\theta$	$v_y = v\sin\theta$
$v_z = v\cos\theta\cos\varphi$	$v_z = v\cos\theta$	$v_z = 0$

Using these velocity components, we substitute into Eqs. (13-2)-(13-4) to obtain the relativistic primed values. For the General Case they are:

$$V_x' = \frac{c\sin\theta\sin\varphi + u}{1 + \sin\theta\sin\varphi u / c}; \qquad (13\text{-}5)$$

$$V_y' = \frac{c\cos\theta}{1 + \sin\theta\sin\varphi u / c}\left(1 - \frac{u^2}{c^2}\right)^{1/2}; \qquad (13\text{-}6)$$

$$V_z' = \frac{c\sin\theta\cos\varphi}{1 + \sin\theta\sin\varphi u / c}\left(1 - \frac{u^2}{c^2}\right)^{1/2}; \qquad (13\text{-}7)$$

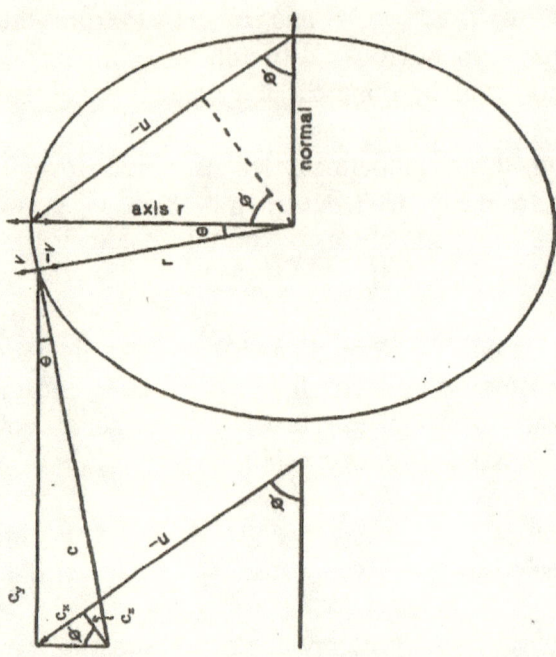

Fig. 13-1. The above is a three dimensional view showing
-u at an angle φ to the normal of the V-v plane orbit, as
well as the breakdown of V, v, and r into the x (u), y, and z
directions relative to –u.

$$v_x' = \frac{v\cos\theta\sin\varphi + u}{1 + \cos\theta\sin\varphi vu / c^2}; \qquad (13\text{-}8)$$

$$v_y' = \frac{v\sin\theta}{1 + \cos\theta\sin\varphi vu / c^2}\left(1 - \frac{u^2}{c^2}\right)^{1/2}; \qquad (13\text{-}9)$$

$$v_z' = \frac{v\cos\theta\cos\varphi}{1 + \cos\theta\sin\varphi vu / c^2}\left(1 - \frac{u^2}{c^2}\right)^{1/2}. \qquad (13\text{-}10)$$

The above equations must be squared, added together, and divided out. $v/c \approx 10^{-22} \approx 0$ simplifies the General Case:

$$V'^2 \approx c^2 + v^2 + u^2, \qquad (13\text{-}11)$$

The reader may do the parallel calculations for the Normal Case and the Parallel Case. But in each case the answer is the same. This is remarkable really. From the Pythagorean Theorem one might expect the result of Eq. (13-11) for the Normal Case, but not for the Parallel Case or the General Case. We are dealing with no averages here over the electrino orbits. Eq. (13-11) holds in the instantaneous case for general θ and φ.

B. Time Dilation

The Lorentz transformation for time is

$$t' = \frac{t - (V/c^2)x}{(1 - V^2/c^2)^{1/2}}. \qquad (13\text{-}12)$$

(See Chapter 3, Eq. (3-35).) The Lorentz transformation for time has an x-component in it. When, however, we take time differences at the same relative location, the x-component drops out and we obtain

$$t' - t_0' = \frac{t - t_0}{\left(1 - V^2 / c^2\right)^{1/2}}. \tag{13-13}$$

Setting t_0' and t_0 equal to 0 and $(1 - V^2/c^2)^{1/2}$ equal to γ, we have

$$t' = \gamma\tau, \tag{13-14}$$

where τ is t in the aether frame. (See Chapter 3, Eq. (3-24).)

We would like to apply Eq. (13-14) to the real mass orbit of electrinos at a general angle φ to the direction of motion relative to the aether. At first this would seem impossible and contradictory, because the electrinos orbit, and in general have x components to their motion. Eq. (13-14) can be applied to the center of the orbit, however, which remains relatively at rest in the prime frame (though not absolutely at rest in the aether frame). We see that the angles φ and θ have the center point of the orbit as the apexes of their angles. That point is unaffected by the angles. Thus Eq. (13-14) is unaffected by the angles φ and θ of the orbit relative to the direction of motion.

We consider, however, the orbiting electrinos to be a clock. Let us derive simply the time transformation from the orbiting electrinos. The orbital velocities of the electrinos at various angles are parameters:

$$V' = \frac{r'd\theta'}{dt'}; \quad V = \frac{rd\theta}{dt}. \tag{13-15}$$

$$dt' = \frac{r'd\theta'}{V'}; \quad dt = \frac{rd\theta}{V}. \tag{13-16}$$

In general, $r' \neq r$, and $d\theta' \neq d\theta$. However, when r' goes down, $d\theta'$ goes up by the same proportion, so that

$$r'd\theta' = k = rd\theta. \qquad (13\text{-}17)$$

From Eqs. (13-1) and (13-11),

$$V'^2 - v'^2 \approx V^2 - v^2 - u^2. \qquad (13\text{-}18)$$

For $u \ll c$, v' and v can be neglected. We then have

$$V'^2 \approx V^2 - u^2; \quad V' \approx V\left(1 - \frac{u^2}{V^2}\right)^{1/2}. \qquad (13\text{-}19)$$

V is almost exactly equal to c. Eq. (13-19) can be expressed approximately

$$V' \approx V\left(1 - \frac{u^2}{c^2}\right)^{1/2}. \qquad (13\text{-}20)$$

Thus dt' can be solved as

$$dt' = \frac{r'd\theta'}{V'} = \frac{rd\theta}{V\left(1 - \dfrac{u^2}{c^2}\right)^{1/2}} = \gamma d\tau. \qquad (13\text{-}21)$$

Since Eqs. (13-1) and (13-11) hold true irrespective of θ and φ, Eqs. (13-14) and (13-21) hold true irrespective of θ and φ.

C. Mass Increase

From Chapter 3, Eqs. (3-59) and (3-60), we see that $M(u) = \gamma M_0$, regardless of time, position, θ, or φ. The only thing that affects the mass increase is the velocity u in the x-direction.   That is because mass and momentum were derived from the y transformation of velocity, which derives from constant y and times depending on u, but not on time, position, θ, or φ.

D. Length Contraction

The Lorentz transformation for x position is

$$x' = \frac{x - Vt}{\left(1 - V^2 / c^2\right)^{1/2}}.$$

(13-22)

(See Chapter 3, Eq. (3-35).) The inverse transformation is

$$x = \gamma\left(x' + \beta ct'\right).$$

(13-23)

(See Chapter 3, Eq. (3-41).)   The transformations have terms in t and t' in them.

$$x_2 = x_2'(t')\gamma + ct'\beta\gamma;$$

(13-24)

$$x_1 = x_1'(t')\gamma + ct'\beta\gamma.$$

(13-25)

However, when we measure lengths simultaneously, the t term drops out and we have

$$x_2 - x_1 = L_0$$

(13-26)

$$= \left[x_2'(t') - x_1'(t')\right]\gamma = L'\gamma.$$

(13-27)

Thus

$$L' = \frac{L_0}{\gamma}. \tag{13-28}$$

(See Chapter 3, Eq. (3-23).)

Eq. (13-28), however, is along the x-axis in relativity, or in the direction of velocity u in this chapter. In this chapter we are interested in the length contraction of radius r of an electrino orbit at general angles θ and φ to the direction of velocity u.

$$L_x' = \frac{L_x}{\gamma}; \quad L_y' = L_y; \quad L_z' = L_z. \tag{13-29}$$

Reading from Fig. 13-1, we have

$$r_x = r\cos\theta\sin\varphi; \tag{13-30}$$

$$r_y = r\sin\theta; \tag{13-31}$$

$$r_z = r\cos\theta\cos\varphi. \tag{13-32}$$

The radius components as observed of the moving frame are as follows:

$$r_x' = r\cos\theta\sin\varphi\left(1 - \frac{u^2}{c^2}\right)^{1/2}; \tag{13-33}$$

$$r_y' = r\sin\theta; \tag{13-34}$$

$$r_z' = r\cos\theta\cos\varphi. \tag{13-35}$$

The squares of the r' components are

$$r_x'^2 = r^2 \cos^2\theta \sin^2\varphi\left(1 - \frac{u^2}{c^2}\right); \qquad (13\text{-}36)$$

$$r_y'^2 = r^2 \sin^2\theta; \qquad (13\text{-}37)$$

$$r_z'^2 = r^2 \cos^2\theta \cos^2\varphi. \qquad (13\text{-}38)$$

Summing the square x', y', and z' components, we find the square of r':

$$r'^2 = r^2[\cos^2\theta \sin^2\varphi(1 - u^2/c^2) + \qquad (13\text{-}39)$$
$$\sin^2\theta + \cos^2\theta \cos^2\varphi].$$

Simplifying,

$$r'^2 = r^2\left[1 - \cos^2\theta \sin^2\varphi\frac{u^2}{c^2}\right], \qquad (13\text{-}40)$$

$$r' = r\left[1 - \cos^2\theta \sin^2\varphi\frac{u^2}{c^2}\right]^{1/2}. \qquad (13\text{-}41)$$

If $\theta = \pi/2$, $\cos^2\theta = 0$, and there is no length contraction in that direction, which we expect. If $\theta = 0$, $\cos^2\theta = 1$, and there is the variable length contraction, which is $r(1 - \sin^2\varphi u^2/c^2)^{1/2}$. If φ is $\pi/2$, there is the maximum length contraction which is $r(1 - u^2/c^2)^{1/2}$, which we expect. Eq. (13-41) works out well, and is the general form for any specific θ and φ.

Problem Set 13

1. Does time dilation and mass increase depend on the direction of the aether wind?

2. What direction relative to the V-v plane of electrino orbits does the aether wind appear to come from in all cases for time dilation and mass increase?

3. Does length contraction depend on the direction of the aether wind? Does it depend on θ? Does it depend on φ?

4. Is length contraction of elementary particles of the right magnitude for special relativity?

5. Why might the outcome of 4 make little difference in macroscopic objects and length contraction?

Chapter 14

INTERSTELLAR RED SHIFT

A. Introduction

The Hubble telescope, heroically repaired by astronauts, has presented a major shock to cosmology theories. How can the Universe appear younger than some local stars?[1] Any reasonable suggestion ought to be considered.

Since the 1920s the red shifting of the spectra of distant galaxies has generally been considered Doppler, or velocity red shifting[2]. That led to a model of an expanding universe and a finite beginning in a Big Bang. The recent use of the Hubble space telescope to re-calibrate the Hubble Constant using the Cepheid variable stars[3,4] has challenged the Big Bang theory by having the Universe appear younger than some local stars.

The Doppler shift is not the only mechanism that can red shift distant galaxies and quasars. The Compton effect of photon collisions with free electrons can red shift light.[5] The absolute particle-wave duality of light may provide another red shift mechanism. If particle-wave duality of photons is correct, then there should be electrical particles in a photon stirring up the wave train. In order to do that a positive and negative particle must orbit about each other as well as travel along the light path axis. If there is a real orbit in the photon, then there must be a finite radius of the orbits of the photon sub-particles. There must also be a finite, non-zero v_t^2 of the particles in orbit at some angle to the axis of photon travel. The v_t^2 of the photon electrinos must obey the equation

$$V^2 \approx u^2 + v_t^2 = c^2 + v_t^2, \qquad (14\text{-}1)$$

(from Chapter 6, Eq. (6-55)). Thus photon' electrinos do not travel at the absolute velocity of light in a vacuum (as whole photons do), but faster.

In all energy photons, v_t is at the identical fixed rate, namely c. Thus different energy photons cannot overtake each other in flight. But different color photons with nearly parallel axes, separated by an appropriate fraction of a wavelength, can have their electrinos approach each other many trillions of times over a light year, with the statistical chance of a collision between the electrinos of different photons. Energy can be exchanged between the photons before they part. We should have already discerned a problem with Doppler red shifting for a model of an expanding universe. A pure velocity Doppler effect should not broaden spectral lines, which we observe in cosmological red shifting. The particle theory above proposed for red shifting would broaden spectral lines, however, statistically with many photon' electrino-photon' electrino collisions.

The cosmological red shift must be broken into two parts (the dual photon, spectral line broadening red shift, and the residual expanding universe term). The spectral line broadening red shift must be subtracted from the total cosmological red shifting to obtain the much smaller expanding universe term. By this method the Universe will appear much older than by the current Hubble hypothesis.

For a reference, the important parameters of the sub-particles (unitons) in the photons are derived below. In the author's research and model, there are two particles in the photon, one positive and one negatively charged, each charged $\pm e$. The structure of a sub-particle (uniton) is a small thin spherical shell of charge. The particles are very small, measured in imaginary radii. Their orbits are much larger, measured in real radii. The quantities of interest are derived below.

B. Mass Due to Charge Alone of Spherical Shell

In this book, oppositely charged unitons ($\pm e$) are picked as the constituent particles of photons. Given that the sub-particles in a photon are charged $\pm e$, finding the other salient physical parameters of the photon sub-particles is relatively easy. The mass M_e (where M_e is the mass of the smallest thin spherical shell of charge distribution of charge $\pm e$ due to charge alone, and not the mass of the electron) of the sub-particles can be found by assuming the particles are the smallest possible for the charge $\pm e$, according to aether general relativity.   Classical electrostatics gives us the capacitance of a spherical surface of radius r:

$$C = 4\pi\varepsilon_0 r. \tag{14-2}$$

For fine structures we multiply by α:

$$C = 4\pi\varepsilon_0 \alpha r. \tag{14-3}$$

The work required to add a charge q to a capacitor with capacitance C is

$$W = 1/2\ q^2\ /\ C. \tag{14-4}$$

Therefore the potential energy of a spherical capacitor in a fine structure, i.e., the energy of its electrostatic field, is

$$E_{pot} = \frac{q^2}{8\pi\varepsilon_0 \alpha r_p} = m_q c^2, \tag{14-5}$$

where m_q is the mass associated with the electric field alone. For the charge mass of the photon sub-particles of charge $\pm e$ (unitons), the energy U is just that of Eq. (14-7), where q = e. We see that for the photon sub-particle

$$c^2 = \frac{-2Gm_e}{r_p} = \frac{U}{m_e} = \frac{e^2}{8\pi\varepsilon_0 \alpha r_p m_e}, \qquad (14\text{-}6)$$

where m_e is the mass of the spherical shell of charge $\pm e$ due to charge alone, not the mass of the electron. Continuing the calculations:

$$m_e^2 = \frac{e^2}{-8\pi\varepsilon_0 \alpha 2G}, \qquad (14\text{-}7)$$

$$m_e = -i\left(\frac{e^2}{4\pi\varepsilon_0 \alpha 4G}\right)^{1/2} = -i\left(\frac{\hbar c}{4G}\right)^{1/2} = M_e, \qquad (14\text{-}8)$$

where

$$\alpha = \frac{e^2}{4\pi\varepsilon_0 \hbar c}. \qquad (14\text{-}9)$$

M_e is one of the parameters we wish to derive for the photon sub-particles.

C. Mass Due to Motion of the Charge

In addition to the electric self potential mass, the uniton or any electrino has kinetic mass in the electrino relative rest frame. The aether field is not static. Every portion of charge in the uniton is traveling at speed c relative to the aether. Therefore m_e or M_e has kinetic energy also. If m_e had a small velocity relative to the aether, its kinetic energy E_{kin} would be $\frac{1}{2}m_e v^2$. But since $v^2 = -c^2$, we take the relativistic form $E_{kin} = -m_e c^2$. The total fundamental mass of the uniton is $M_e + (-M_e) = 0$. The absolute value masses are

$$| M_{kin} | = | M_e | = \frac{1}{2} | M_0 | = \left(\frac{\hbar c}{4G} \right)^{1/2} .$$ (14-10)

D. Total Absolute Values Masses of the Charge Shell

The total absolute value mass of the photon sub-particle is the Planck mass, composed simply of the following constants:

$$| M_{kin} | + | M_e | = | M_0 | = \left(\frac{\hbar c}{G} \right)^{1/2} .$$ (14-11)

Numerically it is

$$| M_0 | \approx 2.177 \ x \ 10^{-8} \ kg.$$ (14-12)

E. Radius of Spherical Shells

The radius of the thin spherical shell photon sub-particle R_0 is

$$R_0 = \frac{2GM_e}{-c^2} = i \left(\frac{\hbar G}{c^3} \right)^{1/2}$$ (14-13)

$$\approx i \ 1.616 \ x \ 10^{-35} \ m.$$ (14-14)

F. Real Mass of Charge Shell Due to Photon Motion

The real mass of the photon sub-particle is easy to calculate. The effective mass of the photon is the energy of the photon E_γ divided by c^2. But a photon sub-particle is half of a photon. Therefore the real mass of the spherical shell photon subparticle is

$$m_{1/2\gamma} = \frac{E_\gamma}{2c^2}. \qquad (14\text{-}15)$$

G. Square of Velocity of Spherical Shells in Orbit

The total spin of a photon is ± 2, not because the photon is not a fine structure, but because $(2)f_p\hbar = (2)(1)\hbar = 2\hbar$. The (2) before f_p arises from the summation of the particles. [See Chapter. 6, Postulate 8.]

The photon is a fine structure, and does not act according to the Coulomb electric force, but by the meso-electric force—equivalent to the strong electric force without the extreme length contracted imaginary radii, or Coulomb force with the multiplication of α in the denominator of the force equation. This means the meso electric force is between the charges of the photon subparticles—the electrinos. This phenomenon is demonstrated in Eq. (14-16).

Now let us make a quick simple calculation of the forces in the outer frame. The meso electric force between the particles is

$$F_{ME} = \frac{e(-e)}{4\pi\varepsilon_0\alpha r^2} = -\frac{\hbar c}{r^2}. \qquad (14\text{-}16)$$

The axis of the photon is just the light path. The orbit of the photon sub-particles is at some angle φ to that (not necessarily perpendicular to it). If, as we will show, the orbit has a finite radius and finite v_t^2, then the photon' electrinos cannot travel at the absolute velocity of light, but faster than c. Their u^2 must obey the relation

$$V'^2 \approx u^2 + v_t^2 = c^2 + v_t^2. \qquad (14\text{-}17)$$

To calculate v_t^2 we first add the meso electric force and inertia to zero. It is important to note that, since the photon sub-particles (unitons and anti-unitons) travel about the speed of light, the mass of one particle is greatly increased in the frame of the other particle. Essentially, it is a one-body problem. There is no reduced mass in the equation. (See Chapter 8, Section G.)

$$0 = F_{ME} + F_I = -\frac{\hbar c}{r^2} + \left(\frac{E_\gamma}{2c^2}\right)\frac{v_t^2 + c^2}{r}. \qquad (14\text{-}18)$$

$$r = \frac{2\hbar c}{E_\gamma\left(1 + \dfrac{v_t^2}{c^2}\right)}; \qquad (14\text{-}19)$$

$$\text{also } r = \frac{\hbar c}{E_\gamma}. \qquad (14\text{-}20)$$

$$\left(1 + \frac{v_t^2}{c^2}\right) = 2; \quad v_t^2 = c^2. \qquad (14\text{-}21)$$

v_t is equal to the speed of light c. In an electron, the sub-particles (electrinos—semions) orbit exactly at the speed of light c, but have no additional linear translational motion. In a photon, the sub-particles (electrinos—unitons) orbit exactly at the speed of light c, but have additional linear translational motion exactly at the speed of light c.

H. Radius of Orbit of Spherical Shells in Photon

If v_t^2 is known, we can solve for r_o, the radius of the sub-particle orbit.

$$r_o = \frac{2\hbar c}{E_\gamma \left(1 + \dfrac{c^2}{c^2}\right)} = \frac{\hbar c}{E_\gamma}. \qquad (14\text{-}22)$$

I. Gravitational Aether Velocity Squared v^2

The gravitational aether velocity transforms away in the frame of the photon sub-particles (unitons and anti-unitons). We either omit it, or count it as 0.

$$v^2 = 0. \qquad (14\text{-}23)$$

J. Square of the Total Velocity of the Photon Electrinos

$$V^2 \approx v^2 + v_t^2 + u^2 = 0 + c^2 + c^2 = 2c^2. \qquad (14\text{-}24)$$

K. Evidences of a Dual Photon

Since the days of Albert Einstein, the model of the photon has been a single particle—the photon or light quanta—and a wave train, which together travel precisely at the speed of light c. Physicists have long overlooked, however, some simple evidences that the Standard Model for the photon is in error.

First there is the common sense cause of the electromagnetic wave train. The simplest cause is a positive and negative electrical particle orbiting about each other as they travel the speed of light linearly. A single particle photon won't do this job. It takes two particles, one positive and one negative.

Second is the matter of polarization. Polarization of light has long been studied.[6] A number of polarization modes are possible: circular polarization; elliptical polarization with a complete range of eccentricities and axes orientations; and linear polarizations with whole range of axes orientations. How can these polarizations of light be accounted for by the single particle model of the photon? Every one of these polarizations of the photon can easily be accounted for by various angular orientations of an orbit of two particles relative to the light path axis. The circular polarization is caused by the orbit of the particles being perpendicular to the light path axis.   The line polarization of sinusoidal waves is when the circle is parallel to the light path axis, and the length along the light path axis is flattened by length contraction of relativity. The elliptical polarization is when the orbital circle is at a general angle to the light path axis.

Third, polarization phenomena bring forth another evidence that puts the Standard Model in disrepute. Polarizations show that there are photon sub-particle motions perpendicular to the light path axis, and that photon sub-particles cannot therefore travel at the absolute velocity c.

L. Dual Photon Red Shift

A great deal of the calculations and evidences of a dual photon have been put in this chapter to elucidate the mathematical photon mechanism for a Particle-Wave Duality Red Shift. Previously there has been a model called "tired light." The Particle-Wave Duality Red Shift makes the "tired light" red shift more scientific, giving it precise formulas to work with. "Tired light" can be seen to result from photon' electrino-photon' electrino collisions along their flight path.   Red shifts will be found to be grainy and statistical. Since a variety of energy photons can collide, the spectra also become broadened with distance. This also is grainy and statistical.

Most of the development in this chapter has been to substantiate the possibility of a dual photon (or a photon-photon collision) red shift.   This possibility is still theoretical. But the Compton effect red shift possibility is not only theoretical, it is measured–experimental. It is well worked out. There are many references on it. (See Google Search Engine.) Probably most of the red shift is Compton, and it should not be neglected. But some of the red shift may be dual photon (or photon-photon collision red shift). This also ought to be explored.

If all red shift were Compton or dual photon red shift, there would be no expansion of the universe at all, and no way to gauge its age. A new feat needs to be attempted: carefully subtract the spectral line broadening effect from the overall cosmic red shift to see if there is a residual difference. If there is, it will be like blowing up a balloon. It will be an expansion of the Universe (true Doppler red shift).

M. The Redshift Goes In Jumps

Recent measurements have given the surprising result that interstellar red shift is quantized and goes in jumps or packets. In 1976 William Tifft of the Steward Observatory at the University of Arizona "claimed that visible-light redshift measurements suggested that galaxies in a cluster in the constellation Coma have redshifts that fall into distinct velocity packets. The velocities, he said, always came out at some multiple of about 72 kilometers per second (km/s)."[7]   In later years Tifft and his colleagues amassed more data from more galaxies, including those closer to home. Later measurements indicated the quanta of redshifts was smaller, as small as 2.6657 km/s, "with different multiples dominating for different types of galaxies."[8]

Few astronomers have taken the notion of "quantized redshifts" seriously in the past, but some galaxy specialists who have seen the new results slated to appear in the journal Astronomy and Astrophysics—are no longer dismissing them out of hand. Says galaxy specialist Mike Disney of the University of Wales at Cardiff, one of the many who have been skeptical in the past, "This paper suggests to me that there is a pretty strong case for getting more of the right kind of data to settle this extremely important debate." Harvard galaxy expert John Huchra, another longtime skeptic, goes further: "My curiosity is now sufficiently whetted that I'm thinking of writing an observing proposal for checking to see if the effect holds up with other galaxies." If it does, standard cosmology might be turned on its ear: "It would mean abandoning a great deal of present research," says Disney.[9]

The majority of interstellar redshifting may not be velocity or Doppler at all—but the result of tired light or photon electrino collisions due to the weak force, or the Compton effect, as we studied earlier in the chapter. The Doppler redshift mechanism may have to be abandoned for the majority of interstellar redshift. So too, the Big Bang theory for the origin of the Universe may have to be discarded. It will be interesting to see how this matter will turn out.

---

[1]Michael D. Lemonick and J. Madeleine Nash, "Unraveling Universe," *Time*, March 6, 1995, pp. 76-84.

[2]Nathan Cohen, Ph.D., *Gravity's Lens* (New York: John Wiley & Sons, Inc., 1988), Forward, p. iv.

[3]*Ibid.* pp. 15, 16.

[4]Lemonick, *op. cit.*, p. 80.

[5]Paul Marmet, Herzberg Institute of Astrophysics, National Research Council, Ottawa, Ontario, Canada, K1A 0R6, *Physics Essays*, Vol. 1, No: 1, p. 24-32, 1988.

[6]"Light," *Encyclopaedia Britannica*, 15th edition (Chicago: Encyclopaedia   Britannica,   Inc.,   1974), Macropaedia, Volume 10, p. 929.

[7]R. Matthews, 'Do Galaxies Fly Through The Universe In Formation?', *Science*, Vol. 271:759, 1996.

[8]T.   Beardsley,   'Quantum   Dissidents',   *Scientific American*, December 1992, pp. 39, 40.

[9]Matthews, *op. cit.*

Problem Set 14

1. What major shock to cosmology theories did the Hubble telescope recently make involving red shift?

2. What type of red shifting has been the model since the 1920s for an expanding Universe?

3. What type of stars are used as a gauge of distance?

4. What type of photon can cause interstellar red shifting without an expanding Universe?

5. How can that type of photon cause interstellar red shifting?

6. Are dual photon red shifts greater or lesser with increasing distance?

7. Should a pure Doppler red shift broaden spectral lines?

8. Are the cosmic red shifts spectra broadened?

9. What two parts must the interstellar red shift be broken into?

10. Will the Universe appear older or younger with dual photon red shifting?

11. Give three evidences the Standard Model for the photon is in error.

12. How much faster than c is a gamma photon' electrino of 1 GeV? of 1 MeV?

13. What is the radius of orbit r_o of a photon of 10 eV?

Chapter 15

APPLICATIONS OF ELECTRINO FUSION

A. Introduction

Currently (December, 2006) there are no facilities in the world designed to produce artificial electrino fusion. (In order to do that you must collide ≥938 MeV electrons or positrons with up axial spins with ≥938 MeV like particles from the opposite direction with down axial spins (so the semions orbit the same direction when collided).) The two semions in orbit in one electron would join the orbit of two semions in the other electron. The four semions in orbit would fuse to two unitons, which would break up gravitons for the additional particles to make anti-protons or anti-neutrons. These would annihilate local protons and neutrons to yield twice the energy as was invested in the production of the anti-matter for those particular particles. Artificial electrino fusion has never been achieved in the history of the world. Natural electrino fusion (such as $\eta'(958) \rightarrow \gamma\gamma$) has been achieved frequently in accelerator laboratories, but not recognized as such because the scientists could not balance the decay schemes with an electrino fusion model since they held to a quark-lepton model. Natural electrino fusion has been right in front of the scientists faces, but not recognized by them.

Experience with electron-positron annihilations would lead us to expect that there would be 1.602×10^{-19} particle collisions/particle for e⁻ e⁻ collisions. Therefore the power we would expect that would result from the collision and fusion of electrons would be:

a. expected calculation of electron fusion power from the fusion of two 1.0 Amp beams of electrons at 2.0 GeV in the Center of Mass Frame:

$$\frac{2 \times 10^9 \; eV}{collision} \; \frac{1.602 \times 10^{-19} \; j}{eV} \; \frac{2.0 \; C}{s} \; \frac{particle}{1.602 \times 10^{-19} \; C} \; x$$

$$\frac{1.602 \times 10^{-19} \; collisions}{particle} \; \frac{2 \; annihilated \; particles}{1 \; fused \; particle}$$

$= 1.282 \times 10^{-9}$ Watts. Since there is 2×10^9 Watts power investment into this fusion, there would never be any hope of electrino fusion power reactors by this calculation. But this calculation is weird. The factor 1.602×10^{-19} appears three times in the equation. This seems unrealistic. A more natural calculation of the electron fusion power from the fusion of two 1.0 Amp beams of electrons is below.

b. natural calculation of electron fusion power from the fusion of two 1.0 Amp beams of electrons at 2.0 GeV in the Center of Mass Frame:

$$\frac{2 \times 10^9 \; eV}{collision} \; \frac{1.602 \times 10^{-19} \; j}{eV} \; \frac{2.0 \; C}{s} \; \frac{1/2 \; collision}{1.602 \times 10^{-19} \; C} \; x$$

$$\frac{2 \; annihilated \; particles}{1 \; fused \; particle} = 4 \times 10^9 \; Watts.$$

Since only 2×10^9 Watts power investment is made with the reaction, two billion Watts (two million kilowatts) net power would be available for recovery and utilization of two 1.0 Amp beams of electrons at 2.0 GeV in the Center of Mass Frame.

B. Electrino Fusion Reactor

If calculation a. is correct, Electrino Fusion Reactors (EFR) are not possible. But if calculation b. is correct, EFR could soon make an absolute glut of electrical power

C. Annihilator of Deadly Wastes

If calculation a. is correct, EFR cannot be used as a device for annihilating deadly radioactive wastes and toxins. But if calculation b. is correct, deadly radioactive wastes and toxins could be used for fuel in EFR, and thus annihilated.

D. Radiation-less Electrical Generator

If calculation b. is correct, EFR can be fuel optimized to primarily hydrogen, thus limiting the radiation emitted to neutrons, which have a twelve minute half-life. Thus radiation emitted can be limited.

E. Antimatter Rocket

If calculation a. is correct, antimatter rockets would not be possible by this means. But if calculation b. is correct, copious quantities of anti-particles could be produced from a self-sustaining reaction for use in anti-matter rockets.

F. Super Nova

In a sun, normally the kinetic energies of the electrons are about 600 MeV. But when a sun collapses the kinetic energies of the electrons can reach 938 MeV. Then suddenly electrons can fuse to antimatter, and antimatter annihilates matter. If calculation a. is corrent, then an insignificant power source would result. But if calculation b. is correct, when the kinetic energies of the electrons reach 938 MeV, then a truly stupendous power source would be made available on the star many times more powerful than hydrogen fusion. The critical electrino fusion process would be so great it would blow the star apart and make it a nebula. The author does not believe such an explosion could come from hydrogen fusion or a calculation a. reaction. Super nova are a strong evidence that calculation b. is correct and EFR and the other listed applications are all possible with a new turn in accelerators, colliders, and reactors. Other than by the super nova evidence, no person on earth should be able to predict a priori which calculation is correct (a. or b.), since the Electrino Fusion Model is new and untested. If super nova are not enough, the matter should be settled through experiment. A safe experiment with EFR can be combined with one for the next chapter.

G. Second Law of Thermodynamics

The applications of electron electrino fusion pale into insignificance in comparison with the positron electrino fusion process's ability to reverse the order to disorder arrow in the second law of thermodynamics, reversing aging, disease, and decay processes. This application is so important, the whole next chapter is devoted to it.

Problem Set 15

1. Balance the decay scheme $\eta'(958) \to \gamma\gamma$. What process does this decay scheme demonstrate? Why have scientists not recognized it?

2. What process is probably the cause of Super Nova?

3. Describe the electrino fusion process for electrons.

Chapter 16

SECOND LAW OF THERMODYNAMICS

A. Introduction

Everything goes from a state of order to more disorder. Brand new automobiles wear out and rust. Objects break or are damaged. A thermos bottle falls off the counter, and the inner glass bottle is shattered. We do not expect the shattered bottle to fall back up to the counter and become whole again. There is a one-way arrow for the events to transpire. That arrow is the Second Law of Thermodynamics.

Houses grow old and fall into decay. Barns fall down. Fruit spoils, people and animals grow old and die. Viruses mutate. People become ill and die. Crime and disorder in society increase. Homes break up. Aborted fetuses disintegrate. Dead people and things decompose. All of these negative occurrences are the outworking of the second law of thermodynamics--that part of which is an arrow making everything go from order to disorder.

Let us consider what other people have written about the second law of thermodynamics.

"Second law of thermodynamics
"An equilibrium macrostate of a system can be characterized by a quantity S (called *entropy*) which has the following properties:

"(i) In any infinitesimal quasi-static process in which the system absorbs heat dQ, its entropy changes by an amount

341

$$dS = \frac{dQ}{T} \qquad (16\text{-}1)$$

where T is a parameter characteristic of the macrostate of the system and is called its *absolute temperature*.

"(ii) In any process in which a thermally isolated system changes from one macrostate to another, its entropy tends to increase, i.e.,

$$\Delta S \geq 0. \qquad (16\text{-}2)$$

"The relation (16-1) is important because it allows one to determine entropy *differences* by measurements of absorbed heat and because it serves to characterize the absolute temperature T of a system. The relation (16-2) is significant because it specifies the direction in which nonequilibrium situations tend to proceed."[1]

The above expression of the second law of thermodynamics is regarding entropy and heat. Other writers include the order to disorder arrow in the second law of thermodynamics.

"It is a matter of common experience that disorder will tend to increase if things are left to themselves. (One has only to stop making repairs around the house to see that!) One can create order out of disorder (for example, one can paint the house), but that requires expenditure of effort or energy and so decreases the amount of ordered energy available.

"A precise statement of this idea is known as the second law of thermodynamics. It states that the entropy of an isolated system always increases, and that when two systems are joined together, the entropy of the combined system is greater than the sum of the entropies of the individual systems. For example, consider a system of gas molecules in a box. The higher the temperature of the gas, the faster the molecules move, and so the more frequently and harder they collide with the walls of the box and the greater the outward pressure they exert on the walls. Suppose that initially the molecules are all confined to the left-hand side of the box by a partition. If the partition is then removed, the molecules will tend to spread out and occupy both halves of the box. At some later time they could, by chance, all be in the right half or back in the left half, but it is overwhelmingly more probable that there will be roughly equal numbers in the two halves. Such a state is less ordered, or more disordered, than the original state in which all the molecules were in one half. One therefore says that the entropy of the gas has gone up. Similarly, suppose one starts with two boxes, one containing oxygen molecules and the other containing nitrogen molecules. If one joins the boxes together and removes the intervening wall, the oxygen and nitrogen molecules will start to mix. At a later time the most probable state would be a fairly uniform mixture of oxygen and nitrogen molecules throughout the two boxes.   This state would be less ordered, and hence have more entropy, than the initial state of two separate boxes."[2]

"The explanation that is usually given as to why we don't see broken cups gathering themselves together off the floor and jumping back onto the table is that it is forbidden by the second law of thermodynamics. This says that in any closed system disorder, or entropy, always increases with time. In other words, it is a form of Murphy's law: Things always tend to go wrong! An intact cup on the table is a state of high order, but a broken cup on the floor is a disordered state. One can go readily from the cup on the table in the past to the broken cup on the floor in the future, but not the other way round.

"The increase of disorder or entropy with time is one example of what is called an arrow of time, something that distinguishes the past from the future, giving a direction to time."[3]

B. Electrino Model and 2nd Law

The natural tendency of leptons in beta decay is that the parent lepton combines with one or more gravitons to produce more particles. In all natural reactions, the order energy of the resultant particles is less than or equal to the order energy of the original particles.

1. Negative Energies. Let us consider antimatter more carefully. "In the Dirac theory also, *the permissible energy values for a free particle range from* $+mc^2$ *to* $+\infty$ *and from* $-mc^2$ *to* $-\infty$. The first of these results is of course just what we expect for a free particle—that its total energy can have any value greater than its rest energy. But the second result is quite puzzling, since it implies the existence of states of *negative total energy*."[4]　Anderson in 1932 discovered positrons in cosmic radiation. These were regarded as Dirac's negative energy particles. "The first two solutions of the Dirac equation . . . clearly describe a free electron of energy E and momentum **p**. The two negative energy electron solutions . . . are to be associated with the antiparticle, the positron."[5]

However, in the annihilation it is not $(+mc^2) + (-mc^2) = 0$, but $2mc^2$ is the result of annihilation.[6]　There is something strange going on with the minus signs in these equations. The calculations are inconsistent.

Maybe there are two kinds of energy considered. One we can call entropy energy E_S. In the annihilation reaction,

$|+mc^2| + |-mc^2| = 2mc^2$. Entropy energy is the higher value. The other energy is order energy E_O. In order energy the same reaction is $(+mc^2) + (-mc^2) = 0$.

Let us consider entropy energy and order energy for particle decay schemes. There are a few decay schemes where no negative order energy (anti-matter) is introduced in the right hand side of the decay schemes. In those few instances, the final order energy is equal to the initial order energy (when kinetic energy is taken into account). But in most cases, a trace of negative order energy (anti-matter) is introduced into the right side of the decay schemes. There is nothing on the left hand sides of the decay schemes to correspond to this addition of a trace of negative order energy on the right sides of the decay schemes. Therefore, total order energy is less on the right hand sides of the decay schemes than on the left hand sides (if only by a trace). A few decay schemes introduce a lot of antimatter (as K^-) on the right side of the decay scheme. The loss of order energy in the systems is greater in those cases. But in every case, for all natural processes, the order energy final is ≤ the order energy initial, or

$$\Delta E_0 \le 0. \tag{16-3}$$

Let us check the order energy for electron electrino fusion reactions. Electrons made energetic by acceleration (as heavy as protons) fuse and form anti-protons. Matter is converted to anti-matter. Entropy energy is conserved, but not so order energy. Order energy is reduced in the extreme from +938 MeV to -938 MeV or more for each electron fused (two electrons are fused in each reaction). The order-disorder arrow for electron electrino fusion points in the usual direction. The system does obey the second law of thermodynamics.

2. Reversing the Order to Disorder Arrow. What would happen if we fused the electrino constituents of positrons instead of the electrino constituents of electrons? Entropy energy E_S would again be conserved. Entropy

would be increased. However, order energy E_O would go from -2 x 938 MeV to +2 x 938 MeV—from disorder to order. The order to disorder arrow would be reversed. This would be a reaction that would be prohibited by the second law of thermodynamics—unless the strong gravitational force that fuses the anti-semions would be stronger than the second law of thermodynamics (which otherwise governs weak interactions). The stronger of the strong gravitational force and the second law of thermodynamics should be determined by experiment. More rides on that one experiment than perhaps on any one other experiment in this generation. If it is found that strong gravity is stronger than the second law of thermodynamics, then order can be restored at first in a small area, then for the whole earth.

Here we see that the entropy arrow of time and the order to disorder arrow of time are separate and distinct, and are not one and the same thing. While all the reactions the author has studied increase entropy, the fusion of positron anti-semions reverse the order to disorder arrow, making more order out of the disorder.

Positron constituent electrino fusion might not only take the electrinos from disorder to order. It could make other physical processes in a local area go from disorder to order. The positron fusion not only violates the second law of thermodynamics, it reverses the order to disorder arrow of that law in a local area, making other processes in that area reverse. Let us consider that process more to see how it might be regulated.

We guess the desired relationships for reversing the order to disorder arrow in the second law of thermodynamics through dimensional analysis. We want to solve for r, the maximum radius in which the reversed law would be effective. There is a way we can obtain a length from combinations of our variables and constants. That way is in the right hand side of Eq. (16-4). The whole expression is the thermodynamic relation we are seeking. The thermodynamic relation is:

$$(\Delta E_o)_t > 0 \ where \ r < \frac{(\Delta E_o)_1 \ c}{ik}, \qquad (16\text{-}4)$$

where E_o is the order energy–the positive or negative energy in the pair production of particles; ΔE_o is the change in the order energy, where $(\Delta E_o)_t$ is the change in the total order energy of the system, and where $(\Delta E_o)_1$ is the change in the order energy for a single source reaction—for a positron fusion reaction it is approximately 2×10^9 eV/collision $\times 1.6 \times 10^{-19}$ joules/eV $= 3.2 \times 10^{-10}$ joules/collision; c is the speed of light—approximately 3.0×10^8 m/s; we shall solve for the effected radius r; i is the beam current in each beam in Coulombs per second (we will solve for 10^{-11}); k is the ratio of particle energy to particle charge. This energy per charge is the accelerated energy of the particle (roughly 1×10^9 ev times 1.6×10^{-19} joules/ev $= 1.6 \times 10^{-10}$ joules) divided by the charge of each positron ($q = 1.6 \times 10^{-19}$ coulombs), which equals 10^9 joules per coulomb. The collision efficiency eff is not needed in this equation, because the result is not in particles, but is already in collisions.

Incredibly, the lower the current, the bigger the radius of the affected area. And the greater the current, the smaller the radius of the effected area. With 10^{-11} A beam currents, the effected radius r solves for 9.6 meters—roughly 10 meters, which describes a small area—less than a tenth of an acre.

To get an idea of the positron beam currents needed to reverse the order to disorder arrow of the second law of thermodynamics in what size of affected radius, see Table 16-1 below.

For an area the size of	r	beam current
House	10 m	10 pA
four football fields	100 m	1 pA
community	1 km	100 fA
city	10 km	10 fA
Israel	160 km	0.6 fA
U.S.	2,400 km	0.04 fA
World	13,000 km	0.008 fA
Sun	1.7E11 m	6E-22 A

Table 16-1. Beam currents versus affected radius for reversal of the order to disorder arrow of the second law of thermodynamics.

Remarkably enough, the affected area of second law reversal calculates to increase with the reduction of positron beam current. Area control is merely a matter of timed gating of the positrons in the positron-positron collider.

---

[1]F. Reif, *Statistical Physics*, Berkeley Physics Course-- Volume 5 (New York: McGraw-Hill Book Company, 1967), p. 283.

[2]Stephen Hawking, *A Brief History of Time*--From the Big Bang to Black Holes (New York: Bantam Books, 1988), pp. 102, 103.

[3]*Ibid.*, pp. 144, 145.

[4]Robert B. Leighton, *Principles of Modern Physics* (New York: McGraw-Hill Book Company, Inc, 1959), p. 665.

[5]Francis Halzen, Alan D. Martin, *Quarks and Leptons* (New York: John Wiley & Sons, 1984), p. 107.

[6]David S. Saxon, *Elementary Quantum Mechanics* (San Francisco: Holden-Day, 1968), p. 386.

Problem Set 16

1. "Humpty-Dumpty sat on a wall. Humpty-Dumpty had a great fall. And all the king's horses and all the king's men couldn't put Humpty-Dumpty back together again." What law of physics does this child's nursery rhyme illustrate? What arrow of time is demonstrated?

2. Is the entropy more or less after two gas containers are opened to each other?

3. You watch a movie. Broken pieces of glass fly up and become a window. Is the movie playing forward or backward? How do you know?

4. What generally happens in beta decay? Is the system going to more or less order?

5. What would the world be like if the order to disorder arrow were reversed?

Chapter 17

INTRODUCTION TO PARTICLE STRUCTURE

I. Introduction

A. The Quest and Approach

The quest of this particle science is to derive the masses, magnetic moments, half lives, and other data of particles from first principles and unite a constant theory. This is not possible without some idea of the structure of particles. To derive, for instance, the masses of particles, we must know what structures are in the particles and what forces are active. For example, protons employ different forces than electrons. Fortunately, some idea of the structure of particles can be gained from the science of chonomics (balancing decay schemes with electrino model tools), without a lot of preliminary information. For the most part, the structures of particles can be induced by balancing all of each particle's decay schemes. To the eye trained in chonomics, the structure of each particle can be obtained by inspection of the various decay products of each particle. This information may be useful in further research and calculations in particle science. Before attempting to derive masses, magnetic moments, and half lives of particles, we shall pause to attempt to induce the structures of known matter particles in this volume.

B. Alternative to the Standard Model

Quark science has become part of what is called the Standard Model of Physics.[1] Quark theory was advanced because it gave more parsimony to particle physics, simplifying the bewildering array of hundreds of "elementary" particles.[2] However, quark-lepton theory requires 61 "elementary" particles to construct known matter.[3] And that does not include gravitons. In contrast, the Electrino Fusion Model of Elementary Particles, advanced in this book, requires only one particle to construct all known matter and gravitons—a pair of oppositely charged octons in orbit. The 3-ness observed in protons in the quark model,[4] attributed to three quarks in the proton, occurs for a different reason in the electrino model for protons. There are three whole particles [uniton, electron, and pion] bound in the proton. But the author discovered that 2-ness (that is 1, 1/2, 1/4, and 1/8) best explained physical properties on a one level deeper model of particle physics derived from first principles. [The uniton is a whole charge, the electron is made up of two half charges orbiting each other, and the pion is made up of four fourth charges orbiting each other.] Attempting to build high level structures like protons and neutrons from bottom level electrinos revealed there were serious errors in the quark model [For instance, there are only two whole particles instead of three quarks in the neutron.], and the electrino model could not be made as a refinement of the quark hypothesis. The reader is therefore asked to consider the electrino model as an alternative to the quark hypothesis and the Standard Model.

II. Summary of Electrino Model Particle Structure

A. Electrinos

a. Creation particles. Scattered throughout the preceeding chapters are various bits of information regarding the structure of matter. On the bottom level are electrinos (unitons, semions, and quartons, which can all be fusion products of octons).[5] Octons themselves are theorized to be the creation particles from which all other electrinos are derived. Quartons, semions and unitons compose all stable and unstable matter, photons, and gravitons.[6]

b. Electrino structure. The structure of each electrino is an infinitesimally thin spherical shell of charge—a whole charge e for a uniton, a half charge for a semion, a fourth charge for a quarton, and an eighth charge for an octon.[7] These are the only charge quantities of electrinos in the model.

c. Electrino radii. The radius of the uniton is the Planck length in imaginary units, $i(\hbar G/c^3)^{1/2}$ or i 1.616 x 10^{-35} m.[8] Due to Special Relativity and aether (boson) velocities at the speed of light at the electrino surface, all electrinos are imaginary in radius. Semions have one half the radius of unitons. Quartons have one fourth the radius of unitons. And octons have one eighth the radius of unitons.

d. Electrino strong mass. The strong mass of the uniton is the Planck mass in minus imaginary units, $-i(\hbar c/G)^{1/2}$ or -i 2.177 x 10^{-08} kg.[9] The strong mass of the semion is half that of the uniton. The strong mass of the quarton is one fourth that of the uniton. And the strong mass of the octon is one eighth that of the uniton. Strong mass is the mass due to forces at or faster than the speed of light.[10]

e. Electrino generated aether motion. Strong mass occurs in electrinos because gravitons are accelerated to the speed of light at the shell/surface of charge in the electrino. The gravitons drift through the particle inner void at the speed of light, and decelerate out the other side of the particle, starting at the speed of light at the particle surface. The Electrino Fusion Model of Elementary Particles is an aether model.[11] The aether is turbulent, caused by gravitational acceleration or electrical force acceleration.

The aether motion at the surface of the electrinos is demonstrated in the strong masses of the electrinos. Half of the strong mass of an electrino is due to its charge, and the other half is due to the charge's motion relative to the aether.[12] The aether is a sea of bosons such as gravitons.[13] The low, but finite imaginary, mass gravitons are accelerated through the electrinos, and decelerate on the other side.

f. Spin. Because all electrinos are smooth, spherically symmetric charge distributions, they cannot have detectable spin.[14] And they do not in the electrino model. The spinless particles can contribute spin to a system, however, when they orbit other particles. They themselves do not spin, but their orbits contribute spins.

B. Echons and Yachons

a. Whole particles. The whole particle concept is important in the Electrino Fusion Model of Elementary Particles.[15] All "elementary" particles in the Standard Model are either, or they are made up of, whole particles in the Electrino Model. This includes matter, photons, and gravitons.

There are two kinds of whole particles—yachons and echons. These particle names are derived from the Hebrew. Yachons are single whole particles—unitons and antiunitons. Echons are whole particles made up of parts—whole particles made up of orbiting electrinos.[16]

Electrinos, other than unitons, are fractional particles. We could not produce and detect free fractional particles unless we could produce 10^{22} times the mass of the electron in our accelerators (over 10^{-8} kg). Fractional particles are bound in a realm where they travel relative to the aether faster than the speed of light, where like charges attract.[17] Whole particles exist in a realm where, relative to the aether, they travel slower than the speed of light, and the slower than the speed of light Coulomb force is active.[18] This is the electric force detected in particle physics, for all detected particles are whole particles or made up of whole particles, and travel slower than the speed of light relative to the aether.

b. Echons. There are two kinds of echons: 1) pions or antipions composed of four fourth charges (quartons or antiquartons, respectively)—two orbiting one way, and two orbiting the opposite way;[19] and 2) electrons or positrons composed of two orbiting half charges (semions or antisemions, respectively).[20] The two types of echons (modified by energy states) plus yachons (unitons and antiunitons) compose all detected matter particles, photons, and gravitons.

c. Radii and masses. The radii and masses of echons have not yet been calculated from first principles. There seems to be a unique radius and mass for a given electrino and aether velocity state. The radii and masses of echons are many orders of magnitude different than the radii and masses of the electrinos because of special relativity. The echons are nearly at rest. The electrinos are nearly at the speed of light (slightly above). The echons and electrinos are in different reference frames. How to calculate echons has been elusive. The mass, radius of echons, the aether velocity at the surface of the echons, and the gamma factor for special relativity for the velocity of the electrinos in the echon is not yet derivable. But all the above variables are related. And if you know one you can calculate all the others. The most fundamental one is probably the echon radius. That may have to be an input into the model—at least for one echon—the electron.

C. Author's Original Particle Structures

A few particle structures were tentatively induced in the first few chapters of this book. There we described three charged leptons with their three associated neutrinos; protons; neutrons; Lambda particles; and several mesons, including charged mesons and neutral mesons, such as K_L^0

and K_S^0. The full chonomic description of those particles, as well as all other known particles, will be listed in Draft Appendix B.

As far as chonomics is concerned, muons and tauons were heavier electrons—occupying higher energy states.

III. Testing the Model

The Electrino Fusion Model of Elementary Particles, though it can induce the structure of each particle, is not advanced enough to derive accurately all the masses of all the particles. But neither is the Standard Model, the Quark Model, or the Superstring Model. More work needs to be done in all those models to achieve the accurate derivation of particle masses. That remains a quest of the Electrino Model. But the masses of charged leptons have been calculated from first principles in Chapter 21. Those are new tests of the model.

But fortunately, two tests of the Electrino Model stand out. One is colliding electrons over 1876 MeV in the center of mass frame, so that the axial spins have the same sense in the center of mass frame, to fuse to produce antiprotons (or antineutrons and antipions) and electron neutrinos and/or antineutrons with electrons. The second test is colliding positrons over 1876 MeV in the center of mass frame, so that the axial spins have the same sense in the center of mass frame, to produce protons (or neutrons and pions) with anti electron neutrinos and/or neutrons with positrons. This second test yields more than particles. This process violates the second law of thermodynamics, and yet it reacts according to the strong force. The second law of thermodynamics and the strong force get in a tug of war. The strong force must win. This affects not only the reaction, but reverses the second law in a calculable radius, which is surprisingly large. For instance, 10 pA of positrons at 1.0 GeV collided and fused with 10 pA of positrons at 1.0 GeV would reverse the order to disorder arrow in the second law of thermodynamics in a radius of 9.6 (approximately 10) meters. Smaller currents would yield larger radii linearly. As long as the fusion process took place, people and things in that radius would be reversed in aging, decay, and disease processes. This is the most exciting of the two tests of the model. Both tests would be desired to evaluate the model.

---

[1]David Griffiths, *Introduction to Elementary Particles* (New York: John Wiley & Sons, Inc., 1987), p. 3.

[2]James S. Trefil, *From Atoms to Quarks: An Introduction to the Strange World of Particle Physics* (New York: Charles Scribner's Sons, 1980), p. 149.

[3]Griffiths, *Op. cit.*, p. 48.

[4]Murray Gell-Mann, *The Quark and the Jaguar* (New York: W.H. Freeman and Company, 1994), p. 181.

[5]Chapter 5, Section I; Chapter 6, Section II.E; Section III.I; Problem Set 6.

[6]Appendix B.

[7]Chapter 6, Section I.D.

[8]Chapter 6, Section II.E.

[9]Chapter 6, Section II.D.

[10]Chapter 6, Section II.E; Section III.I-Section IV.B.

[11]Gordon L. Ziegler, *Electrino Physics*.

[12]Chapter 6, Sections D and E.

[13]Chapter 7, Section D; Chapter 10, Section II.G-Section III.A.

[14]Chapter 6, Section I.B-D.

[15]Chapter 6, Section I.D.

[16]*Ibid.*

[17]Chapter 6, Section III.D.

[18]*Ibid.*

[19]Chapter 6, Section I.D.

[20]*Ibid.*

Problem Set 17.

1. In the electrino fusion model of elementary particles, what are the smallest particles as a class?

2. Counting the anti-particles, how many are there of that class?

3. Name the four basic types.

4. All copies of all particles are fusion products of what particle pair?

5. In the foreseeable future, could man-made machines ionize free electrinos? Why or why not?

6. All detectable particles are, or composed of, what?

7. What is the definition of yachons?

8. What is the definition of echons?

9. Counting up-spin echons separate from down-spin echons, and counting anti-particles as well as particles, how many echons and yachons are there?

10. How can the particles in 9 be succinctly expressed in the science of chonomics?

Chapter 18

INDUCING PARTICLE STRUCTURE

The APPENDIX B: STRUCTURE OF KNOWN PARTICLES is complete now, and it is demonstrated that quarks are not needed to formulate matter. It is not necessary to divide charge into 1/3 e and 2/3 e. All known matter, light, and gravitons can be made up of whole particles (yachons and echons), which are further made up of electrinos (with charges ¼e, ½e, and e), and these electrinos can all be fusion states of octons in octon pairs. The new fusion state system is very parsimonious.

More than one "elementary" particle can have the same quark structure. There is not a unique elementary particle for each quark structure. That shows there must be a deeper level of particle structure than quarks and leptons to differentiate matter into each "elementary particle." By contrast, there is a unique "elementary particle," or chonstruct, for each chonomic structure; and there is a unique chonomic structure for each "elementary particle" or chonstruct. Chonomic structure is adequate to describe all known matter, light, and gravitons; and there are still unused chonomic structures available for further discoveries.

The quark model requires 61 "elementary particles"—quarks, anti-quarks, leptons, anti-leptons, and other necessary particles to compose all known matter. And this does not include gravitons.

By contrast, the chonomic model requires only three whole particles (quarton echons, semion echons, and unitons) and three whole anti-particles (anti-quarton echons, anti-semion echons, and anti-unitons)—a total of six particles—to construct all known matter, light, and gravitons. The semion echons, and the anti-semion echons can be right side up, or upside down. Chonomics treats

359

these cases as though they were separate particles. Thus, in chonomic structures, all known matter, light, and gravitons can be composed of (-, +, o, and ·) and anti-(-, +, o, and ·)–a total of eight whole particles. (Eight is a special number in particle physics.) But these eight whole particles can all be the fusion and spin states of a positive and negative octon pair in orbit. Thus octons could create all known matter, light, and gravitons in the Universe.

By the above we can see that the chonomic model is superior to the quark model. The chonomic model describes particles uniquely, and is far more parsimonious than the quark model. In addition, the chonomic model predicts processes that are not predicted by the quark model—processes which would give whole new powers to mankind—such as converting matter into antimatter or converting antimatter into matter (see Chapter 12), and reversing the order to disorder arrow in the second law of thermodynamics—thereby reversing aging, disease, and decay processes over a wide area (see Chapter 16). Particle physics in accelerators and colliders could affect positively a wide area, and could bless millions. The quark model cannot do that for the human race.

Chonomics can balance decay schemes much more easily and simply than the Standard Model. (See Chapter 10 and Appendix A: LEPTON SUMMARY TABLE.) In the Standard Model decaying particles seem to create a bewildering array of daughter particles. There are few rules to guide in what decay particles are permissible. By contrast, in chonomics what comes in goes out in the same state, unless it is operated on by one or more clearly identifiable processes. Then the charges are conserved and the spins are conserved.

In the Big Bang theory, equal amounts of matter and antimatter should be created in a Big Bang. But the chonomic model shows how to convert antimatter into matter. If this is done on a large scale, an almost totally matter Universe could be constructed by fusing the anti-semion constituents of positrons, but not the semion constituents of electrons.   This makes stable protons, neutrons, and electrons, the constituents of matter. An all matter Universe is safe for habitation and travel.

All these things are brought to light by inducing particle structure in the system of chonomics. This was achieved by carefully, simultaneously following seven criteria in predicting the structure of each particle: particle charge, spin, parity, mass, spin feasibility, preceding particles (to avoid duplication), and decay schemes. This apparently is enough input to uniquely categorize the structure of each "elementary particle" or chonstruct. But it is difficult to keep all seven criteria in focus simultaneously. Thus there may be one or more errors in Appendix B, which is released as a draft for further editing.

Problem Set 18

1. In inducing the structure of matter, is it necessary to divide charge into $2e/3$ and $e/3$, as in the Quark Hypothesis?

2. In the physics model of this book (Electrino Fusion Model of Elementary Particles), what are all whole particles constructed of?

3. These all can be fusion states of what particle-anti-particle pair?

4. On the highest level, what can all matter particles be composed of?

5. Are all quark structures unique?

6. What does that say of the quark model?

7. Do all chonstructs have unique chonomic structures?

8. What does this say of the electrino model?

9. Do known particles use all available structures?

10. Contrast the number of elementary particles the quark model requires to construct all light and matter with the number of elementary particles the electrino model requires to construct all light, matter, and gravitons.

11. How many ways can you show that the electrino model is superior to the quark model of physics?

12. How many criteria were followed simultaneously in arriving at the chonomic structure of each particle in Appendix B? List them.

Chapter 19

MASSES OF ELECTRINOS

A. Introduction

The calculations in this chapter, and referred to in this chapter, are based on earlier chapters in this book (especially Chapters 6 and 14). Those chapters present an aether model of particle physics. The smallest particles in that model are called electrinos. Electrinos are relativistically length contracted spherical shells of charge. Their radii are imaginery on the order of magnitude of 10^{-35} m. They do not have detectable spin. The smallest possible electrino is the octon (1/8 e). Octons can fuse to quartons (1/4 e), which can fuse to semions (1/2 e), which can fuse to unitons (e). Unitons are the largest electrinos possible. Octons precede all other electrinos, and are used in creating the Universe. They do not naturally occur in the Universe. All light, matter, and gravitons are composed of quartons, semions, and unitons. The structures of all known particles are listed in Appendix B: STRUCTURE OF KNOWN PARTICLES, in this book.

Echons are whole particles composed of orbiting electrinos. The masses of electrinos are far easier to calculate than the masses of echons. That is one reason we start with the masses of electrinos. The other reason is that all higher particle masses depend on the foundational masses of electrinos.

B. Masses and Radii of Electrinos

The masses and radii of electrinos are calculated precisely from first principles in Chapters 6 and 14. Those calculations will not be repeated here. The reader is

363

referred to those chapters for the material of this sub-heading.

C. Force Domains.

Before we continue our derivation, we need to clarify a "detail" which makes an enormous difference in the calculations. The detail we need to consider is the domains of forces. In Chapter 8, from a single definition, we derived a system of forces and constants. We had basically two types of forces: ordinary forces (like gravity, inertia, electricity, and magnetism), and strong forces (like strong gravity, strong inertia, strong electric force, and strong magnetism (weak$_0$ force)). The ordinary forces, we agreed, occurred in slower than the speed of light (<c) situations. The strong forces, we agreed, occurred in faster than the speed of light (>c) situations. But what about the case where the velocity of the particle is exactly equal to the speed of light (=c)? Do the ordinary forces and real masses and radii apply? or do strong forces and minus imaginary masses and imaginary radii apply? Or is there half real and half imaginary masses, etc., or is there an entirely new kind of force and mass at that third possible state?

Incredibly enough, the single master definition

$$\frac{0 \, arbitrary \, mass \, unit}{0} \equiv M_0\left(arb.ma.u.\right), \qquad (19\text{-}1)$$

from Chapter 8) does not leave this question to idle speculation or controversy. This question is settled as part of the definition. The uniton is a zero mass particle in the non-relativistic frame, and therefore travels precisely at c. It balances the infinitely high and infinitely thin knife edge of the speed of light barrier. But its strong mass is not said to be real as in <c domain, or half real and half imaginary

as in part of both domains. The strong mass of the uniton is defined to be M_0, which is minus imaginary as in the faster than c region. Then we see precisely that the ordinary forces occur in the domain $v < c$, whereas the strong forces occur in the domain $v \geq c$.

D. Definition of M_0

M_0 is the author's early symbol for the calculated zeroth order of the relativistic strong mass of a uniton—a whole particle. The zeroth order relativistic strong mass of other electrinos, such as quartons and semions, are simple fractions of M_0. From the single master physical definition in Chapter 8, we derived M_0^2 to be $-\hbar c/G$ (Eq. (8-16)) and $-M_0$ to be $(-\hbar c/G)^{1/2}$ (Eq. (8-38). Numerically it is $M_0 \approx -i$ $2.17671407 \times 10^{-08}$ kg.

E. The Transference of Relativistic Strong Mass Among Electrinos

The fact that the zeroth order relativistic strong masses of various electrinos are nominal fractions of M_0 strongly suggests that quartons are fused octons, semions are fused quartons, and unitons are fused semions. We expect to find the true, total, exact relativistic strong masses of the electrinos to be dependant upon the electrino fusion process. The true, total, exact strong masses of quartons will depend on the defined strong masses of octons and the fusion reaction of octons. Similarly with semions and unitons.

F. A More Detailed Scenario of Electrino Fusion

The idea of electrino fusion is introduced in Chapter 6, Section IV.A., and Chapter 9, and is amplified in Chapters

10, 11, 12, 15, and 16. We have considered the joining of particles, reaction formulas, and the prospects that unbalanced artificial electrino fusion could take place with electrons or positrons accelerated to proton-neutron energies, and that the leptons accelerated by one accelerator must have one axial spin orientation, and the other leptons accelerated by the opposite accelerator must have the opposite axial spin orientation as reckoned from frames of the accelerators, or like spin orientations as reckoned in the center of mass frame. However, a detailed fusion scenario has never heretofore been advanced in human history which takes into consideration all of the forces. That is what the author endeavors to present in this section.

The relative strength of forces, and how they affect masses is best illustrated in the constant equal to g/2 of the electron (as described in Chapter 8). The first and predominant force is the strong force (not the strong nuclear force). The strong force is strong gravity equivalent and redundant to the strong electric force. The next prominent effect comes from the non-relativistic Coulomb force electric term. Then comes separate terms for magnetism and weak forces, and last of all comes gravity. All of these forces, except ordinary gravity, play central roles in the electrino fusion processes.

Electrino fusion can occur between single ionized electrinos. But the energies involved for the particles are far greater than man could produce artificially for a long time to come. And we have not observed them naturally. Electrino fusion can occur, however, between electrinos in systems of orbiting electrinos the author calls echons—like electrons, and positrons—at much lower, achievable energies. Such fusions the author has detected in his particle decay schemes he calls chonomics. This is the type of fusions that we would do artificially. Therefore, this is the type of fusion process we will consider in this more detailed scenario.

The repulsive Coulomb electric force is the strongest force an electron senses from another electron in the non-relativistic frame. Like-charged electrons, or like-charged positrons, repel each other in the subrelativistic frame. This repulsive Coulomb force must be overcome by high accelerated velocities, momenta, and energies of the colliding particles if the fusion process is to occur at all. But if and when the Coulomb electric force is overcome, other forces come into play in the fusion process. The ordinary magnetic force then contributes to the process. The centers of mass of the two pairs of orbiting semions in each electron or positron are then magnetically attracted to each other. The electrons or positrons then accelerate toward each other due to magnetism, and kinetic energy is gained which is excess, and the energy can be radiated in an increase of entropy from the fusion reaction. But as the electrons or positrons become extremely close to each other, the semion subparticles of the particles take the predominance, and the system reckons that all particles are in orbit faster than the speed of light. Therefore the system treats itself relativistically instead of subrelativistically. The radii of the particles are considered imaginary instead of real. The reversed strong magnetic force is substituted for the ordinary magnetic force. But since the force is reversed by all the particles, the particle pairs still attract each other axially.

The Coulomb force is reversed to the strong force. The electrons no longer repel each other. The semions in a common orbit all attract each other. The semions attract each other to fusion.

It is important for the sake of the following calculations that we get a clear idea of the initial conditions of the fusion processes, for we have to integrate fusion forces over distances. We need to know what those distances should be. We want to know the typical initial conditions of the fusion process. We can get an idea of that by the fact that the Coulomb force repels the semion pairs as the ordinary magnetic force attracts them. Therefore, as the pairs approach each other polarly, they are strongly aligned at greatest possible separation from each other with particle pair axes at right angles to each other. We expect this effect to be quite precise, but not infinitely precise. Thus there will be some limit to the accuracy of the mass calculations that we can make from this observation. But we expect the calculations to be quite accurate taking this as the initial condition of fusing. When the electrons pass through the speed of light barrier, the semions will be at maximum separation (with $\Delta\theta = \pi/2$ for each of the four semion space intervals). They will then begin their fusion paths from that initial condition.

G. The Origin of Electrino Strong Masses

The strong masses of quartons result from the fusion process of octons. Similarly the strong masses of semions result from the fusion process of quartons; and the strong masses of unitons results from the fusion process of semions. Let us start with the fusion of octons into quartons. The following equations hold true in the greater than or equal to the speed of light region of strong gravity, but not in the less than the speed of light region of ordinary gravity. For instance, the octon strong mass of $1/8\ M_0$ attracts octons to fusion, but the ordinary mass of the octon, 0, slower than the speed of light, cannot attract octons to fusion. The fusion of electrinos is a strong function, not an ordinary function.

The strong mass of the quarton, $M_{1/4}$, is equal to an energy divided by c^2, which is equal to the integral of a minus force times dr divided by c^2. The force we take to be that of the strong mass of octons attracted by strong gravity, where $M_{1/8} = \frac{1}{2}M_{1/4}$ (for consistency of units in the equations). However, we have a two body problem with equal masses. Ordinarily we would employ a reduced mass $\mu = \frac{1}{2}M_{1/8}$ in the inertial side of the equations, and calculate with the unaffected $M_{1/8}$ in the attractive force side of the equations. However, we will not calculate the inertia in this problem. We can account for the effect of the μ in this problem by switching sides of the equation by inverting and multiplying. Since the masses are equal, the $\mu = \frac{1}{2}M_{1/8}$, but $M_{1/8} = \frac{1}{2}M_{1/4}$. Thus the added effect on the equation of the μ is $\frac{1}{4}$. But since when we switched sides of the equation with μ the μ is in the denominator, the effect is multiplying the numerator by 4. But we note that $4(M_{1/8})^2 = (M_{1/4})^2$, which substitution we will make in the following equations.

In the integration the r^2 will be replaced by -r. Two forces must be integrated. The fusing octon feels one force starting at $\sqrt{2}\, R_0$ going to $1/4\, R_0$ as the particle travels. In the opposite direction it feels a force starting at $\sqrt{2}\, R_0$ going to infinity (∞). The associative law lets us separate these forces and integrate them separately. The forces of the two integrals are identical except for minus sign and differing limits of integration. M^2 is negative, so the equation is positive when it appears negative, and vice versa.

Let us now write the described equation and solve for the answer.

$$M_{1/4} = \frac{E}{c^2} = \frac{-\int F dr}{c^2} \tag{19-2}$$

$$= -\int_{\sqrt{2}R_0}^{1/4 R_0} \frac{G\left(M_{1/4}\right)^2 dr}{r^2 c^2} + \int_{\sqrt{2}R_0}^{\infty} \frac{G\left(M_{1/4}\right)^2 dr}{r^2 c^2}. \tag{19-3}$$

$$M_{1/4} = \frac{G\left(M_{1/4}\right)^2}{R_0 c^2}\left(4 - \frac{1}{\sqrt{2}} - \frac{1}{\infty} + \frac{1}{\sqrt{2}}\right). \tag{19-4}$$

$$M_{1/4} = \frac{4G\left(M_{1/4}\right)^2}{R_0 c^2} = \frac{1}{4}\frac{GM_0^2}{R_0 c^2} = \frac{1}{4}M_0. \qquad (19\text{-}5)$$

We see that naturally, through strong mass attraction, quartons have one fourth the strong mass of unitons. Similarly semions have one half the strong mass as unitons, and unitons are whole unitons. In each case the strong mass of the electrino is obtained exactly. The strong mass of each electrino is determined, a step at a time, from the octon strong mass from the original master definition.

The masses of all electrinos are defined above. We need to turn now to the masses of echons (electrinos in orbit).

Problem Set 19

1. Do electrinos have detectable spin?

2. What is the physical shape of electrinos?

3. How are electrinos related to the aether?

4. What is the smallest electrino? What is the largest electrino? What are all the rest of the electrinos?

5. How are all the electrinos related to each other?

6. How can all higher electrinos be created?

7. Do octons now occur naturally?

8. All light, matter, and gravitons are composed of what?

9. How are echons constructed?

10. What is the mass of the uniton? What is the radius of the uniton?

11. What is the mass of the semion? What is the radius of the semion?

12. What is the mass of the quarton? What is the radius of the quarton?

13. What is the mass of the octon? What is the radius of the octon?

14. What if a particle travels exactly at c relative to the aether—do the forces it feels calculate according to <c or >c domains?

15. What is M_0 in terms of $\hbar$, c, and G?

16. How many forces are active in preparing electrinos for fusion? List them.

17. When nucleons fuse, the combined nucleus weighs slightly less than the sum of the parts. The new nucleus has mass defect. Do fused electrinos have mass defect?

Chapter 20

NATURE OF MASSES OF MATTER

How does mass originate? That has long been a mystery in physics. The Higgs boson was theorized to explain it. But the electrino model of physics has another explanation for it. The theory of that is rooted in the single postulate of the Universe. As you recall,

$$\frac{0 \, arbitrary \, mass \, unit}{0} \equiv M_0 \left(arb.ma.u. \right) \qquad (20\text{-}1)$$

(The Fundamental Postulate). The aether rest mass of a uniton is 0; its gamma factor at the speed of light is 1/0, or ∞; and though the product is otherwise undefined, the postulate defines the product as M_0—the minus imaginary Planck's mass. What about non-zero rest mass? It would take more postulates to define the Universe, unless the above postulate would hold true for masses in general, when expressed in more general terms:

$$m\gamma = M_0. \qquad (20\text{-}2)$$

Solving for the mass m, we have:

$$m = \frac{M_0}{\gamma} = M_0 \left(1 - \beta_G^2 \right)^{1/2}, \qquad (20\text{-}3)$$

where β_G is the overall grande, gross, or total beta (velocity divided by c—the speed of light).

$$\beta_G^2 = \beta_o^2 + \beta_r^2 + \beta_p^2 = \beta_x^2 + \beta_y^2 + \beta_{z,}^2 \qquad (20\text{-}4)$$

where o stands for orbital, r stands for radial, and p stands for polar—three dimensions in polar coordinates. For a particle in the relative rest frame,

$$\beta_G^2 = \beta_o^2 + \beta_r^2 .$$
(20-5)

$$m = M_0\left(1 - \beta_o^2 - \beta_r^2\right)^{1/2} .$$
(20-6)

For electron semions,

$$\beta_o = \beta_o^2 = 1.$$
(20-7)

That is, electron semions orbit at the speed of light c. In Eq. (20-6), β_o^2 subtracts out with the 1. That leaves the following equation:

$$m_e = M_0\left(- \beta_r^2\right)^{1/2} = iM_0\beta_r = |M_0|\beta_r .$$
(20-8)

Thus we see that the mass of an electron is proportional to the radial aether velocity through the electron, which is a result in harmony with Chapter 6, Section III.G, derived from different postulates. In our studies, we find that three ratios are equal to each other:

$$\frac{m_e}{|M_0|} = \frac{v_e}{c} = \frac{R_0}{r_e} = \beta_r .$$
(20-9)

Thus we find that if we know any one of m, v, r, β_r, or C (circumference of the particle), or γ, for that matter, we can know all the others. But these are not all independent equations. We cannot solve them simultaneously to obtain a single particle, let alone an infinite series of particles. Will our fundamental postulate suffice for us? or will we need an infinite number of additional inputs to solve for the masses of "elementary particles"?

The photon is not matter. But it is interesting to notice how it fits in the above equations. The photon is composed of a uniton and an anti-uniton orbiting each other at c while traveling in the polar direction at the speed of light. Thus the β_o is 1. The β_r is 0. According to Eq. (20-5) designed for rest particles, β_G^2 equals 1 and, according to Eq. (20-3), m = 0. But particles of zero mass travel at the speed of light c. The polar dimension β_p becomes 1, and β_G^2 becomes 2. Eq. (20-5) now solves to $|M_0|$. This is the effective mass of one uniton in the photon. The other sub-particle in the photon—an anti-uniton—also has effective mass $|M_0|$. Thus the photon has effective mass $2|M_0|$.

The author believes that neutrinos have a velocity just slightly less than the speed of light c. Thus for the neutrino, β_G^2 is slightly less than 1, and m is a small minus imaginary number. This is borne out in recent experiments.[1]

---

[1]The Particle Data Group Authors, "Lepton Summary Table," as quoted in *CRC Handbook of Chemistry and Physics*, 80th Edition, 1999-2000, David R. Lide, Editor-in-Chief (Boca Raton: CRC Press, 1999) p. 11-5 (Mass m note in v_e listing).

Problem Set 20

1. Converting β_r into v/c, what equation in Chapter 6, Section III.G is Eq. (20-8) similar to? The equation in Chapter 6 was deduced from eight postulates. Eq. (20-8) is deduced from one fundamental definition plus the parsimony principle—or two postulates.

2. Why cannot we simultaneously solve v, r, C, β, and γ for m?

3. How many of the above variables do we need to solve for all the rest?

4. What is the effective mass of the photon? What is the mass of the photon?

Chapter 21

PARTICLE ENERGY STATES

A.   Introduction

In preceding chapters the idea was expressed that electrons, muons, and tauons were just energy states of one particle system—and similarly for pions, Kaons, and D-ons as well as other particle sets. The author thought to solve for the various energy states like Niels Bohr solved for the energy states in hydrogen in 1913.[1] Bohr's calculational framework has been very helpful as a guide to the author in solving for the velocities, radii, and masses of particles in elevated states. This chapter will calculate these things. However there are many significant differences in the calculations. These will be pointed out.

B.   The Bohr Atom

Bohr's results followed from algebraic derivations from a few postulates:

"1.   The electrons move in orbits restricted by the requirement that the angular momentum be an integral multiple of h/2π, that is, for circular orbits of radius r, the electron velocity v is restricted by

$$mvr = \frac{nh}{2\pi}$$

(21-1)

and furthermore the electrons in these orbits do not radiate in spite of their acceleration. They were said to be in stationary states."[2]

"2.   Electrons   can   make   discontinuous transitions from one allowed orbit to another, and the change in energy, E-E' will appear as radiation with frequency

$$v = \frac{E - E'}{h} \qquad (21\text{-}2)$$

An atom may absorb radiation by having its electrons make a transition to a higher energy orbit."[3]

3.   Bohr obtained another relevant calculational equation simply by balancing the Coulomb electric force against the centrifugal force.

$$\frac{kq_e^2}{r^2} = \frac{m_e v^2}{r}, \qquad (21\text{-}3)$$

where k = 1/(4πε₀), and q_e is the charge of the electron.[4]

4. "The energy of an electron in an orbit is the sum of its kinetic and potential energies:

$$E = E_{kinetic} + E_{potential} \qquad (21\text{-}4)$$

$$= \frac{1}{2} m_e v^2 - \frac{kq_e^2}{r} .\text{"}[5] \qquad (21\text{-}5)$$

C.   Electron Energy Levels in Hydrogen

Performing simple algebraic operations, Bohr was able to solve for the orbital velocity v, the radius r, and the energy E.

"To begin, multiply both sides of Eq (21-3) by r to see

$$\frac{kq_e^2}{r} = m_e v^2. \tag{21-6}$$

The term on the left hand side is the potential energy. So the equation for the energy becomes

$$E = \frac{1}{2} m_e v^2 - \frac{kq_e^2}{r} = -\frac{1}{2} m_e v^2. \tag{21-7}$$

Now we just need to figure out what the velocity, v is equal to, so solve Eq (21-1) for r,

$$r = \frac{n\hbar}{m_e v}. \tag{21-8}$$

Plug this into Eq (21-6),

$$kq_e^2 \frac{m_e v}{n\hbar} = m_e v^2. \tag{21-9}$$

Then divide both sides by $m_e v$ to see

$$\frac{kq_e^2}{n\hbar} = v. \tag{21-10}$$

Now we can put in this value for v into the equation for energy, and then also plug in the values for k and ħ, and we'll obtain the energy of the different levels of hydrogen:

$$E_n = \frac{-1}{2} m_e \left(\frac{kq_e^2}{n\hbar} \right)^2 \tag{21-11}$$

$$= \frac{-1}{2} m_e \left(\frac{1}{4\pi\varepsilon_0} q_e^2 \frac{2\pi}{nh} \right)^2 \tag{21-12}$$

$$= \frac{-m_e q_e^4}{8h^2 \varepsilon_0^2} \frac{1}{n^2}. \tag{21-13}$$

Or, after substituting values for the constants,

$$E_n = \left(-13.6 \; eV \right) \frac{1}{n^2}. \tag{21-14}$$

Thus, the lowest energy level of hydrogen (n = 1) is about -13.6 eV. The next energy level (n = 2) is -3.4 eV. The third (n = 3) is -1.51 eV, and so on. Note that these energies are less that zero, meaning that the electron is in a bound state with the proton. Positive energy states correspond to the ionized atom where the electron is no longer bound, but is in a scattering state."[6]

D. Three More Quantum Numbers

"Bohr had pictured the electron orbits around the atomic center as being perfectly circular, but this was too simple. There are very few perfect circles in nature, and orbits in atoms are no exception.

"Later, in 1916, the German physicist Arnold Sommerfeld refined Bohr's 'easy' picture with one a bit more complex. In this modified view the electron orbits were not circular, but elliptical. But there are many kinds of ellipses possible (certainly more than one), and this changed the calculations in subtle ways, as each ellipse has a slightly different angular momentum. To take account of the possibility of elliptical orbits, Sommerfeld introduced another number; the **orbital quantum number** (sometimes called the "angular momentum quantum number"), which usually had the symbol "L" [or "l"]. . . .

"Like the principal quantum number, the orbital quantum number can have values of 0, 1, 2, 3, 4, etc., but only up to a whole number value of one **less** than the electrons principal quantum number (i.e. up to a value of n − 1). . . .

"There are two more quantum numbers associated with each electron; the **magnetic quantum number** written as **m**, and the **spin quantum number**, written as **s**. . . .

"To make it easy to picture what is going on, the magnetic quantum number can be thought of as defining the amount of "tilt" there is to the orbit.

"The possible values for **m** follow the same rules as for **L**, except that negative numbers are now allowed (the "tilt" of the orbit can be either "up" or "down"). So for **n** = **2**, the possible values for **m** would be 0, 1, or -1. . . .

"There are only two possible values for **s** [spin] for any value of **n**. These values are usually written as +1/2 and -1/2, meaning either a clockwise spin or an anticlockwise spin.

"But what do these numbers tell us about the electrons?

"Austrian physicist Wolfgang Pauli worked out the significance of these numbers in 1925. He suggested that no two electrons in any given atom could have exactly the same values for all four quantum numbers.

"This became known as the **Pauli exclusion principle – 'No two electrons in any atom may have the same set of quantum numbers'.**"[7]

E.   Accuracies of the Models

The Neils Bohr Model was a close but not an exact fit to the measured data.   The Sommerfeld Model of electron orbital ellipses, taking into account relativistic effects, gave a slightly better fit to the measured data. But as Thayer Watkins[8] demonstrates, no atomic structure model has a perfect fit to the measured data, and the Bohr Model is not much worse than the more advanced models. The fit is best between orbits of low quantum mechanical parameters n, and is worst between orbits of high quantum mechanical parameters n.   At its best, the error can be as low as -0.01234 of 1%. But at its worst, the error can be at least as bad as -1.44546 of 1%.

These also are about the errors of the masses of charged leptons calculated in the next sections. Science has had 97 years to get the energies of the atom perfect, and has not done it. We should not hold back, therefore, until our model of energy states of semion orbits is perfect. We should publish a first cut mass model that is as close to the measured values as Bohr was to his measured values. The model we will advance in the next sections of the energies of semion orbits will be analogous to the Bohr model—not taking into account elliptical orbits, tilted orbits, or varying relativistic effects. There will be room for others to refine the model.

F.   Differences of Semion Orbits with Electron Orbits

There are a number of differences between semion orbits and electron orbits:

1. Semions have e/2 charge. Electrons have whole e charge.
2. Each particle is a miniature black hole. The electron orbits as in Bohr's Model are exterior to black holes. The half charged particles called semions orbiting in charged leptons orbit inside black holes. This makes a difference in the force equation. The force for electrons in their orbit is $e^2/1 \cdot 4\pi\varepsilon_0 \alpha^0 r^2$. The force on semions instead is $e^2/4 \cdot 4\pi\varepsilon_0 \alpha^1 r^2$.
3. Semion orbits are a two body problem instead of a one body problem of electron orbits.
4. Semions orbit faster than the speed of light. Electrons orbit atoms slower than the speed of light.
5. The electron mass in the Bohr Model is the constant m_e. The semion mass in the outer non-relativistic frame is a variable.
6. Angular momentum in Bohr's atom is a function of n. But in our particles, the angular momentum is a function of not only n, but of 1/b.
7. Between the models, standing waves have coincidence under different conditions. Instead of $C = n\lambda$ for electron orbits, $bC = n\lambda$ for semion orbits.
8. It is relatively easy to ionize electrons from orbit. It is virtually impossible to ionize semions from orbit. It is as though the semions are contained in a speed of light boundary which they cannot pass.
9. Because the semion orbits, with the addition of the gravitational aether velocity v, are faster than the speed of light c, the electric force between the semions is reversed in sense, and the sign on the potential energy is changed.

G.   Deriving Particle States

        Deriving particle states is one orbital level deeper than deriving electron orbital states that Niels Bohr did. The calculations are similar, but significantly different. Instead of treating the situation as a one body problem, we must treat the situation as a two body problem, with two equal semions in orbit about each other. This introduces an extra ½ into the expression of the centrifugal force.

        Instead of n standing waves λ working out even in one circumference C, as in Bohr's model of electron orbits, we allow n standing waves λ to work out even in b circumferences C.   This intelligence yields the following equation:

$$bC = n\lambda. \tag{21-15}$$

This reduces to

$$\lambda = \frac{bC}{n} = \frac{2\pi br}{n}. \tag{21-16}$$

The energy of the particle system is reduced by 1/b.

$$E = \frac{h\nu}{b} = \frac{2\pi\hbar\nu}{b} = \frac{2\pi\hbar c}{b\lambda} = \frac{n\hbar c}{b^2 r} = mc^2. \tag{21-17}$$

By the last equation in the above chain of equations, we see

$$r = \frac{n\hbar}{b^2 mc}. \tag{21-18}$$

        To Eq. (21-18) we add the balancing of the force due to charge on the semions with the centrifugal force on the semions. The effective mass of a semion is half the mass of the whole particle in the outer non-relativistic frame. We use this mass of the semion in the centrifugal force along with the ½ from the two body problem. The velocity v_0 is greater than c, and must increase when the energy increases.   In the electric type force side of the equation, the charge of the semion is e/2.

Different particle systems are in different order black holes. The force must depend on the order of black hole the particle system is in. Like the strong force and the electric force differ in strength by a power of $1/\alpha$, the forces in different orders of black holes differ by powers of $1/\alpha$. The electric type force expression, in the right side of Eq. (21-19), we expect to depend on a power of $1/\alpha$. To be in harmony with measured results and Eq. (21-16), we want the power of $1/\alpha$ to be related to n/b. Also, we want the power for the electron to be such that the power of α is 1 when n = 0. We therefore take the power of α for electrons and higher charged leptons to be n/b + 1. Completing the balancing of forces equation, we have

$$\frac{1}{2}\frac{1}{2}\frac{m\,v_o^2}{r} = \left(\frac{e}{2}\right)^2 \frac{1}{4\pi\varepsilon_0\alpha^{(n/b)+1}r^2} . \qquad (21\text{-}19)$$

The first ½ in the equation is from the two body nature of the problem, converting it to a one body problem.

Thanks to the two body nature of the problem, all of the numeral constants in the above equation cancel out. $e^2/4\pi\varepsilon_0\alpha$ can be factored out as $\hbar c$. One r can cancel out of the two sides of the equation. The equation then looks like the following:

$$mv_o^2 = \frac{\hbar c}{\alpha^{n/b}r} . \qquad (21\text{-}20)$$

Combining Eq. (21-20) with Eq. (21-18), we can solve for v_0:

$$v_o^2 = \frac{b^2}{n\alpha^{n/b}} c^2, \qquad (21\text{-}21)$$

$$v_o = \left(\frac{b^2}{n\alpha^{n/b}} \right)^{1/2} c. \qquad (21\text{-}22)$$

H.   Deriving Semion Orbit Energy Levels and Masses

We have solved for v_o in terms of our parameters n and b. We can now plug that formula into the relationship for particle energy to obtain the energy levels of semion orbits, and thus the particle masses. We could blindly continue to employ m in the equations, but it is our desire to solve for m in terms of m_e. We therefore take the special case of m_e in the following equations. The kinetic, potential, and total energies of the semion system can be expressed as

$$Energy_{total} = Energy_{kinetic} + Energy_{potential}. \qquad (21\text{-}23)$$

$$m_T c^2 = +\frac{1}{2} m_e v_o^2 - \left\{ -\frac{b^2 m_e c^2}{n\alpha^{n/b}} \right\}, \qquad (3\text{-}24)$$

where Eq. (21-18) is substituted for r in Eq. (21-20) to obtain the potential energy fraction in Eq. (21-24). Substituting Eq. (21-21), where appropriate, into Eq. (21-24), we obtain

$$m_T c^2 = +\frac{m_e}{2} \frac{b^2}{n\alpha^{n/b}} c^2 + \frac{b^2 m_e c^2}{n\alpha^{n/b}} \qquad (21\text{-}25)$$

$$= \frac{3b^2}{2n\alpha^{n/b}} mc^2. \qquad (21\text{-}26)$$

The measurable mass term of the semion system is

$$m_T = \frac{3b^2}{2n\alpha^{n/b}} m_e. \qquad (21\text{-}27)$$

The expression above in Eq. (21-27), derived from first principles, applies for a term in a series of terms for any charged lepton. But the mass of a particle equals that term plus a series of all previous terms back to that for the electron, where n and b equal zero (see Eq. (21-28)).

$$m_j = \left\{ \frac{3b_j^2}{2n_j\alpha^{n_j/b_j}} + \frac{3b_{j-1}^2}{2n_{j-1}\alpha^{n_{j-1}/b_{j-1}}} + \ldots + \frac{0}{0\alpha^{0/0}} \right\} m_e. \qquad (21\text{-}28)$$

The right most term in Eq. (21-28) times the g/2 factor for the electron is defined as 1.0.

To calculate this in general, we must have a definition of n, b, and j:

j		0	1	2	3	4	5 ....
n		0	1	3	6	10	15 ....
b		0	1	2	3	4	5 ....

Table 21-1

The first three n and b are tested. Higher n and b are calculated. We expect both n and b to increase with j. We expect $n_j - (n_{j-1})$ to be b_j.

Finally, just as the mass is a series of terms, all other force terms are added by multiplying by half the g-factor for the given particle. For the muon, the net mass is the sum of the terms according to Eq. (21-28) times half the g factor for the muon, or 206.5539 m_e times $|g_\mu/2|$, or 206.5539 m_e $|-1.001\ 165\ 912\ 4|$, equals 206.7948. . . m_e, which is 1.000128 times the measured amount, 206.768262 m_e. That is 0.0128 of 1% error, which is almost the same error as the most accurate comparison of the theoretical Bohr Model of orbital differences to the measured values for orbital electrons. With what data we have to work with, our model is quite accurate.

There are only two usable g/2 factors that are available that are measured which can be used in calculating masses—for the electron and the muon. Fortunately, the tie is close enough between particle masses and particle g/2 factors that calculated g/2 factors can be tested by the measured masses of the particles. We will begin employing calculated g/2 factors in this and the next two chapters.

I.   Fathoming the Orbital Velocities

Let us name the electron e_0, the muon e_1, and the tauon e_2. Then, for those particles and higher particles, Eq. (21-22) solves for the orbital semion velocities v_0:

Particle | Semion Orbital Velocity v_0

Particle	Semion Orbital Velocity v_0
e_0	1.0000 c
e_1	11.7062 c
e_2	46.2480 c
e_3	167.8300 c
e_4	593.0600 c
e_5	2070.9000 c

Table 21-2

Compared to Einstein's Special Theory of Relativity, these are very high velocities. But these are velocities in a black hole. Velocities in a black hole should have no limit. Gravity escapes the bounds of a black hole, and communicates the sense of the mass of the black hole.

J.   Theorizing the Radii of Semion Orbits

This model also theorizes the radii of the semion orbits in charged leptons:

Table 21-3

Particle	Radius of Semion Orbit r
e_0	1.0 ℏ/mc
e_1	1.0 ℏ/mc
e_2	3/4 ℏ/mc
e_3	6/9 ℏ/mc
e_4	10/16 ℏ/mc
e_5	15/25 ℏ/mc,

where m is the mass of the given charged lepton—not just the mass of the electron.

K.   Predicted Masses of Charged Leptons

This model theorizes and predicts the masses of any charged leptons where the particles do not stir up pair production. In that case, the mass basis is calculated from which the effects of pair production are subtracted.

Charged Lepton	Mass Term (times m_e)	Predicted Mass	Measured Mass[9]
e_0	1.000 000	1.000 000	1.000 000
e_1	205.553 998	206.793 657	206.768 262
e_2	3 208.351 955	3 418.859 771	3 477
e_3	42 252.446 72	45 720.562 03	
e_4	527 591.661 1	573 928.381 1	

Table 21-4

This model does not predict a limited number of charged leptons (which we now observe). It predicts an infinite number of charged leptons, the next two of which are e_3 and e_4 in the above table.

The calculations in this chapter are for charged leptons. Similar calculations could be made for the quartons in the pion family. Other particle sets could be calculated by taking into account the Chonomic structures of the given particles. All the fundamental whole particles up to state 5 are calculated in *Advanced Electrino Physics* Chapter 4, from which all other particles up to state 5 may be calculated. Accompanying the calculation of fundamental masses in that book Chapter 4, is the calculation of the associated fundamental g/2 factors in Chapter 5. Because of pair production, only the particle masses of charged leptons up to tauons will correspond closely to the calculated values. But the calculated values of higher state particles could be useful as bases for subtracting the effects of pair production from them.

---

[1]Stephen Gasiorowicz, *Quantum Physics* (New York: John Wiley & Sons, 1974), p. 15.

[2]*Ibid.*

[3]*Ibid.*

[4]"Bohr model," *Wikipedia*, the free encyclopedia, http://en.wikipedia.org/wiki/Bohr_model.

[5]*Ibid.*

[6]*Ibid.*

[7]Professor John Blamire, "Atomic Structure—The mystery of . . .—. . .the quantum atom," Exploring Life @

BIOdotEDU,
http://www.brooklyn.cuny.edu/bc/ahp/LAD/C3/C3_elecPo
s_02.html.

[8]Thayer Watkins, "The Relativistic Bohr Model of a
Hydrogen-like Atom," applet-magic.com: Silicon Valley,
Tornado Alley & BB Island USA,
http://www.applet-magic.com/relabohr.htm.

[9]*CRC Handbook of Chemistry and Physics*, 80th
Edition, 1999-2000 (Boca Ratan: CRC Press, 1999), pp, 1-
4, 11-3.

Problem Set 3

1.   How many quantum numbers are there in the
Electrino Model of mass calculations of charged leptons?

2.   Does the author's model account for elliptical or
tilted orbits?

3.   How accurate is the author's prediction of the mass
of the muon?

4.   Why is the author's prediction of the mass of the
tauon so inaccurate?

5.   What charge do semions have?

6.   What is the main difference between the force of
electrons in their orbit and the force of semions in their
orbit?

7.   How many body problem are semions in orbit?

8.   Do semions orbit slower than the speed of light or
faster than the speed of light?

9.   Do charged lepton semion orbital velocities follow Einstein's Special Theory of Relativity?

10.   What difference in the mass is there between Bohr's Model and the author's model?

11.   What different rule for the coincidence of standing waves in the particle does the author's model have as compared to Bohr's Model?

12.   Are semions easy to ionize?

13.   What reverses the sense of the potential energy in semion orbits?

14.   What evidence is there that the mass in Eq. (3-18) is the overall mass of the charged lepton, not just the mass of the electron?

15.   What forces are being balanced in Eq. (3-19)?

16.   What factor occurs in the mass calculations due to both the kinetic and potential energies being positive?

17.   What mass term for the charged lepton is derived from first principles?

18.   How is that mass used in the calculation of the total mass of the particle?

19.   What is n as a function of j? What is b as a function of j?

20.   The Electrino Fusion Model of Elementary Particles predicts there will be how many different kinds of charged leptons in all?

21.   This is a free question: Did you imagine that v_0 should be so much above the speed of light for particles above electrons?

22.   This is a free question: Are the radii of semion orbits a function of α? How?

Appendix A

LEPTON SUMMARY TABLE

The quark and lepton model of particle physics divides charges in quarks to 2e/3 and e/3. The electrino model of particle physics does not do that. Instead, it divides charges in electrinos to e, e/2, e/4, and e/8. The electrino model of particle physics does not hold that the quark and lepton model of particle physics is correct. Nevertheless, to facilitate cross referencing with the existing data, this volume will employ quark model titles and classifications in the subsequent classification of particles.

The chonomic structures and decay schemes contained in the following material are the author's, but the particle data come from SUMMARY TABLES OF PARTICLE PROPERTIES, January 1, 1998, Particle Data Group, as quoted by *CRC Handbook of Chemistry and Physics,* 80th Edition, David R. Lide, Ph.D, Editor in Chief (Boca Raton: CRC Press, 1999), pp. **11**-3 to **11**-5. "In this Summary Table:

"When a quantity has '(S = ...)' to its right, the error on the quantity has been enlarged by the 'scale factor' S, defined as

$$S = \sqrt{\chi^2 / (N - 1)},$$

where N is the number of measurements used in calculating the quantity. We do this when S > 1, which often indicates that the measurements are inconsistent. When S > 1.25, we also show in the Particle Listings an ideogram of the measurements. . . .

"A decay momentum p is given for each decay mode. For a 2-body decay, p is the momentum of each decay product in the rest frame of the decaying particle. For a 3-or-more-body decay, p is the largest momentum any of the products can have in this frame."

To understand how to read and interpret chonomic structures and decay schemes, see Chapter 10, "Introduction to Chonomics."

For gravitons, the mass designations in the chonomic representations is >0i. That means the Electrino Fusion Model of Elementary Particles calls for positive imaginary mass values for gravitons. Experimental values of m^2 for neutrinos is also negative, indicating imaginary mass values for neutrinos (See note at end of this Appendix.). The author's model also predicts the mass values of neutrinos will be positive imaginary. Thus the following chonomic structures give two values for neutrinos: >0i, and <R_i, where R_i is the maximum possible undetected real mass that could be for each respective neutrino type according to 1998 data.

Just as there is more than one way to skin a cat, there is more than one possible way to balance some decay schemes. For instance, the number of required incoming gravitons can vary, in some cases, depending on whether g^o, g^+, and g^- gravitons are employed, without incoming positional-kinetic angular momentum, or g^{o+} and g^{o-} gravitons are employed, with positional-kinetic angular momentum. Also unobserved incoming π^0s could be employed to balance decay schemes, though that is not necessary in this Appendix. There are other flexible parameters in these decay schemes. Often the author just had to pick one example to solve the decay scheme. The following decay schemes illustrate a variety of ways to solve the problems. For a more advanced Summary Table, the various decay modes would have to be summed over the various possible pathways, to determine more fundamental quantities, such as populations of various graviton types.

In Appendix B are listed state numbers with each chonomic structure. There is not room for them in this appendix. The state levels in this work, if not obvious to the investigator, may be obtained by looking up the particles in Appendix B.

The following particles are leptons as counted by the Particle Data Group.

LEPTONS

e $J = \frac{1}{2}$
  Mass m = 0.510998910 ± 0.000000013 MeV
      = (548.57990943 ±0.00000023) x 10^{-6} u
  $|m_{e+}-m_{e-}|/m < 8$ x 10^{-9}, CL 90%
    $|q_{e+}+q_{e-}|/e < 4$ x 10^{-8}
Magnetic moment μ = 1.0011596521811 ± 0.0000000000007 μ_B
    $(g_{e+}-g_{e-})/g_{average} = (-0.5 \pm 2.1)$x 10^{-12}
Electric dipole moment $d = (0.07 \pm 0.07)$x 10^{-26} e cm
    Mean life τ > 4.6 x 10^{26} years, CL = 68%

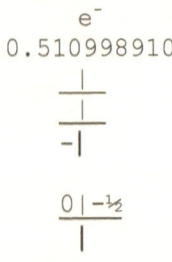

```
              e⁻
        0.510998910

             |
            _|_
           -|

          0|-½
           |
```

μ $J = \frac{1}{2}$
      Mass m = 105.658367 ± 0.000004 MeV
          = 0.1134289256 ± 0.0000000029 u
      Mean life τ = (2.197019 ± 0.000021)x 10^{-6} s (S = 1.1)
      $\tau_{\mu+}/\tau_{\mu-}$ = 1.00002 ± 0.00008
          cτ = 658.650 m
  Magnetic moment μ = 1.0011659208 ± 0.0000000006 e$\hbar$/2m_μ
  $(g_{\mu+}-g_{\mu-})/g_{average}$ = (-0.11 ± 0.12) x 10^{-8}
  Electric dipole moment d = (3.7 ± 3.4) x 10^{-19} e cm
Decay parameters
    ρ = 0.7509 ± 0.0010
    η = -0.001 ± 0.024   (S = 2.0)
    δ = 0.7495 ± 0.0012
    ξP_μ = 1.0007 ± 0.0035
    $\xi P_\mu \delta/\rho$ > 0.99682, CL = 90%
    ξ' = 1.00 ± 0.04

Decay Parameters Continued

$\xi'' = 0.7 \pm 0.4$

$\alpha/A = (0 \pm 4) \times 10^{-3}$

$\alpha'/A = (0 \pm 4) \times 10^{-3}$

$\beta/A = (4 \pm 6) \times 10^{-3}$

$\beta'/A = (1 \pm 5) \times 10^{-3}$

$\overline{\eta} = 0.02 \pm 0.08$

μ^+ modes are charge conjugates of the modes below.

μ^-

105.658367

$-\boxed{}$

$0\,|\,-\tfrac{1}{2}$

$\underline{\mu^-\text{ DECAY MODE}}$	Fraction (Γ_i/Γ)	Confidence level	p (MeV/c)
$e^-\,\overline{\nu}_e\,\nu_\mu$	$\approx 100\%$		53

μ^-		g^{o-}	q.p.	e^-	$\overline{\nu}_e$ >0i	ν_μ >0i
105.6...		>0i	? 0.510...		<.000003	<.19

$$-\boxed{} \;+\; \boxed{}_{-o\,|\,o-} \;\to\; \boxed{-}_{-o\,|\,o-} \;\to\; \boxed{}_{-} \;+\; \boxed{}_{o\,|\,-} \;+\; \boxed{-}_{|\,o}$$

$0\,|\,-\tfrac{1}{2}$	$3\,|\,2$	$3\!-\!1\,|\,\tfrac{1}{2}$	$0\,|\,-\tfrac{1}{2}$	$1\,|\,\tfrac{1}{2}$	$1\,|\,\tfrac{1}{2}$
$0\,|\,-\tfrac{1}{2}$	$-1\,|\,1$	$0\,|\,\tfrac{1}{2}$	$0\,|\,-\tfrac{1}{2}$	$0\,|\,\tfrac{1}{2}$	$0\,|\,\tfrac{1}{2}$

undetected particle

Beta decay followed by straight forward recombinations.

μ⁻ DECAY MODE	Fraction (Γᵢ/Γ)	Confidence level	p (MeV/c)
$e^- \overline{v}_e v_\mu \gamma$	$(1.4 \pm 0.4)\%$		53

This decay mode is the same as the above, except there is an additional unobserved low energy photon on the left of the reaction, which is energized by the decay process and detected on the right side of the decay process.

μ⁻ DECAY MODE	Fraction (Γᵢ/Γ)	Confidence level	p (MeV/c)
$e^- \overline{v}_e v_\mu e^+ e^-$	$(3.4 \pm 0.4) \times 10^{-5}$		53

```
μ⁻              g⁰⁻            g⁻              q·p
105.6...        >0i            >0i

 _|_            _|_            _|_             _|_
 -|      +       |      +      -|       →      -|      →
  |             -o|o-          -|-            --o|o--

0|-½            3|2            -1|-2           3-1|-½
0|-½            0|2            0|-2            0|-½
                |_____|
                    unobserved particles
```

```
 e⁻            v̄ₑ>0i       v_μ>0i        e⁺            e⁻
0.510...     <.000003     <.19        0.510...      0.510...

 _|_           _|_          _|_         _|_           _|_
  |      +      |     +     -|    +      |     +       |
 -|            o|-          |o          |-            -|

0|-½          1|½          1|½         0|-½          0|-½
0|-½          0|½          0|½         0|-½          0|-½
```

Two different unobserved gravitons are simultaneously attracted to the parent particle and contribute subparticles to the quasi particle and subsequent decay products. Beta decay occurs and straight forward recombination of particles.

τ $\qquad\qquad\qquad$ J = ½

Mass m = 1776.84 ± 0.17 MeV

$$\left(m_{\tau^+} - m_{\tau^-} \right) / m_{average} < 2.8x10^{-4}, \; CL = 90\%$$

Mean life τ = (290.6 ± 1.0)x 10^{-15} s

$\qquad$ cτ = 87.11 μm

Magnetic moment anomaly >-0.052 and <0.013, CL = 95%

$\qquad$ τ⁺ modes are charge conjugates of the modes below.

$\qquad$ τ⁺ modes are charge conjugates of the modes below. "h⁺" stands for π⁺ or K⁺, "ℓ" stands for e or μ. "Neutral" means neutral hadron whose decay products include γ's and/or π⁰'s.

τ⁻ DECAY MODE	Fraction (Γ_i/Γ)	Scale factor/ Confidence level	p (MeV/c)
particle⁻ ≥ 0 neutrals ≥0K⁰ₗvτ ("1-prong")	(85.36 ± 0.08)%	S = 1.3	--
particle⁻ ≥ 0 neutrals ≥0K⁰vτ	(84.73 ± 0.08)%	S = 1.4	--
$\mu^-\bar{v}_\mu v_\tau$	(17.36 ± 0.05)%		885

```
  τ⁻        g°⁻       q.p.      μ⁻      V̄_μ >0i    V_τ >0i
1776.84 >0i                  105.6... <.19    <18.2
 -|        |        -|        |        |        -|
 _|   +   _|   →   -|-   →   -|   +    |-   +   _|
  |     -o|o-       o|o       |        o|       |o

0|-½    3|2       3-1|½     0|-½     1|½      1|½
0|-½   -1|1        0|½      0|-½     0|½      0|½
  |_____|
 unobserved particle
```

A graviton composed of a neutrino and an anti-neutrino orbiting about each other gravitationally combine with a tauon, forming a quasi-particle with it. If the - echons in the graviton were not already in an excited state, they are knocked to the 1 state by the tauon, while the minus echon in the tauon retains its state. The echons re-divide straight forwardly.

τ⁻ DECAY MODE	Fraction (Γ_i/Γ)	Scale factor/ Confidence level	p (MeV/c)
$\mu^-\bar{v}_\mu v_\tau\gamma$	(3.6 ± 0.4) x 10⁻³		885

This decay mode is the same as the above, except there is an additional unobserved low energy photon on the left of the reaction, which is energized by the decay process and detected on the right side of the decay process.

	Scale factor/	p
τ⁻ DECAY MODE Fraction (Γᵢ/Γ)	Confidence level	(MeV/c)

$e^-\bar{\nu}_e\nu_\tau$ [i] (17.85 ± 0.05)%   888

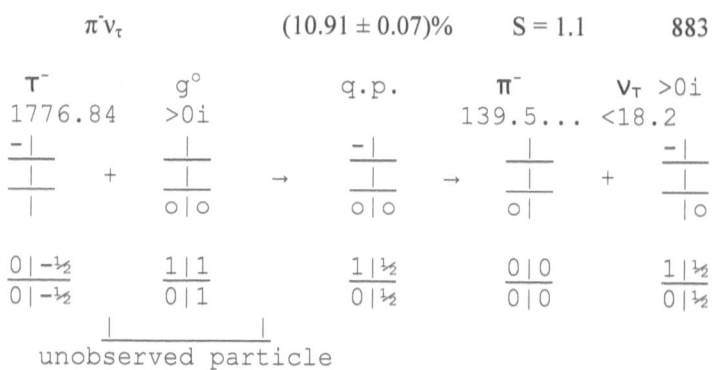

Straight-forward recombinations.

	Scale factor/	p
τ⁻ DECAY MODE Fraction (Γᵢ/Γ)	Confidence level	(MeV/c)

$\pi^-\nu_\tau$ (10.91 ± 0.07)%   S = 1.1   883

Straight-forward recombinations.

τ^- DECAY MODE	Fraction (Γ_i/Γ)	Scale factor/ Confidence level	p (MeV/c)
$K^-\nu_\tau$	$(6.95 \pm 0.23)\times 10^{-3}$	S = 1.1	820

$$\tau^- \qquad g^\circ \qquad\qquad q.p. \qquad\qquad K^- \qquad\qquad \nu_\tau >0i$$
$$1776.84 \quad >0i \qquad\qquad\qquad 493.6... \qquad <18.2$$

(diagram of grids)

$0\mid-\tfrac12$	$1\mid1$	$1\mid\tfrac12$	$0\mid0$	$1\mid\tfrac12$
$0\mid-\tfrac12$	$0\mid1$	$0\mid\tfrac12$	$0\mid0$	$0\mid\tfrac12$

```
 τ⁻         g°            q.p.          K⁻           ν_τ >0i
1776.84   >0i                          493.6...     <18.2
 -|         |            -|            |            -|
 _|_  +    _|_    →     _o|_    →     _o|_   +      _|_
  |       o| o           | o           |             | o

0|-½       1|1          1|½          0|0          1|½
0|-½       0|1          0|½          0|0          0|½
           |_____|
           unobserved
           particle
```

If the g° graviton is not already energized, the tauon - echon knocks the negative o echon of the g° graviton to the 1 state as the tauon and graviton combine to form a quasi-particle. Thereafter there is straight forward recombinations.

τ^- DECAY MODE	Fraction (Γ_i/Γ)	Scale factor/ Confidence level	p (MeV/c)
$h^- \geq 1$ neutrals ν_τ	$(37.08 \pm 0.11)\%$	S = 1.2	--
$h^-\pi^0\,\nu_\tau$	$(25.95 \pm 0.10)\%$	S = 1.1	878
$\pi^-\pi^0\nu_\tau$	$(25.52 \pm 0.10)\%$	S = 1.1	878

```
 τ⁻         g°⁻        q.p.      π⁻          π⁰          ν_τ >0i
1776..     >0i                  139.5...    134.9...    <18.2
 -|          |        -|         |           |          -|
 _|_  +    _____  →  _|_   →   _|_   +    _|_   +     _|_
  |      -o| o-    -o| o-       o|         -|-           | o

0|-½       3|2      3-1|½      0|0         1|0         1|½
0|-½      -1|1        0|½      0|0         0|0         0|½
          |_____|
         unobserved particle
```

A π^0 is a system of a positron and an electron in orbit about one another with the orbital spin oppositely directed to the intrinsic spin of the positron and electron. A π^0 is not composed of the same echons as $\pi^\pm$ particles, though they have similar mass. Decay schemes of π^0 and $\pi^\pm$ will bear this out.

τ DECAY MODE	Fraction (Γ$_i$/Γ)	Scale factor/ Confidence level	p (MeV/c)
π⁻π⁰non-ρ(770)ν$_τ$	(3.0 ± 3.2)x 10⁻³		878
K⁻π⁰ν$_τ$	[i] (4.28 ± 0.15)x 10⁻³		814

τ⁻	g°⁻	q.p.	K⁻	π⁰	ν$_τ$ >0i
1776..	>0i		493.6...	134.9...	<18.2

$$\begin{array}{ccc}
\boxed{} & \boxed{} & \to \boxed{} \to \boxed{} + \boxed{} + \boxed{} \\
\end{array}$$

0│-½	3│2	3-1│½	0│0	1│0	1│½
0│-½	-1│1	0│½	0│0	0│0	0│½

unobserved particle

A g°⁻ graviton gravitationally combines with a tau particle. The negative o echon gets knocked up to the +1 state in the process. Straight forward recombinations result in a K⁻, π⁰, and a τ⁻.

τ DECAY MODE	Fraction (Γ$_i$/Γ)	Scale factor/ Confidence level	p (MeV/c)
h⁻ ≥ 2π⁰ ν$_τ$	(10.84 ± 0.12)%	S = 1.3	--
h⁻2π⁰ν$_τ$	(9.49 ± 0.11)%	S = 1.2	862
h⁻2π⁰ν$_τ$(ex.K⁰)	(9.33 ± 0.12)%	S = 1.2	862
π⁻2π⁰ν$_τ$(ex.K⁰)	(9.27 ± 0.12)%	S = 1.2	862

τ⁻	g°⁻	g⁻	q.p.	π⁻	π⁰	π⁰	ν$_τ$>0i
1776..	>0i	>0i		139...	134...134...<18.2		

0│-½	3│2	-1│-2	4-1│½	0│0	1│0	1│0	1│½
0│-½	1│3	0│-2	0│½	0│0	0│0	0│0	0│½

unobserved particles

The system experiences straight-forward recombinations.

			Scale factor/	p
τ^- DECAY MODE		Fraction (Γ_i/Γ)	Confidence level	(MeV/c)

$K^-2\pi^0\nu_\tau$(ex.K^0)   (6.3 $\pm$ 2.3)x 10^{-4} 796

τ^-	g^-	g^{o-}	q.p.	K^-	π^0	π^0	ν_τ >0i
1776..	>0i	>0i		493...	134...	134...	<18.2

$$\frac{-|\quad}{\quad|\quad} + \frac{|\quad}{-|-} + \frac{|\quad}{-o|o-} \rightarrow \frac{-|\quad}{o|\quad} \rightarrow \frac{|\quad}{o|} + \frac{|\quad}{-|-} + \frac{|\quad}{-|-} + \frac{-|\quad}{|o}$$

$\frac{0|-\frac{1}{2}}{0|-\frac{1}{2}}$	$\frac{-1|-2}{1|-1}$	$\frac{3|2}{0|2}$	$\frac{4-1|\frac{1}{2}}{0|\frac{1}{2}}$	$\frac{0|0}{0|0}$	$\frac{1|0}{0|0}$	$\frac{1|0}{0|0}$	$\frac{1|\frac{1}{2}}{0|\frac{1}{2}}$

|_____|
unobserved particles

One o echon is knocked to the +1 state as a g^{o-} graviton and g^+ graviton gravitationally combine with a τ^- particle. Thereafter it is straight-forward recombinations.

			Scale factor/	p
τ^- DECAY MODE		Fraction (Γ_i/Γ)	Confidence level	(MeV/c)

$h^- \geq 3\pi^0\nu_\tau$	(1.35 $\pm$ 0.07)%	S = 1.1	--
$h^-3\pi^0\nu_\tau$	(1.18 $\pm$ 0.08)%		836
$\pi^-3\pi^0\nu_\tau$(ex.K^0)	(1.04 $\pm$ 0.07)%		836

τ^-	g^{o-}	g^-	g^-	q.p.
1776..	>0i	>0i	>0i	

$$\frac{-|\quad}{\quad|\quad} + \frac{|\quad}{-o|o-} + \frac{|\quad}{-|-} + \frac{|\quad}{-|-} \rightarrow \frac{-|\quad}{---o|o---} \rightarrow$$

$\frac{0|-\frac{1}{2}}{0|-\frac{1}{2}}$	$\frac{3|2}{0|2}$	$\frac{-1|-2}{0|-2}$	$\frac{-1|-2}{0|-2}$	$\frac{3-2|-2\frac{1}{2}}{0|-2\frac{1}{2}}$

|_____|
unobserved particles

```
   π⁻        π⁰        π⁰        π⁰       νₜ  >0i
 139...    134...    134...    134...      <18.2
  |         |         |         |          -|
 _|_   +   _|_   +   _|_   +   _|_   +     _|_
  o|        -|-       -|-       -|-         |o

 0|0       1|0       1|0       1|0        1|½
 ───       ────      ────      ────       ────
 0|0       -1|-1     -1|-1     -1|-1      0|½
```

A g⁰⁻ graviton and two g⁻ gravitons gravitationally combine with the tauon, forming a quasi-particle. Considering the spins, it is straight-forward recombinations.

		Scale factor/	p
τ⁻ DECAY MODE	Fraction (Γᵢ/Γ)	Confidence level	(MeV/c)

$$K^-3\pi^0\nu_\tau(ex.K^0) \quad (4.7\pm2.1)\times10^{-4} \qquad\qquad 765$$

```
   τ⁻       g⁰⁻       g⁻        g⁻        q.p.
 1776..    >0i       >0i       >0i
  -|        |         |         |          -|
 _|_   +   _|_   +   _|_   +   _|_   →     o|      →
  |        -o|o-      -|-       -|-       ---|o---

 0|-½      3|2       -1|-2     -1|-2      3-2|-2½
 ────      ───       ─────     ─────      ──────
 0|-½      0|2       0|-2      0|-2       0|-2½
           |                         |
         unobserved particles
```

```
   K⁻        π⁰        π⁰        π⁰       νₜ  >0i
 493...    134...    134...    134...      <18.2
  |         |         |         |          -|
  o|   +   _|_   +   _|_   +   _|_   +     _|_
  |        -|-       -|-       -|-          |o

 0|0       1|0       1|0       1|0        1|½
 ───       ────      ────      ────       ────
 0|0       -1|-1     -1|-1     -1|-1      0|½
```

A g⁰⁻ graviton and two g⁻ gravitons gravitationally combine with the tauon, forming a quasi-particle. The tauon knocks one o echon to the +1 state. Considering the spins, it is straight-forward recombinations.

τ^- DECAY MODE	Fraction (Γ_i/Γ)	Scale factor/ Confidence level	p (MeV/c)
$h^-4\pi^0\nu_\tau$(ex.K^0)	(1.6 ± 0.4)x 10^{-3}		800
$h^-4\pi^0\nu_\tau$(ex.K^0,η)	(1.0 ± 0.4)x 10^{-3}		800
$K^- \geq 0\pi^0 \geq 0K^0\,\nu_\tau$	(1.57 ± 0.04)%	S=1.1	820
$K^- \geq 1\ (\pi^0$ or $K^0)\,\nu_\tau$	(8.74 ± 0.32)x 10^{-3}		--

Modes with K^0's

K^0(particles)$^-\nu_\tau$	(9.2 ± 0.4) x 10^{-3}	S = 1.4	--
$h^-\overline{K}^0\nu_\tau$	(10.0 ± 0.5)x 10^{-3}	S = 1.8	812
$\pi^-\overline{K}^0$ (non-K*(892)$^-$)ν_τ	(5.4 ± 2.1)x 10^{-3}		812

$\overline{\tau^-}$	g^{o+}	q.p.	π^-	K^0	ν_τ >0i
1776..	>0i		139.5...	497.6...	<18.2

$$\frac{-|}{|}\ + \ \frac{|}{+o|o+} \ \rightarrow \ \frac{-|}{+|+} \ \rightarrow \ \frac{|}{o|} \ + \ \frac{|}{+|+} \ + \ \frac{-|}{|o}$$

$$\frac{0|-\tfrac{1}{2}}{0|-\tfrac{1}{2}} \quad \frac{-3|-2}{0|-2} \quad \frac{-3|-2\tfrac{1}{2}}{0|-2\tfrac{1}{2}} \ \frac{0|0}{-1|-1} \quad \frac{-1|0}{-1|-1} \quad \frac{1|\tfrac{1}{2}}{-1|-\tfrac{1}{2}}$$

unobserved particle

A g^{o+} graviton (a neutrino-anti-neutrino graviton) is gravitationally attracted to a tauon, forming a quasi-particle. One or both + echons are knocked up to the +1 state in the process. Thereafter it is straight-forward recombinations. Anti-K^0s is when both negative and positive + echons are in the +1 state. Anti-K^0_L is when one + echon is in the +1 state, and one is in the 0 state.

τ DECAY MODE	Fraction (Γ$_i$/Γ)	Scale factor/ Confidence level	p (MeV/c)
$K^-\bar{K}^0\nu_\tau$	$(1.58\pm 0.16)\times 10^{-3}$		737

τ^-	g^-	g^0	q.p.	K^-	K^0	$\nu_\tau >0i$
1776..	>0i	>0i		493.6...	497.6...	<18.2

$$
\begin{array}{ccccccc}
\dfrac{-|}{\underline{\quad}|\quad} & + & \dfrac{\quad|\quad}{-|-} & + & \dfrac{\quad|\quad}{o|o} & \to & \dfrac{-|\quad}{-o|-} & \to & \dfrac{\quad|\quad}{o|} & & \dfrac{\quad|\quad}{\quad|\quad} & + & \dfrac{\quad|\quad}{-|-} & + & \dfrac{-|\quad}{\quad|o}
\end{array}
$$

0|-½	-1|-2	1|1	1-1|-1½	0|0	1|0	1|½
0|-½	0|-2	0|1	0|-1½	-1|-1	-1|-1	0|½

unobserved particles

A g⁻ graviton and a g⁰ graviton are gravitationally attracted to a tauon, forming a quasi-particle with it. The tauon knocks one o echon and one or both - echons of the gravitons to the +1 state. Thereafter it is straight forward recombinations.

τ DECAY MODE	Fraction (Γ$_i$/Γ)	Scale factor/ Confidence level	p (MeV/c)
$h^-\bar{K}^0\pi^0\nu_\tau$	$(5.5\pm 0.4)\times 10^{-3}$		794
$\pi^-\bar{K}^0\pi^0\nu_\tau$	$(3.9\pm 0.4)\times 10^{-3}$		794

τ^-	g^{0-}	g^-	q.p.	π^-	K^0	π^0	$\nu_\tau >0i$
1776..	>0i	>0i		139...	497...	134...	<18.2

0|-½	3|2	-1|-2	3-1|-½	0|0	1|0	1|0	1|½
0|-½	0|2	0|-2	0|-½	0|0	0|0	-1|-1	0|½

unobserved particles

A g⁰⁻ graviton and a g⁻ graviton combine with a tauon, forming a quasi-particle. The tauon knocks one or both of the graviton - echons to the +1 state. Thereafter it is straight forward recombinations.

τ^- **DECAY MODE**	Fraction (Γ_i/Γ)	Scale factor/ Confidence level	p (MeV/c)

$$\overline{K}^0 \rho^- v_\tau \qquad (\,2.2 \pm 0.5\,) \times 10^{-3} \qquad\qquad 612$$

τ^-	g^+	g^-	g^o	q.p.	$\overline{K}^0$	$\rho(770)^-$	v_τ
>0i							
1776	>0i	>0i	>0i		497...	775.4	<18.2

```
-|        |        |        |         -|          |          |          -|
 |   +    |   +    |   +    |    →    +|+   →    +|+  +     |    +      |
 |       +|+      -|-      o|o       -o|o-        |         -o|-        |o
```

0	−½	1	2	−1	−2	1	1	2−1	½	−1	0	1−1	−1	1	½
0	−½	0	2	0	−2	0	1	0	½	1	1	0	−1	0	½

```
|                          |
```
unobserved particles

A g^+, a g^-, and a g^o graviton combine with a tauon to form a quasi-particle. The tauon knocks both + echons to the +1 state. Otherwise it is straight-forward recombinations.

τ **DECAY MODE**	Fraction (Γ_i/Γ)	Scale factor/ Confidence level	p (MeV/c)

$$K^- K^0 \pi^0 v_\tau \qquad (\,1.58 \pm 0.20\,) \times 10^{-3} \qquad\qquad 685$$

τ^-	$g-$	g^{o-}	q.p.	K^-	K^0	π^0	v_τ >0i
1776..	>0i	>0i		493...	497...	134...	<18.2

```
-|        |        |         -|          |          |          |         -|
 |   +    |   +    |    →    -o|-   →    o|   +    -|-   +     |    +      |
 |       -|-      -o|o-       -|o-        |          |         -|-        |o
```

0	−½	−1	−2	3	2	3−1	−½	0	0	1	0	1	0	1	½
0	−½	0	−2	0	2	0	−½	0	0	0	0	−1	−1	0	½

```
|                              |
```
unobserved particles

A g^- and a g^{o-} graviton combine with a tauon to form a quasi-particle. The tauon knocks a negative o and both - echons from the gravitons to the +1 state. Otherwise it is straight-forward recombinations.

		Scale factor/	p
τ DECAY MODE	Fraction (Γᵢ/Γ)	Confidence level	(MeV/c)

$$\pi^- \overline{K}^0 \pi^0 \pi^0 \nu_\tau \qquad (\,2.6 \ \pm 2.4\,\,) \text{x} \ 10^{-4} \qquad\qquad 763$$

T⁻	g⁰⁻	g⁻	g⁺	q.p.	
1776...	>0i	>0i	>0i		

```
 T⁻          g⁰⁻          g⁻           g⁺           q.p.
1776...      >0i          >0i          >0i
 -|           |            |            |           -|
_|_    +    _|_    +     _|_    +     _|_    →    _+|+_     →
 |         -o|o-         -|-          +|+         --o|o--

0|-½        3|2         -1|-2        1|2        4-1|1½
----        ---         -----        ---        ------
0|-½        0|2          0|-2        0|2         0|1½
```

|_____|
                unobserved particles

π⁻	$\overline{K}^0$	π⁰	π⁰	νᴛ >0i
139...	497...	134...	134...	<18.2

```
 π⁻          K̄⁰          π⁰           π⁰          νᴛ >0i
139...       497...       134...       134...       <18.2
 |           |            |            |           -|
_|_    +    _+|+_   +    _|_    +     _|_    +     _|_
o|           |          -|-          -|-           |o

0|0        -1|0         1|0          1|0          1|½
---        ----         ---          ---          ---
0|0         0|0         0|0          1|1          0|½
```

The decay mode employs three gravitons combining with the tauon to form a quasi-particle. There is massive particle collisions and reflecting the orbital spins in the quasi-particle. Otherwise it is straight-forward recombinations.

		Scale factor/	p
τ DECAY MODE	Fraction (Γᵢ/Γ)	Confidence level	(MeV/c)

$$\pi^- K^0 \overline{K}^0 \nu_\tau \qquad (\,1.7\pm 0.4\,) \text{ x } 10^{-3} \qquad S = 1.6 \qquad 682$$

T⁻	g⁺	g⁰⁻	q.p.	π⁻	K⁰	$\overline{K}^0$	νᴛ>0i
1776..	>0i	>0i		139...	497...	497...<18.2	

```
 T⁻      g⁺      g⁰⁻         q.p.         π⁻        K⁰         K̄⁰        νᴛ>0i
1776..   >0i     >0i                     139...     497...     497...   <18.2
 -|       |       |          -|           |          |          |         -|
_|_  +  _|_  +  _|_    →  _-+|+-_   →   _|_   +   _-|-_   +   _+|+_  +   _|_
 |      +|+     -o|o-       o|o          o|          |          |         |o

0|-½    1|2     3|2       4-1|2½        0|0        -1|0        1|0        1|½
----    ---     ---       ------        ---        ----        ---        ---
0|-½   -1|1     0|2        0|2½         1|1         1|1        0|0        0|½
```

|_____|
  unobserved particles

A g^+ and a g^{o*} (an upside down g^{o+}) graviton combine with a tauon to form a quasi-particle. The + and - echons from the gravitons are knocked to the +1 state. The + echons trade orbital spin and positional-kinetic angular momentum to form an anti-K^0 particle. Thereafter it is straight-forward recombinations.

			Scale factor/	p
τ^- DECAY MODE		Fraction (Γ_i/Γ)	Confidence level	(MeV/c)

$\pi^- K_S^0 K_S^0 \nu_\tau$ (2.4 ± 0.5)x 10^{-4} 682

unobserved particles

The tauon knocks the graviton - echons to the +1 state. Subsequently there are straight forward recombinations.

τ^- DECAY MODE	Fraction (Γ_i/Γ)	Scale factor/ Confidence level	p (MeV/c)
$\pi^- K^0_S K^0_L \nu_\tau$	$(1.2 \pm 0.4) \times 10^{-3}$	S = 1.7	682

```
 τ-         go-          g-             q.p.
1776..     >0i          >0i
-|         ___|___      __|__          __-|__
_|_  +     ___|___  +   __|__    →     --|-     →
 |         -o|o-         -|-           __o|o-__

0|-½        3|2         -1|-2          3-1|-½
----        ---         -----          ------
0|-½        0|2          0|-2            0|-½
   |_______________________|
    unobserved particles
 π-         K0S         K0L        ντ >0i
139...     497...      497...      <18.2
_|_         _|_         _|_          -|
_|_   +     -|-    +    -|     +    _|_
o|           |           |-          |o

 0|0        1|0         1|0         1|½
-----       ---         ---         ---
-1|-1       0|0         0|0         0|½
```

The tauon knocks all but one of the graviton - echons to the +1 state. Then there are straight forward recombinations.

τ^- DECAY MODE	Fraction (Γ_i/Γ)	Scale factor/ Confidence level	p (MeV/c)
$\pi^- K^0_S K^0_L \pi^0 \nu_\tau$	$(3.1 \pm 1.2) \times 10^{-4}$		614

```
 τ-        go-        g-         g-          q.p.
1776..    >0i        >0i        >0i
-|        ___|___    __|__      __|__       ___-|___
_|_  +    ___|___ +  __|__  +   __|__   →   --|-       →
 |        -o|o-       -|-        -|-        _-o|o--_

0|-½       3|2       -1|-2      -1|-2       3-2|-2½
----       ---       -----      -----       -------
0|-½       0|2        0|-2       0|-2         0|-2½
   |__________________________|
    unobserved particles
```

π^-	K^0_S	K^0_L	π^0	$v_T > 0i$
139...	497...	497...	134...	<18.2

$$\frac{\begin{array}{c} \mid \\ \hline \mid \\ \circ \mid \end{array}}{\begin{array}{c} 0\mid 0 \\ \hline 0\mid 0 \end{array}} \quad + \quad \frac{\begin{array}{c} \mid \\ \hline -\mid - \\ \mid \end{array}}{\begin{array}{c} 1\mid 0 \\ \hline -1\mid -1 \end{array}} \quad + \quad \frac{\begin{array}{c} \mid \\ \hline -\mid \\ \mid - \end{array}}{\begin{array}{c} 1\mid 0 \\ \hline -1\mid -1 \end{array}} \quad + \quad \frac{\begin{array}{c} \mid \\ \hline \mid \\ -\mid - \end{array}}{\begin{array}{c} 1\mid 0 \\ \hline -1\mid -1 \end{array}} \quad + \quad \frac{\begin{array}{c} -\mid \\ \hline \mid \\ \mid \circ \end{array}}{\begin{array}{c} 1\mid \tfrac{1}{2} \\ \hline 0\mid \tfrac{1}{2} \end{array}}$$

Three unobserved gravitons combine with a tauon to form a quasi-particle. The
tauon knocks three of the - echons to the +1 state. The other three graviton -
echons remain in ground state. Subsequently there are straight forward
recombinations.

τ DECAY MODE	Fraction (Γ_i/Γ)	Scale factor/ Confidence level	p (MeV/c)
$K^0 h^+ h^- h^- v_\tau$	$(2.3 \pm 2.0)\times 10^{-4}$		760

Modes with three charged particles

$h^- h^- h^+ \geq 0$neut. v_τ("3-prong")	$(15.18\pm 0.08)\%$	$S = 1.4$	861
$h^- h^- h^+ \geq 0$neutrals v_τ (ex. $K^0_S \rightarrow \pi^+\pi^-$)	$(14.56\pm 0.08)\%$	$S = 1.3$	861
$h^- h^- h^+ v_\tau$	$(9.80\pm 0.08)\%$	$S = 1.4$	861
$h^- h^- h^+ v_\tau$(ex.K^0)	$(9.45\pm 0.07)\%$	$S = 1.3$	861
$h^- h^- h^+ v_\tau$(ex.K^0,ω)	$(9.42\pm 0.07)\%$	$S = 1.3$	861
$\pi^- \pi^+ \pi^- v_\tau$	$(9.32\pm 0.07)\%$	$S = 1.2$	861

τ^-	g°	g°	q.p.	π^-	π^+	π^-	$v_T > 0i$
1776..	>0i	>0i		139...	139...	139...	<18.2

$$\frac{\begin{array}{c} -\mid \\ \hline \mid \\ \mid \end{array}}{\begin{array}{c} 0\mid -\tfrac{1}{2} \\ \hline 0\mid -\tfrac{1}{2} \end{array}} + \frac{\begin{array}{c} \mid \\ \hline \mid \\ \circ\mid\circ \end{array}}{\begin{array}{c} 1\mid 1 \\ \hline 0\mid 1 \end{array}} + \frac{\begin{array}{c} \mid \\ \hline \mid \\ \circ\mid\circ \end{array}}{\begin{array}{c} -1\mid -1 \\ \hline 0\mid -1 \end{array}} \rightarrow \frac{\begin{array}{c} -\mid \\ \hline \mid \\ \circ\circ\mid\circ\circ \end{array}}{\begin{array}{c} 1-1\mid -\tfrac{1}{2} \\ \hline 0\mid -\tfrac{1}{2} \end{array}} \rightarrow \frac{\begin{array}{c} -\mid \\ \hline \mid \\ \circ\mid \end{array}}{\begin{array}{c} 0\mid 0 \\ \hline 0\mid 0 \end{array}} + \frac{\begin{array}{c} \mid \\ \hline \mid \\ \mid\circ \end{array}}{\begin{array}{c} 0\mid 0 \\ \hline -1\mid -1 \end{array}} + \frac{\begin{array}{c} \mid \\ \hline \mid \\ \circ\mid \end{array}}{\begin{array}{c} 0\mid 0 \\ \hline 0\mid 0 \end{array}} + \frac{\begin{array}{c} -\mid \\ \hline \mid \\ \mid\circ \end{array}}{\begin{array}{c} 1\mid \tfrac{1}{2} \\ \hline 0\mid \tfrac{1}{2} \end{array}}$$

unobserved particles

Straight-forward recombinations.

τ DECAY MODE	Fraction (Γ_i/Γ)	Scale factor/ Confidence level	p (MeV/c)
$\pi^-\pi^+\pi^-\nu_\tau(ex.K^0)$	(9.03± 0.06)%	S = 1.2	861
$\pi^-\pi^+\pi^-\nu_\tau(ex.K^0,\omega)$	(8.99± 0.06)%	S = 1.2	861
$h^-h^-h^+\geq 1$neutrals ν_τ	(5.38± 0.07)%	S = 1.2	--
$h^-h^-h^+\geq 1\pi^0\nu_\tau(ex.~K^0)$	(5.08± 0.06)%	S = 1.1	--
$h^-h^-h^+\pi^0\nu_\tau$	(4.75± 0.06)%	S = 1.2	834
$h^-h^-h^+\pi^0\nu_\tau(ex.K^0)$	(4.56± 0.06)%	S = 1.2	834
$h^-h^-h^+\pi^0\nu_\tau(ex.K^0,\omega)$	(2.79± 0.08)%	S = 1.2	834
$\pi^-\pi^+\pi^-\pi^0\nu_\tau$	(4.61± 0.06)%	S = 1.1	834

```
T⁻       g°⁻      g°              q.p.
1776..   >0i      >0i
-|        |        |               -|
_|_  +   _|_  +   _|_  →          _|_      →
 |       -o|o-    o|o             -oo|oo-

0|-½     3|2      -1|-1           3-1|½
0|-½     0|2      0|-1            0|½
         |                   |
         unobserved particles


π⁻      π⁺       π⁻       π⁰       ν_T  >0i
139...  139...   139...   134...   <18.2
 |        |        |        |        -|
_|_  +   _|_  +   _|_  +   _|_  +   _|_
o|        |o      o|       -|-       |o

0|0      0|0      0|0      1|0      1|½
0|0      0|0      0|0      0|0      0|½
```

A g°⁻ graviton and a g° graviton are gravitationally attracted to a tauon. They form a quasi-particle with it. Then there are straight-forward recombinations.

τ DECAY MODE	Fraction (Γ_i/Γ)	Scale factor/ Confidence level	p (MeV/c)
$\pi^-\pi^+\pi^-\pi^0\nu_\tau(\text{ex.K}^0)$	(4.48± 0.06)%	S = 1.1	834
$\pi^-\pi^+\pi^-\pi^0\nu_\tau(\text{ex.K}^0,\omega)$	(2.70± 0.08)%	S = 1.2	834
$h^-(\rho\pi)^0\nu_\tau$	(2.88± 0.35)%	--	--

$(\rho^-\pi^+)^0$ is equivalent to ρ^0--echon wise.

τ^- 1776..	g^- >0i	g^0 >0i	g^0 >0i	q.p.	π^- 139...	$\rho(770)^0$ 775.4	ν_T >0i <18.2

$$\frac{-|}{\frac{|}{|}} + \frac{|}{\frac{|}{-|-}} + \frac{|}{\frac{|}{o|o}} + \frac{|}{\frac{|}{o|o}} \rightarrow \frac{-|}{\frac{|}{-oo|oo-}} \rightarrow \frac{|}{\frac{|}{o|}} + \frac{|}{\frac{|}{-o|o-}} + \frac{-|}{\frac{|}{|o}}$$

$\frac{0|-\frac{1}{2}}{0|-\frac{1}{2}}$	$\frac{-1|-2}{0|-2}$	$\frac{1|1}{0|1}$	$\frac{-1|-1}{0|-1}$	$\frac{1-2|-2\frac{1}{2}}{0|-2\frac{1}{2}}$	$\frac{0|0}{-1|-1}$	$\frac{1-1|-1}{-1|-2}$	$\frac{1|\frac{1}{2}}{0|\frac{1}{2}}$

unobserved particles

Straight-forward recombinations involving a $\rho(770)^0$.

τ DECAY MODE	Fraction (Γ_i/Γ)	Scale factor/ Confidence level	p (MeV/c)
$h^-\rho\pi^0\nu_\tau$	(1.35± 0.20)%		--

τ^- 1776..	g^0 >0i	g^0 >0i	g^- >0i	g^- >0i	q.p.

$$\frac{-|}{\frac{|}{|}} + \frac{|}{\frac{|}{o|o}} + \frac{|}{\frac{|}{o|o}} + \frac{|}{\frac{|}{-|-}} + \frac{|}{\frac{|}{-|-}} \rightarrow \frac{-|}{\frac{|}{--oo|oo--}} \rightarrow$$

$\frac{0|-\frac{1}{2}}{0|-\frac{1}{2}}$	$\frac{1|1}{0|1}$	$\frac{1|1}{0|1}$	$\frac{-1|-2}{0|-2}$	$\frac{-1|-2}{0|-2}$	$\frac{2-2|-2\frac{1}{2}}{0|-2\frac{1}{2}}$

unobserved particles

```
   π⁻         ρ⁰          π⁰       vᴛ >0i
 139...     775...     134...     <18.2

   |           |          |         -|
  ___         ___        ___       ___
   |     +     |     +    |     +    |
  ___         ___        ___       ___
  o|        -o|o-        -|-        |o

 0|0       1-1|-1        1|0       1|½
 ___       ____         ____       ___
-1|-1       0|-1        -1|-1      0|½
```

Two g° and two g⁻ gravitons gravitationally combine with a tauon to form a quasi-particle. The - echons and o echons collide and exchange their orbital spins. Subsequently there are straight forward recombinations.

τ⁻ DECAY MODE	Fraction (Γᵢ/Γ)	Scale factor/ Confidence level	p (MeV/c)
h⁻ρ⁺h⁻vₜ	(4.5 ± 2.2)x 10⁻³		--

```
   τ⁻        g⁰⁻      g⁰      q.p.          π⁻        ρ⁺        π⁻  vᴛ >0i
 1776.. >0i   >0i     >0i                 139...    775.4   139...<18.2

  -|          |        |        -|          |          |         |     -|
 ___         ___      ___      ____        ___        ___       ___   ___
  |     +     |    +   |   →    |     →     |     +     |    +    |  +  |
 ___         ___      ___      ____        ___        ___       ___   ___
  |         -o|o-     o|o    -oo|oo-        o|        -|o-       o|    |o

 0|-½       3|2      -1|-1    3-1|½         0|0  1-1|-1         0|0    1|½
 ___        ___      ____     ____         ___  ____           ___    ___
 0|-½       0|2       0|-1     0|½         1|1   0|-1          0|0    0|½
            |_____|
           unobserved particles
```

A g⁰⁻ and a g⁰ graviton combine with a tauon to form a quasi-particle. There are straight-forward recombinations. The resultant ρ⁺ has net zero orbital spin and -1 net particle spin, due to the intrinsic spin of the - echons. One π⁻ particle carries off +1 positional-kinetic angular momentum.

τ DECAY MODE	Fraction (Γ$_i$/Γ)	Scale factor/ Confidence level	p (MeV/c)
h⁻ρ⁻h⁺ν$_τ$	(1.17± 0.23)%		--

τ⁻	g°⁻	g°	q.p.	π⁻	ρ⁻	π⁺	ν$_τ$ >0i
1776..	>0i	>0i		139...	775.4	139...	<18.2

$$\frac{-|}{|} \; + \; \frac{|}{-o|o-} \; + \; \frac{|}{o|o} \; \rightarrow \; \frac{-|}{-oo|oo-} \; \rightarrow \; \frac{|}{o|} \; + \; \frac{|}{-o|-} \; + \; \frac{|}{|o} \; + \; \frac{-|}{|o}$$

0│-½	3│2	-1│-1	3-1│½	0│0	1-1│-1	0│0	1│½
0│-½	0│2	0│-1	0│½	1│1	0│-1	0│0	0│½

|_____|
        unobserved particles

Same as previous decay mode, except the ρ⁻ has the negative o echon, and the π⁺ has the positive o echon.

τ DECAY MODE	Fraction (Γ$_i$/Γ)	Scale factor/ Confidence level	p (MeV/c)
h⁻h⁻h⁺2π⁰ν$_τ$	(5.04 ± 0.32)x 10⁻³		797
h⁻h⁻h⁺2π⁰ν$_τ$(ex.K⁰)	(4.94 ± 0.32)x 10⁻³		797
h⁻h⁻h⁺2π⁰ν$_τ$(ex.K⁰,ϖ,η)	(9 ± 4)x 10⁻⁴		797

τ⁻	g°	g°	g⁻	g⁻	q.p.
1776..	>0i	>0i	>0i	>0i	

$$\frac{-|}{|} \; + \; \frac{|}{o|o} \; + \; \frac{|}{o|o} \; + \; \frac{|}{-|-} \; + \; \frac{|}{-|-} \; \rightarrow \; \frac{-|}{--oo|oo--} \; \rightarrow$$

0│-½	1│1	1│1	-1│-2	-1│-2	2-2│-2½
0│-½	0│1	0│1	0│-2	0│-2	0│-2½

|____ unobserved particles ____|

π⁻	π⁻	π⁺	π⁰	π⁰	ν$_τ$ >0i
139...	139...	139...134...	134...		<18.2

$$\frac{|}{o|} \; + \; \frac{|}{o|} \; + \; \frac{|}{|o} \; + \; \frac{|}{-|-} \; + \; \frac{|}{-|-} \; + \; \frac{-|}{|o}$$

0│0	0│0	0│0	1│0	1│0	1│½
0│0	-1│-1	0│0	-1│-1	-1│-1	0│½

Two g^o gravitons and two g^- gravitons gravitationally combine with the tauon, forming a quasi-particle with it. The graviton echons collide and reflect their orbital spins. Subsequently there are straight-forward recombinations.

τ DECAY MODE	Fraction (Γ_i/Γ)	Scale factor/ Confidence level	p (MeV/c)
$h^-h^-h^+3\pi^0\nu_\tau$	$(2.3 \pm 0.6) \times 10^{-4}$	S = 1.2	749

τ^-	g^{o-}	g^{o-}	g^-	q.p.
1776..	>0i	>0i	>0i	

$$
\begin{array}{c|c|c|c} \text{-|} & \text{|} & \text{|} & \text{|} \\ \hline \text{|} & \text{|} & \text{|} & \text{|} \\ \hline \text{|} & \text{-o|o-} & \text{-o|o-} & \text{-|-} \end{array}
$$

0 | -½	3 | 2	3 | 2	-1 | -2	6-1 | 1½
0 | -½	0 | 2	0 | 2	0 | -2	0 | 1½

|unobserved particles |

π^-	π^-	π^+	π^0	π^0	π^0	ν_τ >0i
139..	139..	139..	134..	134..	134..	<18.2

0 | 0	0 | 0	0 | 0	1 | 0	1 | 0	1 | 0	1 | ½
0 | 0	0 | 0	0 | 0	1 | 1	0 | 0	0 | 0	0 | ½

Outgoing positional-kinetic angular momentum makes this decay mode possible. Three gravitons join the tauon in forming a quasi-particle. With graviton echon collisions and reflecting of orbital spins, the particles have straight forward recombinations.

τ DECAY MODE	Fraction (Γ_i/Γ)	Scale factor/ Confidence level	p (MeV/c)
$K^-h^+h^-\geq 0$neutrals ν_τ	$(6.24 \pm 0.24) \times 10^{-3}$	S = 1.5	794
$K^-\pi^+\pi^-\geq 0$neutrals ν_τ	$(4.78 \pm 0.21) \times 10^{-3}$	S = 1.3	794
$K^-\pi^+\pi^-\nu_\tau$	$(3.41 \pm 0.16) \times 10^{-3}$	S = 1.8	794
$K^-\pi^+\pi^-\nu_\tau(ex.K^0)$	$(2.87 \pm 0.16) \times 10^{-3}$	S = 2.1	794

--

τ DECAY MODE	Fraction (Γ_i/Γ)	Scale factor/ Confidence level	p (MeV/c)
$K^-\pi^+\pi^-\pi^0\nu_\tau$	(1.35 $\pm$ 0.14)x 10^{-3}		763
$K^-\pi^+\pi^-\pi^0\nu_\tau$(ex.$K^0$)	(8.1 $\pm$ 1.2)x 10^{-4}		763

```
   τ⁻           g°⁻           g°            q.p.
 1776...        >0i           >0i
  -|             |             |             -|
  ─┼─    +     ─┼─    +      ─┼─     →       ─┼─     →
   |           -o|o-          o|o          -o|oo-

 0|-½          3|2          -1|-1          3-1|½
 ─────         ───          ─────          ─────
 0|-½          0|2           0|-1           0|½
  |                                      |
            unobserved particles

  K⁻          π⁺           π⁻           π⁰        ν_τ >0i
 493...       139...        139...        134...       <18.2
  |            |            |             |            -|
 ─┼─    +    ─┼─    +     ─┼─    +      ─┼─    +      ─┼─
 o|           |o          o|           -|-            |o

 0|0         0|0          0|0          1|0           1|½
 ───         ───          ───          ───           ───
 0|0         0|0          0|0          0|0           0|½
```

A o echon is knocked to the +1 state in the formation of the quasi-particle. Straight-forward recombinations.

τ DECAY MODE	Fraction (Γ_i/Γ)	Scale factor/ Confidence level	p (MeV/c)
$K^-K^+\pi^-\geq$0neut. ν_τ	(1.46 $\pm$ 0.06)x 10^{-3}	S = 1.6	685
$K^-K^+\pi^-\nu_\tau$	(1.40 $\pm$ 0.05)x 10^{-3}	S = 1.7	685

τ^-	g^o	g^o	q.p.	K^-	K^+	π^-	$v_\tau > 0i$
1776..	>0i	>0i		493...	493...	139...	<18.2

$$\frac{-|}{\underset{|}{\underline{\quad|\quad}}} + \frac{|}{\underset{o|o}{\underline{\quad|\quad}}} + \frac{|}{\underset{o|o}{\underline{\quad|\quad}}} \rightarrow \frac{-|}{\underset{o|o}{\underline{o|o}}} \rightarrow \frac{|}{\underset{|}{\underline{o|}}} + \frac{|}{\underset{|}{\underline{|o}}} + \frac{|}{\underset{o|}{\underline{\quad|\quad}}} + \frac{-|}{\underset{|o}{\underline{\quad|\quad}}}$$

| $\frac{0\,|-\frac{1}{2}}{0\,|-\frac{1}{2}}$ | $\frac{1\,|\,1}{0\,|\,1}$ | $\frac{-1\,|-1}{0\,|-1}$ | $\frac{1-1\,|-\frac{1}{2}}{0\,|-\frac{1}{2}}$ | $\frac{0\,|\,0}{-1\,|-1}$ | $\frac{0\,|\,0}{0\,|\,0}$ | $\frac{0\,|\,0}{0\,|\,0}$ | $\frac{1\,|\frac{1}{2}}{0\,|\frac{1}{2}}$ |

|_____|
unobserved particles

A pair of o echons from a g^o graviton are knocked to the +1 state as the gravitons combine with the tauon to form a quasi-particle. Straight-forward recombinations.

τ DECAY MODE	Fraction (Γ_i/Γ)	Scale factor/ Confidence level	p (MeV/c)
$K^-K^+\pi^-\pi^0 v_\tau$	$(6.1 \pm 2.5) \times 10^{-5}$	$S = 1.4$	618

τ^-	g^{o-}	g^o	q.p.
1776..	>0i	>0i	

$$\frac{-|}{\underset{|}{\underline{\quad|\quad}}} + \frac{|}{\underset{-o|o-}{\underline{\quad|\quad}}} + \frac{|}{\underset{o|o}{\underline{\quad|\quad}}} \rightarrow \frac{-|}{\underset{-o|o-}{\underline{o|o}}} \rightarrow$$

| $\frac{0\,|-\frac{1}{2}}{0\,|-\frac{1}{2}}$ | $\frac{3\,|\,2}{0\,|\,2}$ | $\frac{-1\,|-1}{0\,|-1}$ | $\frac{3-1\,|\frac{1}{2}}{0\,|\frac{1}{2}}$ |

|_____|
unobserved particles

K^-	K^+	π^-	π^0	$v_\tau > 0i$
493...	493...	139...	134...	<18.2

$$\frac{|}{\underset{|}{\underline{o|}}} + \frac{|}{\underset{|}{\underline{|o}}} + \frac{|}{\underset{o|}{\underline{\quad|\quad}}} + \frac{|}{\underset{-|-}{\underline{\quad|\quad}}} + \frac{-|}{\underset{|o}{\underline{\quad|\quad}}}$$

| $\frac{0\,|\,0}{0\,|\,0}$ | $\frac{0\,|\,0}{0\,|\,0}$ | $\frac{0\,|\,0}{0\,|\,0}$ | $\frac{1\,|\,0}{0\,|\,0}$ | $\frac{1\,|\frac{1}{2}}{0\,|\frac{1}{2}}$ |

A negative and a positive o echon are knocked to the +1 state as a g^{o-} and a g^o graviton combine with a tauon to form a quasi-particle. Then there are straight-forward recombinations.

τ DECAY MODE	Fraction (Γ_i/Γ)	Scale factor/ Confidence level	p (MeV/c)
$e^-e^-e^+\bar{v_e}v_\tau$	$(2.8 \pm 1.5) \times 10^{-5}$		888

τ^-	g^{o-}	g^-	q.p.
1776..	>0i	>0i	

$$ -| \qquad | \qquad | \qquad -| $$
$$ = + = + = \rightarrow = \rightarrow $$
$$ | \qquad -o|o- \qquad -|- \qquad --o|o-- $$

$0|-\frac{1}{2}$	$3|2$	$-1|-2$	$3-1|-\frac{1}{2}$
$0|-\frac{1}{2}$	$0|2$	$0|-2$	$0|-\frac{1}{2}$

| unobserved particles |

e^-	e^-	e^+	$\bar{v}_e$	v_τ >0i
.510..	.510..	.510..	<.000010	<18.2

$0|-\frac{1}{2}$	$0|-\frac{1}{2}$	$0|-\frac{1}{2}$	$1|\frac{1}{2}$	$1|\frac{1}{2}$
$0|-\frac{1}{2}$	$0|-\frac{1}{2}$	$0|-\frac{1}{2}$	$0|\frac{1}{2}$	$0|\frac{1}{2}$

A g^{o-} graviton and a g^- graviton combine with a tauon to form a quasi-particle. There are straight-forward recombinations. The o echons go out in the neutrino and anti-neutrino.

τ DECAY MODE	Fraction (Γ_i/Γ)	Scale factor/ Confidence level	p (MeV/c)
Modes with five charged particles			
$3h^-2h^+\geq0$ neutrals v_τ (ex. $K^0_s \rightarrow \pi^-\pi^+$) ("5-prong")	$(1.02 \pm 0.04) \times 10^{-3}$	S = 1.1	794
$3h^-2h^+v_\tau$(ex.K^0)	$(8.39 \pm 0.35) \times 10^{-4}$	S = 1.1	794

```
  τ⁻        g°        g°        g°         q.p.
1776..     >0i       >0i       >0i
 -|         |         |         |           -|
 _|_   +   _|_   +   _|_   +   _|_    →     _|_        →
  |        o|o       o|o       o|o        ooo|ooo
```

```
0|-½      1|1       -1|-1      1|1        2-1|½
0|-½      0|1        0|-1      0|1         0|½
  |_____|
            unobserved particles
```

```
  π⁻        π⁻        π⁻        π⁺        π⁺        vₜ >0i
139...     139...     139...    139...    139...    <18.2
  |         |         |         |         |          -|
 _|_   +   _|_   +   _|_   +   _|_   +   _|_   +    _|_
 o|        o|        o|         |o        |o         |o
```

```
0|0       0|0       0|0        0|0        0|0        1|½
0|0       1|1       -1|-1      0|0        0|0        0|½
```

Orbital spins of g° gravitons go out as positional-kinetic angular momentum for two of five pions and orbital spin of a tauon neutrino.

τ DECAY MODE	Fraction (Γᵢ/Γ)	Scale factor/ Confidence level	p (MeV/c)
$3h⁻2h⁺\pi⁰v_\tau(ex.K⁰)$	(1.78 ± 0.27)x 10⁻⁴		746

```
  τ⁻        g°⁻        g°        g°         q.p.
1776..     >0i        >0i       >0i
 -|         |          |         |           -|
 _|_   +   _|_   +    _|_   +   _|_    →     _|_        →
  |       -o|o-       o|o       o|o       -ooo|ooo-
```

```
0|-½      3|2       -1|-1      -1|-1      3-2|-½
0|-½      0|2        0|-1      0|-1        0|-½
  |_____|
            unobserved particles
```

π^-		π^-		π^-		π^+		π^+		π^0		ν_τ >0i
139...		139...		139...		139...		139...		134...		<18.2

										-		-
+		+		+		+		+		+		
o		o		o			o		o		-	-		o

0	0	0	0	0	0	0	0	0	0	1	0	1	½
0	0	0	0	0	0	0	0	0	0	-1	-1	0	½

Straight-forward recombinations.

τ^- DECAY MODE	Fraction (Γ_i/Γ)	Scale factor/ Confidence level	p (MeV/c)
	Miscellaneous other allowed modes		
$(5\pi)^-\nu_\tau$	$(7.6 \pm 0.5)\times 10^{-3}$		800
$K^*(892)^-\geq 0h^0\geq 0K^0_L\nu_\tau$	$(1.42\pm 0.18)\%$	S = 1.4	665
$K^*(892)^-\nu_\tau$	$(1.20 \pm 0.07)\%$	S = 1.8	665

τ^-		g^-		g^0		q.p.		$K^*(892)^-$		ν_τ >0i
1776..		>0i		>0i				895.5		<18.2

-						-				-
+		+		→		→		+		
		-	-	o	o	o		o		o
						-	o-	-	-	

0	-½	-1	-2	1	1	1-1	-1½	1-1	-1	1	½
0	-½	0	-2	0	1	0	-1½	-1	-2	0	½

unobserved particles

The tauon knocks a o echon from the g^0 graviton to the +1 state as they and a g^- graviton combine to form a quasi-particle. Straight-forward recombinations occur. (For further information on the structure of the $K^*(892)^-$ particle, see Appendix B, Structure of Known Particles.)

	Scale factor/	p
τ DECAY MODE Fraction (Γ$_i$/Γ) Confidence level	(MeV/c)	

	Scale factor/ Confidence level	p (MeV/c)

$K^*(892)^0 K \geq 0$neutrals v_τ (3.2 ± 1.4) x 10^{-3} 542
$K^*(892)^0 K^- v_\tau$ (2.1 ± 0.4) x 10^{-3} 542

τ⁻	g°	g°	g⁻	q.p.	K*(892)⁰	K⁻	v$_T$ >0i
1776.. >0i	>0i	>0i			895.5	493...	<18.2

$$-| \qquad | \qquad | \qquad | \qquad -| \qquad | \qquad | \qquad -|$$

(recombination diagram)

$\frac{0\,|-½}{0\,|-½}$	$\frac{1\,|1}{0\,|1}$	$\frac{1\,|1}{0\,|1}$	$\frac{-1\,|-2}{0\,|-2}$	$\frac{2-1\,|-½}{0\,|-½}$	$\frac{1-1\,|-1}{0\,|-1}$	$\frac{0\,|0}{0\,|0}$	$\frac{1\,|½}{0\,|½}$

                unobserved particles

A g⁻ graviton and two g° gravitons combine with a tauon to form a quasi-particle. A pair of o echons is knocked up to the +1 state in the process. The tauon neutrino gets the +1 spin from one g° graviton. The K*(892)⁰ particle receives the +1 and -1 spins of the other two gravitons. There are straight-forward recombinations.

	Scale factor/	p
τ DECAY MODE Fraction (Γ$_i$/Γ) Confidence level		(MeV/c)

$\overline{K}^*(892)^0 \pi \geq 0$neutrals v_τ (3.8 ± 1.7) x 10^{-3} 655

$\overline{K}^*(892)^0 \pi^- v_\tau$ (2.2 ± 0.5) x 10^{-3} 655

τ⁻	g°	g°	g⁻	q.p.	$\overline{K}$*(892)⁰	π⁻	v$_T$ >0i
1776.. >0i	>0i	>0i			895.5	139...	<18.2

(recombination diagram)

$\frac{0\,|-½}{0\,|-½}$	$\frac{1\,|1}{0\,|1}$	$\frac{1\,|1}{0\,|1}$	$\frac{-1\,|-2}{0\,|-2}$	$\frac{2-1\,|-½}{0\,|-½}$	$\frac{1-1\,|-1}{0\,|-1}$	$\frac{0\,|0}{0\,|0}$	$\frac{1\,|½}{0\,|½}$

                unobserved particles

This is very similar to the previous mode, except only one o echon is bumped to the +1 state.

			Scale factor/	p
τ DECAY MODE		Fraction (Γ$_i$/Γ)	Confidence level	(MeV/c)

$(\overline{K}*(892)\pi)^-v_\tau \rightarrow$ (1.0 ± 0.4)x 10^{-3} --

$\pi^-\overline{K}^0\pi^0 v_\tau$

τ⁻	g⁺	g°	g⁻	q.p.
1776..	>0i		>0i	

```
  -|          |           |           |           -|
 _|_    +    _|_    +    _|_    +    _|_    →    ___|___   →
   |         +|+         o|o         -|-          -+o|o+-
```

0|-½	1|2	1|1	-1|-2	2-1|½
0|-½	0|2	0|1	0|-2	0|½

```
         |_____|
```
                 unobserved particles

$(\overline{K}*(892)\pi)^-$ v_τ >0i   π⁻ $\overline{K}^0_S$ π⁰ v_τ >0i
                      <18.2   139.. 497..   134..   <18.2

```
     |          -|           |           |          |           -|
   __o|    +    _|_    →    _|_    +   _|_   +    _|_    +    _|_
   -+|+-         |o         o|         +|+        -|-          |o
```

1-1|0	1|½	0|0	-1|0	1|0	1|½
0|0	0|½	0|0	0|0	0|0	0|½

Three gravitons of different varieties combine with a tauon to form a quasi-particle. The tauon knocks one o echon from one graviton to the +1 state. The quasi-particle first divides into a tau neutrino and a net negative composite particle composed of an anti K*(892)⁻ and a π⁰ particle. The ensemble further divides into a π⁻, an anti K⁰, a π⁰, and a tau neutrino when the o echon and two + echons trade energy states.

τ DECAY MODE	Fraction (Γ_i/Γ)	Scale factor/ Confidence level	p (MeV/c)
$K_1(1270)^-\nu_\tau$	(4.7 ± 1.1)x 10^{-3}		433

```
  τ⁻        g⁺      g°          q.p.1      q.p.2    K₁(1270)⁻  ν_T  >0i
1776..      >0i     >0i                             1272       <18.2
 -|          |       |           -|         -|         |        -|
 _|    +    _|      _|     →    O|    →    O|    →   O|    +    _|
  |        +|+      o|o         +|o+       -|o+      -|+        |o

0|-½      1|2     -1|-1       2-1|1½      2|1½       1|1       1|½
0|-½      0|2      1|0        0|1½        0|1½       0|1       0|½
 |          _____|
   unobserved
   particles
```

A g^+ and a g° graviton are gravitationally attracted to a tauon, combine with it, and form a quasi-particle. The tauon echon knocks the negative o echon up to the +1 state. The -1 orbital spin in the incoming g° graviton knocks over a + echon of the other graviton to a - echon. Thus the second quasi-particle is formed, which redivides into a $K_1(1270)^-$ particle and a tauon neutrino.

τ DECAY MODE	Fraction (Γ_i/Γ)	Scale factor/ Confidence level	p (MeV/c)
$K_1(1400)^-\nu_\tau$	(1.7 ± 2.6)x 10^{-3}	S = 1.7	335

```
  τ⁻        g⁻       g°              q.p.      K₁(1400)⁻   ν_T  >0i
1776..      >0i      >0i                       1403        <18.2
 -|          |        |               -|          |         -|
 _|    +    _|   +   _|     →       -O|    →    -O|    +    _|
  |        -|-       o|o             |o-         |-          |o

0|-½      -1|-2     1|1          1-1|-1½      1-1|-1       1|½
0|-½      0|-2      0|1          0|-1½        -1|-2        0|½
 |          _____|
   unobserved
   particles
```

A g^- and g° graviton combine with a tauon to form a quasi-particle, which redivides into a $K_1(1400)^-$ and a tauon neutrino. Straight-forward recombinations.

			Scale factor/	p
τ^- **DECAY MODE**		Fraction (Γ_i/Γ)	Confidence level	(MeV/c)

$\eta\pi^-\pi^0\nu_\tau$ $\qquad$ (1.81 ± 0.24)x 10^{-3} $\qquad\qquad$ 778

τ^-	g^+	g^0	g^-	g^-	q.p.
1776..	>0i	>0i	>0i	>0i	

$$
\begin{array}{c}-|\\ \hline \\ \hline | \end{array} + \begin{array}{c}|\\ \hline \\ \hline +|+ \end{array} + \begin{array}{c}|\\ \hline \\ \hline \circ|\circ \end{array} + \begin{array}{c}|\\ \hline \\ \hline -|- \end{array} + \begin{array}{c}|\\ \hline \\ \hline -|- \end{array} \rightarrow \begin{array}{c}-|\\ \hline \\ \hline --+\circ|\circ+-- \end{array} \rightarrow
$$

$\dfrac{0\,|-\frac{1}{2}}{0\,|-\frac{1}{2}}$	$\dfrac{1\,|\,2}{0\,|\,2}$	$\dfrac{1\,|\,1}{0\,|\,1}$	$\dfrac{-1\,|-2}{0\,|-2}$	$\dfrac{-1\,|-2}{0\,|-2}$	$\dfrac{2-2\,|-1\frac{1}{2}}{0\,|-1\frac{1}{2}}$

unobserved particles

η	π^-	π^0	ν_τ >0i
547.85	139...	134...	<18.2

$$
\begin{array}{c}|\\ \hline \\ \hline -+|+- \end{array} + \begin{array}{c}|\\ \hline \\ \hline \circ| \end{array} + \begin{array}{c}|\\ \hline \\ \hline -|- \end{array} + \begin{array}{c}-|\\ \hline \\ \hline |\circ \end{array}
$$

$\dfrac{1-1\,|\,0}{-1\,|-1}$	$\dfrac{0\,|\,0}{0\,|\,0}$	$\dfrac{1\,|\,0}{-1\,|-1}$	$\dfrac{1\,|\,\frac{1}{2}}{0\,|\,\frac{1}{2}}$

Straight forward recombinations.

			Scale factor/	p
τ^- **DECAY MODE**		Fraction (Γ_i/Γ)	Confidence level	(MeV/c)

$\eta\pi^-\pi^0\pi^0\nu_\tau$ $\qquad$ (1.4 ± 0.7)x 10^{-4} $\qquad\qquad$ 746

τ^-	g^{0-}	g^-	g^-	g^+	q.p.
1776..	>0i	>0i	>0i	>0i	

$$
\begin{array}{c}-|\\ \hline \\ \hline | \end{array} + \begin{array}{c}|\\ \hline \\ \hline -\circ|\circ- \end{array} + \begin{array}{c}|\\ \hline \\ \hline -|- \end{array} + \begin{array}{c}|\\ \hline \\ \hline -|- \end{array} + \begin{array}{c}|\\ \hline \\ \hline +|+ \end{array} \rightarrow \begin{array}{c}-|\\ \hline \\ \hline ---+\circ|\circ+--- \end{array} \rightarrow
$$

$\dfrac{0\,|-\frac{1}{2}}{0\,|-\frac{1}{2}}$	$\dfrac{3\,|\,2}{0\,|\,2}$	$\dfrac{-1\,|-2}{0\,|-2}$	$\dfrac{-1\,|-2}{0\,|-2}$	$\dfrac{1\,|\,2}{0\,|\,2}$	$\dfrac{4-2\,|-\frac{1}{2}}{0\,|-\frac{1}{2}}$

| unobserved particles |

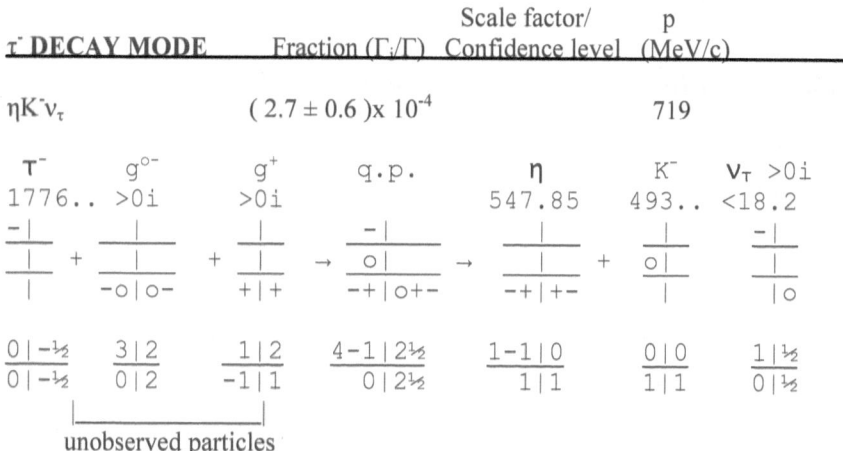

Mostly, there are straight forward recombinations.

τ^- DECAY MODE	Fraction (Γ_i/Γ)	Scale factor/ Confidence level	p (MeV/c)
$\eta K^- \nu_\tau$	$(2.7 \pm 0.6) \times 10^{-4}$		719

unobserved particles

A g^{o-} graviton and a g^+ graviton combine with a tauon to form a quasi-particle. There are straight-forward recombinations.

τ^- DECAY MODE	Fraction (Γ_i/Γ)	Scale factor/ Confidence level	p (MeV/c)
$\eta \pi^- \pi^+ \pi^- \nu_\tau$	$(2.3 \pm 0.5) \times 10^{-4}$		744

unobserved particles

η	π⁻	π⁺	π_	ν_T >0i

$$\frac{\eta}{547.85} \quad \frac{\pi^-}{139...} \quad \frac{\pi^+}{139...} \quad \frac{\pi_-}{139...} \quad \frac{\nu_T >0i}{<18.2}$$

│		│		│		│		−│
│ │	+	│ │	+	│ │	+	│ │	+	│ │
−+│+−		○│		│○		○│		│○

$$\frac{1-1│0}{0│0} \qquad \frac{0│0}{0│0} \qquad \frac{0│0}{-1│-1} \qquad \frac{0│0}{0│0} \qquad \frac{1│½}{0│½}$$

Four gravitons of various types combine with a tauon to form a quasi-particle.
There are straight-forward recombinations.

		Scale factor/	p
τ⁻ **DECAY MODE**	Fraction (Γ_i/Γ)	Confidence level	(MeV/c)

$f_1(1285)\pi^-\nu_\tau$	$(4.1 \pm 0.8) \times 10^{-4}$		408
$f_1(1285)\pi^-\nu_\tau \rightarrow$	$(1.3 \pm 0.4) \times 10^{-4}$		--
$\pi^-\pi^+\pi^-\nu_\tau$			

$$\frac{\tau^-}{1776..} \quad \frac{g^{o-}}{>0i} \quad \frac{g^{o+}}{>0i} \quad q.p.$$

−│		│		│		−│
│	+	│	+	│	→	│
│		−○│○−		+○│○+		−+○○│○○+−

$$\frac{0│-½}{0│-½} \qquad \frac{3│2}{0│2} \qquad \frac{-3│-2}{0│-2} \qquad \frac{3-3│-½}{0│-½}$$

|_____|

                   unobserved particles

$$\frac{f_1(1285)}{1281.8} \quad \frac{\pi^-}{139...} \quad \frac{\nu_T >0i}{<18.2}$$

│		│		−│
│	+	│	+	│	→
−+○│○+−		○│		│○

$$\frac{2-1│1}{0│1} \qquad \frac{0│0}{-1│-1} \qquad \frac{1│½}{-1│-½}$$

(Continued on next page.)

η		π⁻		π⁺		π_		ν_T >0i
547.85		139...		139...		139...		<18.2

$$\frac{\quad|\quad}{-+|+-} \;+\; \frac{\quad|\quad}{\circ\;|} \;+\; \frac{\quad|\quad}{|\;\circ} \;+\; \frac{\quad|\quad}{\circ\;|} \;+\; \frac{-\;|}{|\;\circ}$$

1-1 | 0	0 | 0	0 | 0	0 | 0	1 | ½
0 | 0	1 | 1	0 | 0	-1 | -1	-1 | -½

Two different gravitons combine with a tauon to form a quasi-particle, which first divides into a $f_1(1285)$, a pi minus, and a tauon neutrino. The $f_1(1285)$ further divides into an eta particle, a pi minus, and a pi plus.

τ⁻ DECAY MODE	Fraction (Γ_i/Γ)	Scale factor/ Confidence level	p (MeV/c)
h⁻ω≥0neutrals v_τ	(2.40 ± 0.09)%	S = 1.2	708
h⁻ωv_τ	(1.99 ± 0.08)%	S = 1.3	708

τ⁻		g⁺		g°		q.p.1		q.p.2		π⁻		ω		ν_T>0i
1776.. >0i		>0i		>0i						139...		782.65		<18.2

$$\frac{-|}{|} \;+\; \frac{\quad|\quad}{+|+} \;+\; \frac{\quad|\quad}{\circ\;|\;\circ} \;\rightarrow\; \frac{-|}{+\circ\;|\;\circ+} \;\rightarrow\; \frac{-|}{-\circ\;|\;\circ+} \;\rightarrow\; \frac{\quad|\quad}{\circ\;|} \;+\; \frac{\quad|\quad}{-\;|\;+} \;+\; \frac{-|}{|\;\circ}$$

0 | -½	1 | 2	-1 | -1	2-1 | 1½	2 | 1½	0 | 0	1 | 1	1 | ½
0 | -½	0 | 2	1 | 0	0 | 1½	0 | 1½	0 | 0	0 | 1	0 | ½

|unobserved particles|

Two gravitons combine with a tauon to form a quasi-particle. The minus orbital spin of the g° graviton knocks over a + echon to a - echon, forming the second quasi-particle. Subsequently there are straight-forward recombinations, including the ω particle.

τ⁻ DECAY MODE	Fraction (Γ_i/Γ)	Scale factor/ Confidence level	p (MeV/c)
h⁻ωπ⁰v_τ	(4.1 ± 0.4)x 10⁻³		684
h⁻ω2π⁰v_τ	(1.4 ± 0.5)x 10⁻⁴		644

These two modes are similar to the above decay scheme, except there are additional one or two g⁺ gravitons with -1 positional-kinetic angular momentum coming in, and one or two π⁰s with +1 positional-kinetic angular momentum going out.

Neutrinos

ν_e $J = \frac{1}{2}$

   The following results are obtained using neutrinos associated with e^+ or e^-.
Mass $m < 3$ eV. Interpretation of tritium beta decay experiments is
   complicated by anomalies near the endpoint, and the limits are not without
   ambiguity.
Mean life/mass, $\tau/m_\nu > 7 \times 10^9$ s/eV   (solar)
Mean life/mass, $\tau/m_{ve} > 300$ s/eV, CL $= 90\%$ (reactor)
Magnetic moment $\mu < 1.0 \times 10^{-10}$ μ_B, CL $= 90\%$

```
         Ve                                   Ve
      <.000003                             <.000003
         |                                    |
       __|__                                __|__
       - | o                                + | o  ,

       1 | ½                                -1 | -½
         |                                    |
```
                              or

depending on whether the electron neutrino has up or down spin. The down
spin alternative will be taken as valid as well as the up spin alternative in all the
data. For brevity, however, only the net positive spin alternatives are presented
in most cases.

ν_μ $\qquad$ $J = \frac{1}{2}$

Mass $m < 0.19$ MeV, CL = 90%
Mean life/mass, $\tau/m_\nu > 15.4$ s/eV, CL = 90%
Magnetic moment $\mu < 6.8 \times 10^{-10}$ μ_B, CL = 90%

$$\nu_\mu$$
$$<0.19$$

$$\frac{|}{\frac{-|}{|\;o}}$$

$$\frac{1\;|\;\frac{1}{2}}{|}$$

ν_τ $\qquad\qquad\qquad\qquad$ $J = \frac{1}{2}$

Mass $m < 18.2$ MeV, CL = 95%
Magnetic moment $\mu < 3.9 \times 10^{-7}$ μ_B, CL = 90%
Electric dipole moment $d < 5.2 \times 10^{-17}$ e cm, CL = 95%

$$\nu_\tau$$
$$<18.2$$

$$\frac{-|}{\frac{|}{|\;o}}$$

$$\frac{1\;|\;\frac{1}{2}}{|}$$

Appendix B

STRUCTURE OF KNOWN PARTICLES

The quark and lepton model of particle physics divides charges in quarks to ±2e/3 and ±e/3. The electrino model of particle physics does not do that. Instead, it divides charges in electrinos to ±e, ±e/2, ±e/4, and ±e/8. The electrino model of particle physics does not hold that the quark and lepton model of particle physics is correct. Nevertheless, to facilitate cross referencing with the existing data, this volume will employ quark model titles and classifications in the subsequent classification of particles.

The chonomic structures contained in the following material are the author's, but the particle data come from C. Amsler *et al.* (Particle Data Group), PL **B667**, 1 (2008) (URL: http//pdg.lbl.gov). and S. Eidelman *et al.* (Particle Data Group), Phys Lett B **592** (2004) (URL: http://pdg.lbl.gov). The author's chonomic structures in this appendix are induced from the following eight criteria: particle charge, spin, parity, mass, spin feasibility, preceding particles (to avoid duplication), decay schemes, and the Pauli Exclusion Principle. The use of isospin in the precursor data instead of the simple charge made the author's work difficult; so too the convention of listing any charge π as π, and decay products of baryons as N . . ., where N can stand for many different baryons. For accurate results, please change to precise reporting conventions. These results are highly valuable, and worth doing right.

This appendix is a draft, and may contain some errors. We hope to find and correct those errors when the masses of all the particles in this appendix are solved for. This appendix proves that all known matter, light, and gravitons can better be constructed of electrinos rather than quarks and leptons. All particles may be formulated with yachons and echons, or with +, -, o's, and ●'s. For the most part, in this appendix, n replaces a ● and a +. The key to understanding chonomics is in Chapter 10.

431

Particle state levels are listed at the left side of the chonomic structures. The bottom state level is called 0 or ground state.

In the higher chonomic structures there can appear the symbols no, which do not mean the word no, but neutron and pion.

This document, "Structure of Known Particles," is a necessary adjunct to *Prediction of the Masses of Every Known Particle, Step 2 and Step 3*. The investigator needs to know the particle structures he/she is calculating the masses for.

GAUGE AND HIGGS BOSONS

γ $I(J^{PC}) = 0,1(1^{--})$

Mass m < 1 x 10^{-18} eV
Charge q < 5 x 10^{-30} e
Mean life τ_γ = Stable

```
             γ
             0
   2         |
   1         |
   0       · | ·

           1 | 1
             |
```

g
or gluon

$$I(J^P) = 0(1^-)$$

Mass m = 0
SU(3) color octet

No electrino formulation of gluon in system.

W

J = 1
Charge = ±1 e
Mass m = 80.398 ± 0.025 GeV
$m_Z - m_W$ = 10.4 ± 1.6 GeV
$m_{W^+} - m_{W^-}$ = -0.2 ± 0.6 GeV
Full width Γ = 2.141 ± 0.041 GeV

```
                    W
                    ?
        2           |
        1          _|_
        0        - | o - ,

        1 -1 | -1
             |
```

The listed mass corresponds to energy at which system ionizes. Particle mass
is much less.

Z

J = 1
Charge = 0
Mass m = 91.1876 ± 0.0021 GeV
Full width Γ = 2.4952 ± 0.0023 GeV

```
                  Z
                  ?
        2         |
        1        _|_
        0      - | - ,

        1 -1 | -1
             |
```

The listed mass corresponds to energy at which system ionizes. Particle mass
is much less.

Higgs Bosons – H^0 and $H^\pm$, Searches for

H^0 Mass m > 114.4 GeV, CL = 95%

H_1^0 In Supersymmetric Models $\left(m_{H_1^0} < m_{H_2^0}\right)$

Mass m > 92.8 GeV, CL = 95%

A^0 Pseudoscalar Higgs Boson in Supersymmetric Models
Mass m > 93.4 GeV, CL = 95%   tanβ>0.4
$H^\pm$ Mass m > 79.3 GeV, CL = 95%

No formulation in system.

Unknown heavy or light bosons, searches for, are not covered in this appendix. This appendix covers known particles.

LEPTONS

e

$J = \frac{1}{2}$

Mass m = 0.510998910 ± 0.000000013 MeV
= (548.57990943 ± 0.00000023) x 10^{-6} u

$$\left(m_{e^+} - m_{e^-}\right)/m < 8 \times 10^{-8}, \; CL = 90\%$$

$$\left|q_{e^+} + q_{e^-}\right|/e < 4 \times 10^{-8}$$

Magnetic moment μ = 1.0011596521811 ± 0.0000000000007 μ_B

$$\left(g_{e^+} - g_{e^-}\right)/g_{average} = \left(-0.5 \pm 2.1\right) \times 10^{-12}$$

Electric dipole moment d = (0.07 ± 0.07) x 10^{-26} e cm
Mean life τ > 4.6 x 10^{26} yr, CL = 90%

```
              e
     0.510 998 910 MeV
  2        |
  1       _|_
  0       -|

        0|-½
          |
```

μ $J = \frac{1}{2}$
Mass m = 105.658367 ± 0.000004 MeV
      = 0.1134289256 ± 0.0000000029 u
Mean life τ = (2.197019 ± 0.000021) x 10^{-6} s   (S = 1.1)
Magnetic moment μ = 1.0011659208 ± 0.0000000006 eℏ/2m$_\mu$

```
            μ
     105.671 929 MeV
   2       |
   1      -|
   0       |

      0 | -½
         |
```

τ $J = \frac{1}{2}$
Mass m = 1776.84 ± 0.17 MeV

```
            τ
      1747.03 MeV
   2      -|
   1       |
   0       |

      0 | -½
         |
```

Neutrinos

ν_e $J = \frac{1}{2}$
Mass m >0i MeV
          < 0.000003 MeV
              ν_e >0i
              0
   2 |
   1 |
   0   - | o

      1 | ½
         |
```
```

ν_μ $J = \frac{1}{2}$
Mass m < 0.19 MeV, CL = 90%

ν_μ >0i
0

2 |
1       $-$|
0 | o

1 | ½
|

ν_τ $J = \frac{1}{2}$
Mass m < 18.2 MeV, CL = 95%

ν_τ >0i
0

2       $-$|
1 |
0 | o

1 | ½
|

QUARKS

No formulation of quarks in the electrino system.

ELECTRINOS

See Gordon L. Ziegler and Iris I Koch, *Prediction of the Masses of Every Known Particle (as of 2008) Step 2, Part 1* (http://benevolententerprises.org Book List), pp. 17, 18.

LIGHT UNFLAVORED MESONS
(S = C = B = 0)

$\pi^{\pm}$

$$I^G(J^P) = 1^-(0^-)$$

Mass m = 139.57018 ± 0.00035 MeV   (S = 1.2)

```
              π⁺
         137.002 202 MeV
      2        |
      1        |
      0        | o

           0 | 0
             |
```

π^0

$$I^G(J^{PC}) = 1^-(0^{-+})$$

Mass m = 134.9766 ± 0.0006 MeV   (S = 1.1)

```
              π⁰
         137.002 202 MeV
      2        |
      1        |
      0      - | -

           1 | 0
             |
```

η

$$I^G(J^{PC}) = 0^+(0^{-+})$$

Mass m = 547.853 ± 0.024 MeV

```
              η
         548.008 806 MeV
      2        |
      1        |
      0     -+ | +-

        1 -1 | 0
             |
```

---

f₀(600) $I^G(J^{PC}) = 0^+(0^{++})$
or σ

        Mass m = (400-1200) MeV
                f₀(600)
           548.008 807 MeV
        2 |
        1   _|_
        0   o | o

           1-1 | 0
               |

ρ(770)± $I^G(J^{PC}) = 1^+(1^{--})$
            Mass m = 775.49 ± 0.34 MeV

                    ρ⁺
           758.084 547 1 MeV
        2 |
        1   _|_
        0   - | o -

           1-1 | -1
               |

ρ(770)⁰ $I^G(J^{PC}) = 1^+(1^{--})$
        Mass m = 775.49 ± 0.34 MeV   (S = 1.8)

                    ρ⁰
           1096.017 616 MeV
        2 |
        1   _|_
        0   - o | o -

           1-1 | -1
               |

$\omega(782)$ $I^G(J^{PC}) = 0^-(1^{--})$

Mass m = 782.65 ± 0.12 MeV   (S = 1.9)

```
        ω(782)
          ?
  2      |
  1      |
  0     -|+

       -1|-1
         |
```

$\eta'(958)$ $I^G(J^{PC}) = 0^+(0^{-+})$
Mass m = 957.66 ± 0.24 MeV

```
        η'(958)
          ?
  2     |
  1     |
  0    -o|o-

      2-1|0
        |
```

$f_0(980)$ $I^G(J^{PC}) = 0^+(0^{++})$
Mass m = 980 ± 10 MeV

```
        f₀(980)
          ?
  2      |
  1     o|o
  0     o|o

      1-1|0
        |
```

$a_0(980)$ $\qquad$ $I^G(J^{PC}) = 1^-(0^{++})$

Mass $m = 984.7 \pm 1.2$ MeV $\qquad$ $(S = 1.5)$

```
        a₀(980)
          ?
2      __|__
1      __|__
0      -+|o+-

      1-1|0
        |
```

$\varphi(1020)$ $\qquad$ $I^G(J^{PC}) = 0^-(1^{--})$

Mass $m = 1019.455 \pm 0.020$ MeV $\qquad$ $(S = =1.1)$

```
        φ(1020)
          ?
2        _|_
1      -+o|o+
0       o|o-

      2-3|-1
        |
```

$h_1(1170)$ $\qquad$ $I^G(J^{PC}) = 0^-(1^{+-})$

Mass $m = 1170 \pm 20$ MeV

```
        h₁(1170)
          ?
2      __|__
1      __|__
0      -o|o-

       2|1
        |
```

---

b₁(1235) $I^G(J^{PC}) = 1^+(1^{+-})$

Mass m = 1229.5 ± 3.2 MeV   (S = 1.6)

```
         b₁(1235)
            ?
  2    ___|___
  1    ___|___
  0    -|o+

       -1|-1
         |
```

---

a₁(1260) $I^G(J^{PC}) = 1^-(1^{++})$
              Mass m = 1230 ± 40 MeV

```
         a₁(1260)
            ?
  2    ___|___
  1    ___|o__
  0    -o|o-

      1-1|-1
         |
```

---

f₂(1270) $I^G(J^{PC}) = 0^+(2^{++})$
              Mass m = 1275.1 ± 1.2 MeV   (S = 1.1)

```
         f₂(1270)
            ?
  2    ___|___
  1    -o|o-
  0    -o|o-

      2-2|-2
         |
```

$f_1(1285)$ $I^G(J^{PC}) = 0^+(1^{++})$

Mass m = 1281.8 ± 0.6 MeV   (S = 1.6)

```
          f₁(1285)
            ?
2          |
1         o | o
0        -o | o-

         2-2 | -1
            |
```

$\eta(1295)$ $I^G(J^{PC}) = 0^+(0^{-+})$

Mass m = 1294 ± 4 MeV   (S = 1.6)

```
          η(1295)
            ?
2          |
1          |
0        -o | o-

         2-1 | 0
            |
```

$\pi(1300)$ $I^G(J^{PC}) = 1^-(0^{-+})$

Mass m = 1300 ± 100 MeV

```
          π(1300)
            ?
2          |
1         | o
0        -o | o-

         2-1 | 0
            |
```

$a_2(1320)$ $\qquad$ $I^G(J^{PC}) = 1^-(2^{++})$

Mass m = 1318.3 ± 0.6 MeV   (S = 1.2)

```
         a₂(1320)
            ?
   2    ____|____
   1      - | o-
   0      -o | o-

        2-2 | -2
            |
```

$f_0(1370)$ $\qquad$ $I^G(J^{PC}) = 0^+(0^{++})$

Mass m = 1200 to 1500 MeV

```
         f₀(1370)
            ?
   2    ____|____
   1      o | o
   0     -+o | o+-

        2-2 | 0
            |
```

$\pi_1(1400)$ $\qquad$ $I^G(J^{PC}) = 1^-(1^{-+})$

Mass m =1351 ± 30 MeV   (S = 2.0)

```
         π₁(1400)
            ?
   2    ____|____
   1     ____|____
   0      -+ | o+-

        2-1 | 1
            |
```

$\eta(1405)$

$$I^G(J^{PC}) = 0^+(0^{-+})$$
$$\text{Mass } m = 1409.8 \pm 2.5 \text{ MeV} \quad (S = 2.2)$$

$\eta(1405)$

?

```
2      |
1     o│o
0     ─│─

    2-1│0
      |
```

$f_1(1420)$

$$I^G(J^{PC}) = 0^+(1^{++})$$
$$\text{Mass } m = 1426.4 \pm 0.9 \text{ MeV} \quad (S = 1.1)$$

$f_1(1420)$

?

```
2      |
1     o│o
0    ─+│+─

   1-2│-1
      |
```

$\omega(1420)$

$$I^G(J^{PC}) = 0^-(1^{--})$$

$$\text{Mass } m = 1400 \pm 1450 \text{ MeV}$$

$\omega(1420)$

?

```
2      |
1      |
0    ─o│o+

   1-2│-1
      |
```

$a_0(1450)$ $\qquad$ $I^G(J^{PC}) = 1^-(0^{++})$
$$\text{Mass m} = 1474 \pm 19 \text{ MeV}$$

```
          a₀(1450)
             ?
  2          |
  1        o | o
  0       -+ | +-

        2-2 | 0
             |
```

$\rho(1450)$ $\qquad$ $I^G(J^{PC}) = 1^+(1^{--})$

$$\text{Mass m} = 1465 \pm 25 \text{ MeV}$$

```
          ρ(1450)
             ?
  2          |
  1        - | -
  0          | o

        1-1 | -1
             |
```

$\eta(1475)$ $\qquad$ $I^G(J^{PC}) = 0^+(0^{-+})$

$$\text{Mass m} = 1476 \pm 4 \text{ MeV} \qquad (S = 1.3)$$

```
          η(1475)
             ?
  2          |
  1        o | o
  0        + | +

        -2+1 | 0
             |
```

$f_0(1500)$ $\qquad$ $I^G(J^{PC}) = 0^+(0^{++})$

Mass m = 1505 ± 6 MeV   (S = 1.3)

$f_0(1500)$

?

```
2        |
1      o | o
0    -+o|o+-

    2-2 | 0
        |
```

$f_2'(1525)$ $\qquad$ $I^G(J^{PC}) = 0^+(2^{++})$

Mass m = 1525 ± 5 MeV

$f_2'(1525)$

?

```
2      o | o
1      - | -
0    -o | o-

   2-2 | -2
        |
```

$\pi_1(1600)$ $\qquad$ $I^G(J^{PC}) = 1^-(1^{-+})$

Mass m = 1662^{+15}_{-11} MeV   (S = 1.2)

$\pi_1(1600)$

?

```
2        |
1      + | +
0    + | o+

   1-2 | 1
        |
```

$\eta_2(1645)$ $I^G(J^{PC}) = 0^+(2^{-+})$
Mass $m = 1617 \pm 5$ MeV $(S = 1.6)$

$\eta(1645)$

```
        ?
2       |
1     o | o
0     + | +

    2-1 | 2
        |
```

$\omega(1650)$ $I^G(J^{PC}) = 0^-(1^{--})$

Mass $m = 1670 \pm 30$ MeV

$\omega(1650)$

```
        ?
2       |
1     o | o
0    -o | o+

   1-2 | -1
        |
```

$\omega_3(1670)$ $I^G(J^{PC}) = 0^-(3^{--})$

Mass $m = 1667 \pm 4$ MeV

$\omega_3(1670)$

```
        ?
2       |
1     o | o
0    -o | o+

   -3 | -3
        |
```

$\pi_2(1670)$ $\qquad$ $I^G(J^{PC}) = 1^-(2^{-+})$
Mass m = 1672.4 ± 3.2 MeV

```
        π₂ (1670)
            ?
2           |
1         ○ | ○
0        ─ ○ | ○ ─

         1─2 | ─2
            |
```

$\varphi(1680)$ $\qquad$ $I^G(J^{PC}) = 0^-(1^{--})$

Mass m = 1680 ± 20 MeV

```
        φ (1680)
            ?
2           |
1        ○ | ○
0        ─ | ─

          2 | 1
            |
```

$\rho_3(1690)$ $\qquad$ $I^G (J^{PC}) = 1^+(3^{--})$
Mass m = 1688.8 ± 2.1 MeV

```
        ρ₃(1690)
            ?
2         ─ | ─
1           |
0         ○ | ○

         1─3 | ─3
            |
```

ρ(1700) $I^G(J^{PC}) = 1^+(1^{--})$
                   Mass m = 1720 ± 20 MeV

                        ρ(1700)
                          ?
            2   - | -
            1 |
            0   o | o

            1-1 | -1
                 |

---

f_j(1710) $I^G(J^{PC}) = 0^+(0^{++})$

               Mass m = 1724 ± 7 MeV   (S = 1.5)

                        f_j(1710)
                          ?
            2 |
            1   o | o
            0   -+o | o+-

            2-2 | 0
                 |

π(1800) $I^G(J^{PC}) = 1^-(0^{-+})$
                   Mass m = 1816 ± 14 MeV   (S = 2.3)

                        π(1800)
                          ?
            2 | o
            1   o | o
            0   -+o | o+-

            2-2 | 0
                 |

φ₃(1850) $I^G(J^{PC}) = 0^-(3^{--})$
Mass m = 1854 ± 7 MeV

```
        φ₃(1850)
          ?
 2        |
 1       o | o
 0       - | -

       -2 | -3
          |
```

π₂(1880) $I^G(J^{PC}) = 1^-(2^{-+})$
Mass m = 1895 ± 16 MeV

```
        φ₃(1880)
          ?
 2          | o
 1        o | o
 0      -+o | o+-

       4-2 | 2
          |
```

f₂(1950) $I^G(J^{PC}) = 0^+(2^{++})$

Mass m = 1944 ± 12 MeV   (S = 1.5)

```
        f₂(1950)
          ?
 2          |
 1       -o | o-
 0       -o | o-

       2-2 | -2
          |
```

$f_2(2010)$ $I^G(J^{PC}) = 0^+(2^{++})$

Mass $m = 2011^{+60}_{-80}$ MeV

```
        f₂(2010)
           ?
  2        |
  1      +o|o+
  0      +o|o+

        2-2|2
           |
```

$a_4(2040)$ $I^G(J^{PC}) = 1^-(4^{++})$
Mass $m = 2020 \pm 16$ MeV

```
        a₄(2040)
           ?
  2        |o
  1      -o|o-
  0      -o|o-

        1-3|-4
           |
```

$f_4(2050)$ $I^G(J^{PC}) = 0^+(4^{++})$
Mass $m = 2018 \pm 2.1$ MeV $(S = 2.1)$

```
        f₄(2050)
           ?
  2        |
  1       o|o
  0     -+o|o+-

        -4|-4
           |
```

$f_2(2300)$ $I^G(J^{PC}) = 0^+(2^{++})$
Mass m = 2297 ± 28 MeV
$f_2(2300)$
?

```
2    ___|___
1       o | o
0     -+o | o+-

     1-3 | -2
        |
```

$f_2(2340)$ $I^G(J^{PC}) = 0^+(2^{++})$
Mass m = 2339 ± 60 MeV
?

```
2    ___|___
1       o | o
0     -+o | o+-

     1-3 | -2
        |
```

This appears to be the identical particle as $f_2(2300)$. It has the same characteristics, the same decay products, and the mass of $f_2(2300)$ is within the error bar of the mass of $f_2(2340)$.

STRANGE MESONS
(S = ± 1, C = B = 0)

$K^\pm$ $I(J^P) = \frac{1}{2}(0^-)$
Mass m = 493.677 ± 0.016 MeV   (S = 2.8)

```
       K+
        ?
2     __|__
1      _|_o
0       |

      0 | 0
        |
```

K_S^0 $I(J^P) = \frac{1}{2}(0^-)$
                     Mass $m = 497.614 \pm 0.024$ MeV $(S = 1.6)$

$$K_S^0$$
                          ?
        2 |
        1   - | -
        0 |

                 1 | 0
                   |

K_L^0 $I(J^P) = \frac{1}{2}(0^-)$
                     Mass $m = 497.614 \pm 0.024$ MeV $(S = 1.6)$

$$K_L^0$$
                          ?
        2 |
        1   - |
        0 | -

                 1 | 0
                   |

$K^*(892)^\pm$ $I(J^P) = \frac{1}{2}(1^-)$
                     Mass $m = 891.66 \pm 0.26$ MeV

                     K* (892) $^+$
                          ?
        2 |
        1 | o
        0   - | -

              1 -1 | -1
                   |

K*(892)^0 I(J^P) = ½(1$^-$)

         Mass m = 896.00 ± 0.25 MeV   (S = 1.4)

```
        K*(892) 0
           ?
 2         |
 1         _|o
 0         o|

        -1|-1
           |
```

K$_1$(1270)   I(J^P) = ½(1$^+$)
        Mass m = 1272 ± 7 MeV

```
        K₁(1270)
           ?
 2         |
 1         _|o
 0        -o|+

        1-2|-1
           |
```

K$_1$(1400)   I(J^P) = ½(1$^+$)
        Mass m = 1403 ± 7 MeV

```
        K₁(1400)
           ?
 2         |
 1         _|o
 0        -o|-
        1-1|-1
           |
```

K*(1410)   I(J^P) = ½(1$^-$)

Mass m = 1414 ± 15 MeV   (S = 1.3)

```
            K* (1410)
               ?
      2        |
      1       o | o
      0       o | o

           1-2 | -1
               |
```

K_0^* (1430)   I(J^P) = ½(0$^+$)

Mass m = 1425 ± 50 MeV

```
          K_0^* (1430)
               ?
      2        |
      1        | o
      0       o |

           1-1 | 0
               |
```

K_2^*(1430)$^+$ I(J^P) = ½(2$^+$)

Mass m = 1425.6 ± 1.5 MeV   (S = 1.1)

```
          K_2^* (1430)^+
               ?
      2        |
      1       - | o-
      0       - | -

          1-1 | -2
               |
```

$K_2^*(1430)^0$ $I(J^P) = \frac{1}{2}(2^+)$

Mass m = 1432.4 ± 1.3 MeV

$$K_2^*(1430)^0$$

```
        ?
        |
2       |
1     - | o -
0     - o | -

    2 - 2 | -2
        |
```

K*(1680) $I(J^P) = \frac{1}{2}(1^-)$
Mass m = 1717 ± 27 MeV   (S = 1.4)

K* (1680)

```
        ?
        |
2     ──┼──
1      o | o
0     - o | o -

    1 - 1 | -1
        |
```

K₂(1770) $I(J^P) = \frac{1}{2}(2^-)$
Mass m = 1773 ± 8 MeV

K₂ (1770)

```
        ?
        |
2     ──┼──
1      o | o
0     - o | o -

    1 - 2 | -2
        |
```

$K_3^*(1780)$ $I(J^P) = \frac{1}{2}(3^-)$

Mass m = 1776 ± 7 MeV   (S = 1.1)

$K_3^*(1780)$

?

```
2        |
1        |   o
0      -o | -

      -2 | -3
         |
```

$K_2(1820)$ $I(J^P) = \frac{1}{2}(2^-)$

Mass m = 1816 ± 13 MeV

$K_2(1820)$

?

```
2        |
1        |   o
0      -o | -

     1-2 | -2
         |
```

$K_4^*(2045)$ $I(J^P) = \frac{1}{2}(4^+)$

Mass m = 2045 ± 9 MeV   (S = 1.1)

$K_4^*(2045)$

?

```
2        |
1     -o | o-
0     -o | o-

      -2 | -4
         |
```

CHARMED MESONS
(C = ± 1)

$D^\pm$ $I(J^P) = \frac{1}{2}(0^-)$
                        Mass m = 1869.62 ± 0.20 MeV   (S = 1.1)

D^+
?
2 | ○
1 |
0 |

0 | 0
|

D^0 $I(J^P) = \frac{1}{2}(0^-)$
                        Mass m = 1864.84 ± 0.17 MeV   (S = 1.1)

D^0
?
2   — | —
1 |
0 |

1 | 0
|

$D*(2007)^0$ $I(J^P) = \frac{1}{2}(1^-)$ I, J, P need confirmation
                        Mass m = 2006.97 ± 0.19 MeV   (S = 1.1)
D* (2007)^0
?
2   — | —
1 |
0   — | —

1 | −1
|

D*(2010)$^\pm$ I(J^P) = ½(1⁻)   I, J, P need confirmation.

Mass m = 2010.27 ± 0.17 MeV   (S = 1.1)

```
         D* (2010) +
            ?
2          |o
1          _|_
0         -|-

         1-1|-1
            |
```

D$_1$(2420)^0 I(J^P) = ½(1⁺)   I, J, P need confirmation.

Mass m = 2422.3 ± 1.3 MeV   (S = 1.2)

```
         D₁ (2420)⁰
            ?
2          -|-
1           |
0          o|o

         1-1|-1
            |
```

$D_2^*(2460)^0$ I(J^P) = ½(2⁺) $J^P = 2^+$ assignment strongly
favored.

Mass m = 2461.1 ± 1.6 MeV   (S = 1.3)

```
         D₂*(2460)⁰
            ?
2          -|-
1           |
0          -|-

         1-1|-2
            |
```

$D_2^*(2460)^+$ $I(J^P) = \frac{1}{2}(2^+)$ $J^P = 2^+$ assignment strongly favored.

Mass m = $2460.1^{+2.6}_{-3.5}$ MeV   (S = 1.5)

$$D_2^*(2460)^+$$

```
        ?
2      | o
1      |
0    - | -

     -1 | -2
        |
```

CHARMED, STRANGE MESONS
$(C = S = \pm 1)$

$D_S^\pm$ $I(J^P) = 0(0^-)$
was $F^\pm$

Mass m = 1968.49 ± 0.34 MeV   (S = 1.3)

$$D_S^+$$

```
        ?
2      | o
1    - | -
0      |

    2-1 | 0
        |
```

$D_S^{*\pm}$ $I(J^P) = 0(?^?)$ J^P is natural, width and decay modes consistent with 1^-

Mass m = 2112.3 ± 0.5 MeV   (S = 1.1)

$$D_S^{*+}$$

```
        ?
2      | o
1    - | -
0    - | -

    2-1 | -1
        |
```

$D_{s0}^{*}(2317)^{\pm}$ $I(J^{P}) = 0(0^{+})$ J, P need confirmation.

Mass m = 2317.8 ± 0.6 MeV   (S = 1.1)

$$D_{s0}^{*}(2317)^{+}$$

```
        ?
2      | o
1      |
0      o |
```

$\underline{1-1 \mid 0}$
$\mid$

$D_{s1}(2460)^{\pm}$ $I(J^{P}) = 0(1^{+})$

Mass m = 2459.6 ± 0.6 MeV   (S = 1.1)

$$D_{s0}^{*}(2460)^{+}$$

```
        ?
2      | o
1      - | -
0      |
```

$\underline{1-1 \mid 0}$
$\mid$

$D_{s1}(2536)^{\pm}$ $I(J^{P}) = 0(1^{+})$ J, P need confirmation.

Mass m = 2535.35 ± 0.34 MeV

$$D_{s1}(2536)^{+}$$

```
        ?
2      | o
1      - | -
0      |
```

$\underline{2-2 \mid -1}$
$\mid$

$D_{s2}(2573)^{\pm}$ $I(J^P) = 0(?^?)$ J^P is natural, width and decay

Mass m = 2572.6 ± 0.9 MeV   modes consistent with 2^+.

```
D_sJ(2573 )^+
        ?
2      | o
1      o | o
0       |

      -2 | -2
        |
```

BOTTOM MESONS
(B = ±1)

$B^{\pm}$ $I(J^P) = \frac{1}{2}(0^-)$ I, J, P need confirmation. Quantum

Mass m = 5279.15 ± 0.31 MeV   numbers shown are quark-

B⁺   model predictions.

```
        B^+
        ?
3      | o
2       |
1       |
0       |

       0 | 0
        |
```

B^0 $I(J^P) = \frac{1}{2}(0^-)$ I, J, P need confirmation.   Quantum numbers shown are
          Mass m = 5279.53 ± 0.33 MeV   quark-model predictions.

$$B^0$$
$$?$$

3	$-\mid-$
2	$\mid$
1	$\mid$
0	$\mid$

1 | 0
$\mid$

$B*$ $I(J^P) = \frac{1}{2}(1^-)$ I, J, P need confirmation.   Quantum numbers shown are
          Mass m = 5325.1 ± 0.5 MeV   quark-model predictions.

$$B*$$
$$?$$

3	$\mid o$
2	$\mid$
1	$\mid$
0	$-\mid-$

1-1 | -1
$\mid$

$B_1(5721)^0$ $I(J^P) = \frac{1}{2}(1^+)$ I, J, P need confirmation.
          Mass m = 5720.7 ± 0.5 MeV

$$B_1(5721)^0$$
$$?$$

3	$o \mid o$
2	$\mid$
1	$\mid$
0	$-\mid-$

1-1 | -1
$\mid$

$B_2^*(5747)^0$ $I(J^P) = 1/2(2+)$ I, J, P need confirmation.

Mass m = 5746.9 ± 2.9 MeV

$$B_2^*(5747)^0$$

?

3   o | o

2   o | o

1 |

0   − | −

3 | 2

|

BOTTOM, STRANGE MESONS
(B = ±1, S = ∓1)

B_S^0 $I(J^P) = 0(0^-)$ I, J, P need confirmation. Quantum numbers
Mass m = 5366.3 ± 0.6 MeV   shown are quark-model

B_S^0 predictions.

?

3 | o
2 |
1   o |
0 |

0 | 0
|

B_s^* $I(J^P) = 0(1^-)$ I, J, P need confirmation. Quantum numbers
Mass m = 5412.8 ± 1.3 MeV   shown are quark-model

B_S^* predictions.

?

3   - | -
2 |
1   - | -
0   - | -

2 | -1
|

$B_{s1}(5830)^0$ $I(J^P) = \tfrac{1}{2}(1^+)$ I, J, P need confirmation.
Mass m = 5829.4 ± 0.7 MeV

$B_{s1}(5830)^0$

?

3   o | o
2 |
1   - | -
0   - | -

2-1 | -1
|

$B_{s2}^{*}(5840)^{0}$ $I(J^{P}) = \frac{1}{2}(2^{+})$ I, J, P need confirmation.

Mass m = 5839.7 ± 0.6 MeV

$B_{s1}(5840)^{0}$

?

```
3    o | o
2      |
1    - | -
0    - | -
```

```
2 - 2 | - 2
       |
```

BOTTOM, CHARMED MESONS
(B = C = ± 1)

$B_{c}^{\pm}$ $I(J^{P}) = 0(0^{-})$ I, J, P need confirmation. Quantum numbers shown

Mass m = 6276 ± 4 MeV   are quark-model predictions.

B_{c}^{+}

?

```
3     | o
2   - | -
1     |
0     |
```

```
1 | 0
   |
```

$\overline{c}c$ **MESONS**

$\eta_c(1S)$ $\qquad\qquad$ $I^G(J^{PC}) = 0^+(0^{-+})$
$\qquad\qquad$ Mass m = 2980.3 ± 1.2 MeV $\quad$ (S = 1.7)

$$\eta_c(1S)$$
$$?$$
```
2    o | o
1      |
0      |
```
$$\underline{1-1\,|\,0}$$
$$|$$

$J/\psi(1S)$ $\qquad\qquad$ $I^G(J^{PC}) = 0^-(1^{--})$
$\qquad\qquad$ Mass m = 3096.916 ± 0.011 MeV

$$J/\psi(1S)$$
$$?$$
```
2    o | o
1      |
0    - | +
```
$$\underline{1-2\,|-1}$$
$$|$$

$\chi_{c0}(1P)$ $\qquad\qquad$ $I^G(J^{PC}) = 0^+(0^{++})$
$\qquad\qquad$ Mass m = 3414.75 ± 0.31 MeV

$$\chi_{c0}(1P)$$
$$?$$
```
2    o | o
1      |
0    o | o
```
$$\underline{1-1\,|\,0}$$
$$|$$

$\chi_{c1}(1P)$ $\qquad$ $I^G(J^{PC}) = 0^+(1^{++})$

$\qquad$ Mass m = 3510.66 ± 0.07 MeV

```
        Xc1 (1P)
          ?
2        o | o
1         |
0       -o | o-

      2-2 | -1
         |
```

$h_c(1P)$ $\qquad$ $I^G(J^{PC}) = ?^?(1^{+-})$

$\qquad$ Mass m = 3525.93 ± 0.27 MeV $\qquad$ (S = 1.5)

```
        h_c (1P)
          ?
2        o | o
1         |
0        - | -

      1-1 | -1
         |
```

$\chi_{c2}(1P)$ $\qquad$ $I^G(J^{PC}) = 0^+(2^{++})$

$\qquad$ Mass m = 3556.20 ± 0.09 MeV

```
        Xc2 (1P)
          ?
2        o | o
1         |
0       -- | --

      2-2 | -2
         |
```

$\eta_c(2S)$ $I^G(J^{PC}) = 0^+(0^{-+})$

Mass m = 3637 ± 4 MeV   (S = 1.7)

$$\eta_c(2S)$$

```
           ?
2        o | o
1          |
0        -+ | +-

         2-2 | 0
           |
```

$\psi(2S)$ $I^G(J^{PC}) = 0^-(1^{--})$

Mass m = 3686.09 ± 0.04 MeV   (S = 1.6)

$$\psi(2S)$$

```
           ?
2        o | o
1          |
0        -o | o+

         1-2 | -1
           |
```

$\psi(3770)$ $I^G(J^{PC}) = 0^-(1^{--})$

Mass m = 3772.92 ± 0.35 MeV   (S = 1.1)

$$\psi(3770)$$

```
           ?
2        o | o
1          |
0        +o | o-

         1-2 | -1
           |
```

$\chi(3872)$

$I^G(J^{PC}) = 0^?(?^{?+})$

Mass m = 3872.2 ± 0.8 MeV   (S = 2.5)

$\chi(3872)$

?

```
2     o | o
1       |
0    -o | o+


    -3 | -3
       |
```

$\psi(4040)$

$I^G(J^{PC}) = 0^-(1^{--})$
Mass m = 4039 ± 1 MeV

$\psi(4040)$
?

```
2    -+ | +-
1       |
0       |


   1-2 | -1
       |
```

$\psi(4160)$

$I^G(J^{PC}) = 0^-(1^{--})$
Mass m = 4159 ± 20 MeV

$\psi(4160)$
?

```
2    -+ | +-
1       |
0    -o | o-


   2-2 | -1
       |
```

$\chi(4260)$ $\qquad$ $I^G(J^{PC}) = ?^?(1^{--})$

Mass m = 4263^{+8}_{-9} MeV $\quad$ (S = 1.1)

$\chi(4260)$

$?$

```
2     -+ | +-
1        |
0      o | o

    1-2 | -1
        |
```

$\psi(4415)$ $\qquad$ $I^G(J^{PC}) = 0^-(1^{--})$

Mass m = 4421 ± 4 MeV

$\psi(4415)$

$?$

```
2     -+ | +-
1        |
0      o | o

    2-1 | 1
        |
```

$$b\bar{b} \text{ MESONS}$$

Y(1S)

$$I^G(J^{PC}) = 0^-(1^{--})$$
Mass m = 9460.30 ± 0.26 MeV   (S = 3.3)

Y(1S)
?

```
3        o | o
2          |
1          |
0        -o | o+

       1-2 | -1
           |
```

χ_{b0}(1P)

$$I^G(J^{PC}) = 0^+(0^{++})$$ J needs confirmation.
Mass m = 9859.44 ± 0.42 MeV

χ_{b0}(1P)
?

```
3        o | o
2          |
1          |
0        o | o

       1-1 | 0
           |
```

$\chi_{b1}(1P)$ $I^G(J^{PC}) = 0^+(1^{++})$ J needs confirmation.

Mass m = 9892.78 ± 0.26 MeV

$\chi_{b1}(1P)$
?

```
3        o | o
2          |
1          |
0        — o | o —

         2 – 2 | – 1
            |
```

$\chi_{b2}(1P)$ $I^G(J^{PC}) = 0^+(2^{++})$ J needs confirmation.

Mass m = 9912.21 ± 0.26 MeV

$\chi_{b2}(1P)$
?

```
3        o | o
2          |
1          |
0        — o | o —

         1 – 2 | – 2
            |
```

Y(2S) $I^G(J^{PC}) = 0^-(1^{--})$

Mass m = 10.02326 ± 0.00031 GeV

Y(2S)
?

```
3        o | o
2          |
1        — | —
0        — | —

         2 – 1 | – 1
            |
```

$\chi_{b0}(2P)$ $\qquad$ $I^G(J^{PC}) = 0^+(0^{++})$ J needs confirmation.

Mass m = 10.2325 ± 0.0005 GeV

$\chi_{b0}(2P)$

?

```
3      o | o
2        |
1        |
0      -+ | +-

      2-2 | 0
          |
```

$\chi_{b1}(2P)$ $\qquad$ $I^G(J^{PC}) = 0^+(1^{++})$ J needs confirmation.

Mass m = 10.25546 ± 0.00022 GeV

$\chi_{b1}(2P)$

?

```
3      o | o
2        |
1        |
0      - | -

      1-1 | -1
          |
```

$\chi_{b2}(2P)$ $\qquad$ $I^G(J^{PC}) = 0^+(2^{++})$ J needs confirmation.

Mass m = 10.26865 ± 0.00022 GeV

$\chi_{b2}(2P)$

?

```
3      o | o
2        |
1      - | -
0      - | -

      2-2 | -2
          |
```

Y(3S) $I^G(J^{PC}) = 0^-(1^{--})$

Mass m = 10.3553 ± 0.0005 GeV

Y(3S)

?

```
3        o | o
2          |
1          |
0        - + | + -

      1 - 2 | - 1
          |
```

Y(4S)
or Y(10580) $I^G(J^{PC}) = 0^-(1^{--})$

Mass m = 10.5794 ± 0.0012 GeV

Y(4S)

?

```
3        o | o
2          |
1          |
0        - | +

      1 - 2 | - 1
          |
```

Y(10860) $I^G(J^{PC}) = 0^-(1^{--})$

Mass m = 10.865 ± 0.008 GeV   (S = 1.1)

Y(10860)

?

```
3        o | o
2          |
1          |
0        - o | o +

      1 - 2 | - 1
          |
```

Y (11020) $I^G(J^{PC}) = 0^-(1^{--})$
                  Mass m = 11.019 ± 0.008 GeV
                      Y (11020)
                           ?
          3   o | o
          2 |
          1 |
          0 |

                     1 | 1
                       |

_____

                        N BARYONS
                      (S = 0, I = ½)

    p $I(J^P) = \frac{1}{2}(\frac{1}{2}^+)$

          Mass m = 938.27203 ± 0.00008 MeV
                           p
                           ?
          2 |
          1 |
          0 | no

                   1-1 | ½
                       |

_____

    n $I(J^P) = \frac{1}{2}(\frac{1}{2}^+)$
              Mass m = 939.56536 ± 0.00008 MeV
                           n
                   933.948 3545 MeV
          2 |
          1 |
          0 | n

                    -1 | -½
                       |

N(1440) P_{11} $I(J^P) = \frac{1}{2}(\frac{1}{2}^+)$

  Breit-Wigner mass = 1420 to 1470 ($\approx$ 1440) MeV

```
           N(1440)  P₁₁
               ?
    2         |
    1         | o
    0        o | no

        1-1 | ½
             |
```

N(1520) D_{13} $I(J^P) = \frac{1}{2}(3/2^-)$

  Breit-Wigner mass = 1515 to 1525 ($\approx$ 1520) MeV

```
           N(1520)  D₁₃
               ?
    2         |
    1         | o
    0        o | no

        2-1 | 1½
             |
```

N(1535) S_{11} $I(J^P) = \frac{1}{2}(\frac{1}{2}^-)$

  Breit-Wigner mass = 1525 to 1545 ($\approx$ 1535) MeV

```
           N(1535)  S₁₁
               ?
    2         |
    1         | o
    0      +o | no+
```

$$\frac{1-2 \mid \frac{1}{2}}{\mid}$$

N(1650) S_{11} $I(J^P) = \frac{1}{2}(\frac{1}{2}^-)$

   Breit-Wigner mass = 1645 to 1670 ($\approx$ 1655) MeV

```
        N(1650)  S₁₁
             ?
   2        |
   1        _|o
   0         |n

        1-1 | ½
            |
```

N(1675) D_{15} $I(J^P) = \frac{1}{2}(5/2^-)$

   Breit-Wigner mass = 1670 to 1680 ($\approx$ 1675) MeV

```
        N(1675)  D₁₅
             ?
   2        |
   1        _|o
   0        o|no

        2 | 2½
          |
```

N(1680) F_{15} $I(J^P) = \frac{1}{2}(5/2^+)$

   Breit-Wigner mass = 1680 to 1690 ($\approx$ 1685) MeV

```
        N(1680) F₁₅
             ?
   2        |
   1        +|+
   0        +|no+

        3-3 | 2½
            |
```

N(1700) D$_{13}$ I(J^P) = ½(3/2$^-$)

Breit-Wigner mass = 1650 to 1750 (≈ 1700) MeV

$$N(1700)D_{13}$$

```
            ?
2           |
1           |
0           | no

        1 | 1½
          |
```

N(1710) P$_{11}$ I(J^P) = ½(½$^+$)

Breit-Wigner mass = 1680 to 1740 (≈ 1710) MeV

$$N(1710)\ P_{11}$$

```
            ?
2           |
1           | o
0         o | no

       1 − 1 | ½
            |
```

N(1720) P$_{13}$ I(J^P) = ½(3/2$^+$)

Breit-Wigner mass = 1700 to 1750 (≈ 1720) MeV

$$N(1720)P_{13}$$

```
            ?
2           |
1           | o
0         + | n +

       1 − 1 | 1½
            |
```

N(2190) G$_{17}$ I(J^P) = ½(7/2$^-$)

    Breit-Wigner mass = 2100 to 2200 (≈ 2190) MeV

```
              N(2190)G₁₇
                 ?
   2             |
   1             |
   0           + | no+

               2 | 3½
                 |
```

N(2220) H$_{19}$ I(J^P) = ½(9/2$^+$)

    Breit-Wigner mass = 2200 to 2300 (≈ 2250) MeV

```
              N(2220)H₁₉
                 ?
   3             |
   2             |
   1           + | o+
   0          +o | no+

               2 | 4½
                 |
```

N(2250) G$_{19}$ I(J^P) = ½(9/2$^-$)

    Breit-Wigner mass = 2200 to 2350 (≈ 2275) MeV

```
              N(2250)G₁₉
                 ?
   2             |
   1             | o
   0          +o | no+

               3 | 4½
                 |
```

$N(2600)$ $I_{1,11}$ $I(J^P) = \frac{1}{2}(11/2^-)$

Breit-Wigner mass = 2550 to 2750 ($\approx$ 2600) MeV

```
            N(2600)l₁,₁₁
                 ?
  2              |
  1            + | +
  0            + | no+

             3 | 5½
                 |
```

Δ BARYONS
(S = 0, I = 3/2)

$\Delta(1232)$ P_{33} $I(J^P) = 3/2(3/2^+)$

Breit-Wigner mass (mixed charges) = 1231 to 1233 ($\approx$ 1232) MeV

```
            Δ(1232) P₃₃
                 ?
  2              |
  1              | o
  0              | no

             1 | 1½
                 |
```

$\Delta(1600)$ P_{33} $I(J^P) = 3/2(3/2^+)$

Breit-Wigner mass = 1550 to 1700 ($\approx$ 1600) MeV

```
            Δ(1600) P₃₃
                 ?
  2            | o
  1            | o
  0          +o | no+

           2-2 | 1½
                 |
```

$\Delta(1620)\,S_{31}$ $I(J^P) = 3/2(1/2^-)$

Breit-Wigner mass = 1600 to 1660 ($\approx$ 1630) MeV

```
        Δ(1620) S₃₁
            ?
2            |
1           | o
0           | no

    1-1 | ½
         |
```

$\Delta(1700)\,D_{33}$ $I(J^P) = 3/2(3/2^-)$

Breit-Wigner mass = 1670 to 1750 ($\approx$ 1700) MeV

```
        Δ(1700) D₃₃
            ?
2            |
1           | o
0           | no

    2-1 | 1½
         |
```

$\Delta(1905)\,F_{35}$ $I(J^P) = 3/2(5/2^+)$

Breit-Wigner mass = 1865 to 1915 ($\approx$ 1890) MeV

```
        Δ(1905) F₃₅
            ?
2            |
1           | o
0           | no

    2 | 2½
         |
```

$\Delta(1910)$ P_{31} $I(J^P) = 3/2(1/2^+)$

Breit-Wigner mass = 1870 to 1920 ($\approx$ 1910) MeV

$$\Delta(1910)\, P_{31}$$

```
        ?
2        |
1        | o
0        | no

   1-1 | ½
        |
```

$\Delta(1920)$ P_{33} $I(J^P) = 3/2(3/2^+)$

Breit-Wigner mass = 1900 to 1970 ($\approx$ 1920) MeV

$$\Delta(1920)\, P_{33}$$

```
        ?
2        |
1        | o
0      + | no +

   2-2 | 1½
        |
```

$\Delta(1930)$ D_{35} $I(J^P) = 3/2(5/2^-)$

Breit-Wigner mass = 1900 to 2020 ($\approx$ 1960) MeV

$$\Delta(1930)\, D_{35}$$

```
        ?
2        | o
1        | o
0       o | no

   3-1 | 2½
        |
```

$\Delta(1950)$ F_{37} $I(J^P) = 3/2(7/2^+)$

Breit-Wigner mass = 1915 to 1950 ($\approx$ 1930) MeV
$\Delta(1950)$ F_{37}

```
        ?
2       |
1       | o
0       | no

    3 | 3½
        |
```

$\Delta(2420)$ $H_{3,11}$ $I(J^P) = 3/2(11/2^+)$

Breit-Wigner mass = 2300 to 2500 ($\approx$ 2420) MeV
$\Delta(2420)$ $H_{3,11}$

```
         ?
2        |
1        | o
0      + | no+

    4 | 5½
        |
```

Λ BARYONS
(S = -1, I = 0)

Λ $I(J^P) = 0(\frac{1}{2}^+)$
Mass m = 1115.683 $\pm$ 0.006 MeV

```
        Λ
        ?
2       |
1       | n
0       |

   -1 | -½
       |
```

$\Lambda(1405)\ S_{01}$ $\qquad$ $I(J^P) = 0(1/2^-)$

Mass m = 1406 ± 4 MeV

$$\Lambda(1405)\ S_{01}$$

$$?$$

2		
1		n
0	o	o

$$\underline{1-2\ |-\tfrac{1}{2}}$$
$$|$$

$\Lambda(1520)\ D_{03}$ $\qquad$ $I(J^P) = 0(3/2^-)$

Mass m = 1519.5 ± 1.0 MeV

$$\Lambda(1520)\ D_{03}$$

$$?$$

2		
1	o	n
0		o

$$\underline{-2\ |-1\tfrac{1}{2}}$$
$$|$$

$\Lambda(1600)\ P_{01}$ $\qquad$ $I(J^P) = 0(1/2^+)$

Mass m = 1560 to 1700 ($\approx$ 1600) MeV

$$\Lambda(1600)\ P_{01}$$

$$?$$

2		
1	o	n
0		o

$$\underline{1-2\ |-\tfrac{1}{2}}$$
$$|$$

$\Lambda(1670)\ S_{01}$ $I(J^P) = 0(1/2^-)$

Mass m = 1660 to 1680 ($\approx$ 1670) MeV

$$\Lambda(1670)\ S_{01}$$

```
          ?
2         |
1       o | n
0       + | o +

     1-2 | ½
          |
```

$\Lambda(1690)\ D_{03}$ $I(J^P) = 0(3/2^-)$

Mass m = 1685 to 1695 ($\approx$ 1690) MeV

$$\Lambda(1690)\ D_{03}$$

```
          ?
2         |
1       o | n o
0       o | o

     2-1 | 1½
          |
```

$\Lambda(1800)\ S_{01}$ $I(J^P) = 0(1/2^-)$

Mass m = 1720 to 1850 ($\approx$ 1800) MeV

$$\Lambda(1800)\ S_{01}$$

```
          ?
2         |
1       o | n
0         | o

     1-1 | ½
          |
```

$\Lambda(1810)\,P_{01}$ $\qquad I(J^P) = 0(1/2^+)$

Mass m = 1750 to 1850 ($\approx$ 1810) MeV

$$\Lambda(1810)\,P_{01}$$

```
        ?
        |
2       |
1      o | no
0      o | o

     2-2 | ½
        |
```

$\Lambda(1820)\,F_{05}$ $\qquad I(J^P) = 0(5/2^+)$

Mass m = 1815 to 1825 ($\approx$ 1820) MeV

$$\Lambda(1820)\,F_{05}$$

```
          ?
          |
2         |
1     +o | n+
0      + | o+

     2-2 | 2½
        |
```

$\Lambda(1830)\,D_{05}$ $\qquad I(J^P) = 0(5/2^-)$

Mass m = 1810 to 1830 ($\approx$ 1830) MeV

$$\Lambda(1830)\,D_{05}$$

```
        ?
        |
2       |
1      o | n
0        | o

      2 | 2½
        |
```

$\Lambda(1890)$ P_{03} $I(J^P) = 0(3/2^+)$

Mass m = 1850 to 1910 ($\approx$ 1890) MeV

$$\Lambda(1890)\ P_{03}$$

```
          ?
2         |
1       o | n
0       + | o +

      2 - 2 | 1½
          |
```

$\Lambda(2100)$ G_{07} $I(J^P) = 0(7/2^-)$

Mass m = 2090 to 2110 ($\approx$ 2100) MeV

$$\Lambda(2100)\ G_{07}$$

```
          ?
2         |
1       o | n
0       - | o -

     - 3 | - 3½
          |
```

$\Lambda(2110)$ F_{05} $I(J^P) = 0(5/2^+)$

Mass m = 2090 to 2140 ($\approx$ 2110) MeV

$$\Lambda(2110)\ F_{05}$$

```
          ?
2         |
1     + o | n +
0       + | o +

      2 - 2 | 2½
          |
```

$\Lambda(2350)$ H_{09} $I(J^P) = 0(9/2^+)$

Mass m = 2340 to 2370 ($\approx$ 2350) MeV

$\Lambda(2350)$ H_{09}

```
          ?
3        |
2        |
1      +o|n+
0      +|o+

       2|4½
        |
```

Σ BARYONS
(S = -1, I = 1)

Σ^+ $I(J^P) = 1(½^+)$

Mass m = 1189.37 $\pm$ 0.07 MeV   (S = 2.2)

Σ^+

```
        ?
2      |
1      |n
0      |o

     1-1|½
        |
```

Σ^0 $I(J^P) = 1(½^+)$

Mass m = 1192.642 $\pm$ 0.024 MeV

Σ^0

```
        ?
2      |
1      |n
0     o|o

     1-1|½
        |
```

Σ^- $I(J^P) = 1(1/2^+)$

Mass m = 1197.449 ± 0.030 MeV   (S = 1.2)

Σ^-
?
```
2     |
1     | n
0     o |
```
1-1 | ½
|

$\Sigma(1385)^+ P_{13}$ $I(J^P) = 1(3/2^+)$

Mass m = 1382.8 ± 0.4 MeV   (S = 2.0)

$\Sigma(1385)^+ P_{13}$
?
```
2     |
1     | n
0   + | o+
```
1-1 | 1½
|

$\Sigma(1385)^0 P_{13}$ $I(J^P) = 1(3/2^+)$

Mass m = 1383.7 ± 1.0 MeV   (S = 1.4)

$\Sigma(1385)^0 P_{13}$
?
```
2     |
1     | n
0   +o | o+
```
2-2 | 1½
|

$\Sigma(1385)^- P_{13}$ $I(J^P) = 1(3/2^+)$

     Mass m = 1387.2 ± 0.5 MeV   (S = 2.2)

$$\Sigma(1385)^- P_{13}$$

```
            ?
2         __|__
1            |no
0         +o|o+

        2-2|1½
           |
```

$\Sigma(1660) P_{11}$ $I(J^P) = 1(1/2^+)$

     Mass m = 1630 to 1690 (≈ 1660) MeV

$$\Sigma(1660) P_{11}$$

```
            ?
2         __|__
1          o|no
0           |o

        1-1|½
           |
```

$\Sigma(1670) D_{13}$ $I(J^P) = 1(3/2^-)$

     Mass m = 1665 to 1685 (≈ 1670) MeV

$$\Sigma(1670) D_{13}$$

```
            ?
2         __|__
1          o|no
0           |o

        2-1|1½
           |
```

___

$\Sigma(1750)\, S_{11}$ $I(J^P) = 1(1/2^-)$

Mass m = 1730 to 1800 ($\approx$ 1750) MeV

$$\Sigma(1750)\, S_{11}$$

```
        ?
2       |
1      |no
0      |o

     2-2|½
        |
```

$\Sigma(1775)\, D_{15}$ $I(J^P) = 1(5/2^-)$

Mass m = 1770 to 1780 ($\approx$ 1775) MeV

$$\Sigma(1775)\, D_{15}$$

```
        ?
2       |
1     o|no
0     +|o+

     2-1|2½
        |
```

$\Sigma(1915)\, F_{15}$ $I(J^P) = 1(5/2^+)$

Mass m = 1900 to 1935 ($\approx$ 1915) MeV

$$\Sigma(1915)\, F_{15}$$

```
        ?
2       |
1    +o|no+
0     +|o+

     2-2|2½
        |
```

APP. B: STRUCTURE OF KNOWN PARTICLES

$\Sigma(1940)$ D_{13} $I(J^P) = 1(3/2^-)$

   Mass m = 1900 to 1950 ($\approx$ 1940) MeV

```
                   Σ(1940) D₁₃
                       ?
         2             |
         1          o | no
         0          + | o+

                  -3 | -1½
                       |
```

$\Sigma(2030)$ F_{17} $I(J^P) = 1(7/2^+)$

   Mass m = 2025 to 2040 ($\approx$ 2030) MeV

```
                   Σ(2030) F₁₇
                       ?
         2           + | +
         1          +o | no+
         0           + | o+

                  3-3 | 3½
                       |
```

$\Sigma(2250)$ $I(J^P) = 1(?^?)$

   Mass m = 2210 to 2280 ($\approx$ 2250) MeV

```
                   Σ(2250)
                       ?
         2             |
         1           + | n+
         0           + | o+

                   3 | 5½      Unknown
                       |
```

A493

APP. B: STRUCTURE OF KNOWN PARTICLES

Ξ BARYONS
$(S = -2, I = \frac{1}{2})$

Ξ^0 $I(J^P) = \frac{1}{2}(\frac{1}{2}^+)$ P is not yet measured; + is the quark model prediction.

Mass m = 1314.86 ± 0.20 MeV

```
              Ξ⁰
              ?
2           | n
1           |
0           |

         -1 | -½
            |
```

Ξ^- $I(J^P) = \frac{1}{2}(\frac{1}{2}^+)$ P is not yet measured; + is the quark model prediction.

Mass m = 1321.71 ± 0.07 MeV

```
             Ξ⁻
             ?
2          | n
1          |
0        o |

       1-1 | ½
           |
```

$\Xi(1530)^0\, P_{13}$ $I(J^P) = \frac{1}{2}(3/2^+)$

Mass m = 1531.80 ± 0.32 MeV   (S = 1.3)

$\Xi(1530)^0\, P_{13}$

```
             ?
2          | n
1          |
0     +o | o+

      2-2 | 1½
          |
```

APP. B: STRUCTURE OF KNOWN PARTICLES

$\Xi(1530)^- P_{13}$ $I(J^P) = \frac{1}{2}(3/2^+)$

   Mass m = 1535.0 ± 0.6 MeV

$$\Xi(1530)^- P_{13}$$

```
            ?
2        __|n_
1        __|__
0        +o|+

        1-1|1½
          |
```

$\Xi(1690)$ $I(J^P) = \frac{1}{2}(?^?)$

   Mass m = 1690 ± 10 MeV

$$\Xi(1690)$$

```
            ?
2        __|n_
1         o|
0          |o

        1-1|½        Unknown
          |
```

$\Xi(1820) D_{13}$ $I(J^P) = \frac{1}{2}(3/2^-)$

   Mass m = 1823 ± 5 MeV

$$\Xi(1820) D_{13}$$

```
            ?
2        __|n_
1         o|
0          |o

        2-1|1½
          |
```

A495

APP. B: STRUCTURE OF KNOWN PARTICLES

$\Xi(1950)$ $I(J^P) = \frac{1}{2}(?^?)$

Mass m = 1950 ± 15 MeV

```
              Ξ(1950)
                 ?
    2          | n
    1          o |
    0            |

          1 | 1½    Unknown
               |
```

$\Xi(2030)$ $I(J^P) = \frac{1}{2}(\geq 5/2^?)$
Mass m = 2025 ± 5 MeV

```
              Ξ(2030)
                 ?
    2          | n
    1          o |
    0            | o

          2 | 2½    Unknown
               |
```

Ω BARYONS
(S = - 3, I = 0)

Ω^- $I(J^P) = 0(3/2^+)$ J^P is not yet measured; $3/2^+$ is the quark model prediction.
Mass m = 1672.45 ± 0.29 MeV

```
               Ω⁻
               ?
    2          | n
    1          o |
    0          + | +

          1-1 | 1½
                |
```

A496

$\Omega(2250)^-$ $I(J^P) = 0(?^?)$

Mass m = 2252 ± 9 MeV

$\Omega(2250)^-$

```
            ?
2         | n
1         o |
0         o | o

        2 | 2½      Unknown
          |
```

CHARMED BARYONS
(C = + 1)

Λ_c^+ $I(J^P) = 0(\tfrac{1}{2}^+)$ J not confirmed; ½ is the quark model
prediction.

Mass m = 2286.46 ± 0.14 MeV

Λ_c^+

```
            ?
2         | no
1         |
0         |

      1-1 | ½
          |
```

$\Lambda_c(2595)^+$ $I(J^P) = 0(\tfrac{1}{2}^-)$ The spin-parity follows from the fact
that $\Sigma_c(2455)\pi$ decays, with little available space, are
dominant.

Mass m = 2595.4 ± 0.6 MeV

$\Lambda_c(2595)^+$

```
            ?
2         | no
1         |
0         o | o

      1-1 | ½
          |
```

$\Lambda_c(2625)^+$ $I(J^P) = 0(3/2^-)$ J^P is expected to be $3/2^-$.

Mass m = 2628.1 ± 0.6 MeV   (S = 1.5)

$$\Lambda_c(2625)^+$$
?

```
2      | no
1       |
0      o | o
```

2−1 | 1½
|

$\Lambda_c(2880)^+$ $I(J^P) = 0(5/2^+)$ There is some good evidence that
indeed $J^P = 5/2^+$.
   Mass m = 2881.53 ± 0.35 MeV

$$\Lambda_c(2880)^+$$
?

```
2      | no
1      + | +
0     +o | o+
```

2−2 | 2½
|

$\Lambda_c(2940)^+$ $I(J^P) = 0(?^?)$

Mass m = $2939.3^{+1.4}_{-1.5}$ MeV

$$\Lambda_c(2940)^+$$
?

```
2      | no
1      + | +
0     +o | o+
```

3−1 | 4½   Unknown
|

A498

$\Sigma_c(2455)^{++}$ $I(J^P) = 1(1/2^+)$ J^P not confirmed; $1/2^+$ is the quark model
prediction.
Mass m = 2454.02 ± 0.18 MeV

$\Sigma_c(2455)^{++}$
?

```
2        | no
1         |
0        | o
```

1−1 | ½
|

$\Sigma_c(2455)^{+}$ $I(J^P) = 1(1/2^+)$ J^P not confirmed; $1/2^+$ is the quark model
prediction.
Mass m = 2452.9 ± 0.4 MeV

$\Sigma_c(2455)^{+}$
?

```
2        | no
1         |
0       o | o
```

1−1 | ½
|

$\Sigma_c(2455)^{0}$ $I(J^P) = 1(½^+)$ J^P not confirmed; $1/2^+$ is the quark model
prediction.
Mass m = 2453.76 ± 0.18 MeV

$\Sigma_c(2455)^{0}$
?

```
2        | no
1         |
0       o |
```

1−1 | ½
|

$\Sigma_c(2520)^{++}$ $I(J^P) = 1(3/2^+)$

Mass m = 2518.4 ± 0.6 MeV

$$\Sigma_c(2520)^{++}$$
$$?$$

```
2        |no
1         |
0       + | o+
```

$$\underline{2-2\,|\,1\tfrac{1}{2}}$$
$$|$$

$\Sigma_c(2520)^0$ $I(J^P) = 1(3/2^+)$

Mass m = 2518.0 ± 0.5 MeV

$$\Sigma c(2520)^0$$
$$?$$

```
2        |no
1         |
0       +o | +
```

$$\underline{2-2\,|\,1\tfrac{1}{2}}$$
$$|$$

$\Sigma_c(2800)$ $I(J^P) = 1(?^?)$ Unknown structures.

$\Sigma_c(2800)^{++}$ mass m = 2801^{+4}_{-6} MeV

$\Sigma_c(2800)^+$ mass m = 2792^{+14}_{-5} MeV

$\Sigma_c(2800)^0$ mass m = 2802^{+4}_{-7} MeV

Ξ_c^+ $I(J^P) = \frac{1}{2}(\frac{1}{2}^+)$ $I(J^P)$ not confirmed; $\frac{1}{2}(\frac{1}{2}^+)$ is the quark
                                                                    model prediction.

Mass m = 2467.9 ± 0.4 MeV

Ξ_c^+

?

```
2          | no
1        o |
0          | o
```

1-1 | ½

Ξ_c^0 $I(J^P) = \frac{1}{2}(\frac{1}{2}^+)$ $I(J^P)$ not confirmed; $\frac{1}{2}(\frac{1}{2}^+)$ is the quark model
                                                                    prediction.

Mass m = 2471.0 ± 0.4 MeV

Ξ_c^0

?

```
2          | no
1        o |
0        o | o
```

2-2 | ½

$\Xi_c'^+$ $I(J^P) = \frac{1}{2}(\frac{1}{2}^+)$ $I(J^P)$ not confirmed; $\frac{1}{2}(\frac{1}{2}^+)$ is the quark

model prediction.

Mass m = 2575.7 ± 3.1 MeV

Ξ_c^0

?

```
2        o | no
1        o | o
0        o | o
```

2-2 | ½

A501

Ξ'^{0}_{c} $I(J^P) = \frac{1}{2}(\frac{1}{2}^+)$ $I(J^P)$ not confirmed; $\frac{1}{2}(\frac{1}{2}^+)$ is the
quark model prediction.
Mass m = 2578.0 ± 2.9 MeV

$$\Xi^{0}_{c}$$
?

```
2        o|no
1        o | o
0        o | o

        2-2 | ½
          |
```

$\Xi_c(2645)^+$ $I(J^P) = \frac{1}{2}(3/2^+)$

Mass m = 2646.6 ± 1.4 MeV   (S = 1.6)

$$\Xi_c(2645)^+$$
?

```
2         | no
1        o |
0        + | o+

        2-2 | 1½
          |
```

$\Xi_c(2645)^0$ $I(J^P) = 1/2(3/2^+)$

Mass m = 2646.1 ± 1.2 MeV
$$\Xi_c(2645)^0$$
?

```
2         | no
1        o |
0        + | +

        2-2 | 1½
          |
```

$\Xi_c(2790)^+$ $I(J^P) = 1/2(1/2^-)$

J^P has not been measured; ½⁻ is the quark model prediction.

Mass $m = 2789.2 \pm 3.2$ MeV

$\Xi_c(2790)^0$

?

```
2      | no
1      o |
0      + | o+
```

```
1-2 | ½
  |
```

$\Xi_c(2790)^0$ $I(J^P) = 1/2(1/2^-)$

J^P has not been measured; ½⁻ is the quark model prediction.

Mass $m = 2791.9 \pm 3.3$ MeV

$\Xi_c(2790)^0$

?

```
2      | no
1      o |
0      +o | o+
```

```
1-2 | ½
  |
```

$\Xi_c(2815)^+$ $I(J^P) = 1/2(3/2^-)$

J^P has not been measured; ½⁻ is the quark model prediction.

Mass $m = 2816.5 \pm 1.2$ MeV

$\Xi_c(2815)^0$

?

```
2      | no
1      o |
0      o | o
```

```
2-1 | 1½
  |
```

A503

$\Xi_c(2815)^0$ $I(J^P) = 1/2(3/2^-)$

J^P has not been measured; $\frac{1}{2}^-$ is the quark model prediction.

Mass m = 2818.2 ± 1.2 MeV

$\Xi_c(2815)^0$

?

```
2          | no
1         o |
0         o | o
```

2-1 | 1½

$\Xi_c(2980)$ $I(J^P) = 1/2(?^?)$ Unknown structures.

$\Xi_c(2980)^+$ m = 2974 ± 5 MeV   (S = 2.3)

$\Xi_c(2980)^0$ m = 2974 ± 4 MeV

$\Xi_c(3080)$ $I(J^P) = 1/2(?^?)$ Unknown structures.

$\Xi_c(3080)^+$ m = 3077.0 ± 0.4 MeV

$\Xi_c(3080)^0$ m = 3079.9 ± 1.4 MeV

Ω_c^0 $I(J^P) = 0(\frac{1}{2}^+)$ $I(J^P)$ not confirmed; $0(\frac{1}{2}^+)$ is the quark
                                     model prediction.

Mass m = 2697.5 ± 2.6 MeV   (S = 1.2)

Ω_c^0

?

```
2          | no
1         o |
0         o | o
```

2-2 | ½

$\Omega_c(2770)^0$ $I(J^P) = 0(3/2^+)$ $I(J^P)$ not confirmed; $0(\frac{1}{2}^+)$ is the quark
model prediction.

Mass $m = 2768.3 \pm 3.0$ MeV $(S = 1.2)$

$$\Omega_c^0$$

```
        ?
2      |no
1     o|
0     +o|o+

    2-2|1½
      |
```

BOTTOM BARYONS
(B = - 1)

Λ_b^0 $I(J^P) = 0(1/2^+)$ $I(J^P)$ not yet measured; $0(1/2^+)$ is the
quark model prediction.

Mass $m = 5620.2 \pm 1.6$ MeV

$$\Lambda_b^0$$

```
        ?
3      |n
2      |
1      |
0      |

    -1|-½
      |
```

Σ_b^+ $I(J^P) = 1(1/2^+)$ I, J, P need confirmation.

Mass $m = 5807.8 \pm 2.7$ MeV

$$\Sigma_b^+$$

```
        ?
3      |n
2      |
1      |
0      |o

    1-1|½
      |
```

A505

Σ_b^- $I(J^P) = 1(1/2^+)$ I, J, P need confirmation.

Mass m = 5815.2 ± 2.0 MeV

$$\Sigma_b^-$$

?

3 | n
2 |
1 |
0   o |

1−1 | ½
|

Σ_b^{*+} $I(J^P) = 1(3/2^+)$ I, J, P need confirmation.

Mass m = 5829.0 ± 3.4 MeV

$$\Sigma_b^{*+}$$

?

3 | n
2 |
1 |
0   + | o+

1−1 | 1½
|

Σ_b^{*-} $I(J^P) = 1(3/2^+)$ I, J, P need confirmation.

Mass m = 5836.4 ± 2.8 MeV

$$\Sigma_b^{*-}$$

?

3 | n
2 |
1 |
0   +o | +

1−1 | 1½
|

A506

Ξ_b^0 $I(J^P) = 1/2(1/2^+)$ I, J, P need confirmation.

Mass m = 5792.4 ± 3.0 MeV

$$\Xi_b^0$$

```
        ?
3      |n
2      |
1      |
0     o|o

   1-1|½
      |
```

Ξ_b^- $I(J^P) = 1/2(1/2^+)$ I, J, P need confirmation.

Mass m = 5792.4 ± 3.0 MeV

$$\Xi_b^-$$

```
        ?
3      |n
2      |
1     o|
0     o|o

   1-1|½
      |
```

GRAVITONS

$g^\pm$ $I(J) = 0(2)$

Mass m undetermined; expect positive imaginary component.

```
    g+                    g-
   >0i                   >0i
2   |                 2   |
1   |                 1   |
0  +|+                0  -|-

  1|2                  -1|-2
   |                     |
```

Higher energy state orbiting particle pairs can also compose $g^\pm$ gravitons.

g^{o+} $I(J) = 0(2)$
                    Mass m undetermined; expect positive imaginary component.

```
            g^o+
            >0i
   2     ___|___
   1     ___|___
   0     +o | o+

         -3 | -2
          ___|___
```

A negative + echon orbits a positive o echon with -1 orbital spin, forming a neutrino. A positive + echon orbits a negative o echon with -1 orbital spin, forming an anti-neutrino. The neutrino and anti- neutrino orbit each other with an additional -1 orbital spin. The result is a g^{o+} graviton with -2 spin.

---

g^{o-} $I(J) = 0(2)$ The spin conjugate graviton g^{o-} is also possible.

```
            g^o-
            >0i
   2     ___|___
   1     ___|___
   0     -o | o-

          3 | 2
          ___|___
```

According to chonomic decay schemes, a 1 spin graviton also exists.

g^{o} $I(J) = 0(1)$
                    Mass m undetermined; expect positive imaginary component.

```
            g^o
            >0i
   2     ___|___
   1     ___|___
   0      o | o

          1 | 1
          ___|___
```

MAGNETONS

The magnetic moments cancel out in gravitons. Not so in magnetons. The $\omega(782)$ particle is a magneton, as are other ω particles.

$\omega(782)$ $\qquad\qquad$ $I^G(J^{PC}) = 0^-(1^{--})$

$\qquad$ Mass $m = 781.94 \pm 0.12$ MeV $\quad$ (S = 1.5)

$$\begin{array}{cl} & \omega(782) \\ & ? \\ 2 & | \\ 1 & \overline{|} \\ 0 & -\,|\,+ \\ \\ & \underline{1\,|\,1} \\ & | \end{array}$$

PROBLEM SOLUTIONS

Problem Set 1

1. One eye can see a flat projection, but two eyes can see depth of field. One ear can hear sound, but two ears can estimate the direction of the sound. One hand can grasp an object, but the second hand can manipulate the object in a great variety of ways, making work and industry possible. One foot can balance the weight of the body, but two feet can walk or run.

2. If electrons acted solely as particles in atoms, they would spiral in to the nucleus and atoms would be impossible. The existence of atoms and molecules depend on the particle-wave duality.

3. A general planetary orbit has two foci. Such an orbit is an ellipse.

4. Hyperfine splitting of atomic spectra is on account of two spin orientations of elementary particles in magnetic fields.

5. Mysteries are not always dualities. Any number of examples might be cited of mysteries that compose three or more aspects, such as:
    a. +1, 0, -1 charges for π, ρ, and Σ particles.
    b. Three cords compose the ideal rope.
    c. Four organic bases are employed in the DNA molecule: adenine, thymine, guanine, and cytosine.
    d. Three kinds of multiplication of vectors: cross product, dot product, and scaler product.
    e. Three dimensions in a crystal lattice.
    f. Three particles as building blocks of matter: proton, neutron, and electron.
    g. etc.

6. Wording of the "mystery resolution principle" may vary. One drafted by the author is as follows: "The secrets of wisdom are paradoxical dualities or multi-faceted truths where singular solutions are unsuccessfully sought by men."

Problem Set 2

1. James Clerk Maxwell (1831-1879) interpreted his laws in an aether with "displacement currents," "electric displacements" in a vacuum, and "aether strains."

2. [This is a no score question depending on the student. At this point the student could possibly explain static electric charge through polarization of the aether. Electric composite aether particles may have magnetic moments with their spin, which may be aligned to make magnetic fields. Other aether forces, such as gravity and inertia, will be explained in chapters 4 and 5.]

3. No. Contrary to popular scientific opinion, the Michelson-Morley experiment did not prove there is no aether. Even Einstein did not say that an aether was incompatible with relativity.

4. The Michelson-Morley experiment proved that the speed of light is constant to all observers, with or without an aether.

5. No. Though Einstein's Special Theory of Relativity was ascribed to the principle of relativity, his two postulates themselves do not require the absolute principle of relativity. Those two postulates are 1) The speed of light

is constant to all observers; 2) The laws of physics are covariant with respect to uniform motion.

6. Yes, surprisingly enough. High precision atomic clocks and rockets are required.

7. Yes, surprisingly enough. High precision atomic clocks and rockets are required.

8. Aether momentum. Aether molecules composed of opposite charges, which can be ionized.

9. Relativity in an aether.

Problem Set 3

1. The permittivity of free space is reduced by the division of γ in the transverse direction and reduced by the division of γ^2 in the longitudinal direction. The clocks run faster by the factor γ, the length is expanded by the factor γ, and the mass is decreased by the division of γ for the self values of the moving observer. We do not observe these effects because the system renormalizes the permittivity, clock speed, length, and mass of the object to be equal to the values at rest in the self frame of the moving observer.

2. A moving observer sees a constant permittivity of free space, clock speed, length, and mass in his own self frame at all relative constant velocities. Because the system has an equal and opposite reaction to the earlier action, the system renormalizes the above values to a constant rest permittivity, rest length, rest clock speed, and rest mass in the self frame of the moving observer.

3. In normalizing down the length to the rest length of the moving observer, the system has to scale down the length

of the moving object to an observer at rest. In slowing down the clocks to the rest clock speed in the self frame of the moving observer, the system has to scale down the clock speed of the moving object as seen by an observer at rest, etc.

4. See pages 60-62.

5. See page 62-64.

6. According to Einstein, the given transformations of length and time are valid between any two objects with a relative velocity, without respect to any particular absolute space. According to the author, the given transformations of length and time are valid between the aether frame and an object moving in that aether. For correct, non-contradictory results, the aether speed and direction must be known in the system. [See Chapter 3, Section F.]

7. Assuming the aether velocity relative to the earth laboratory frame is small relative to the 0.99 c of the cosmic particle, the particle can be assumed to travel at 0.99 c relative to the aether, which is at rest relative to the laboratory frame. Then the Lorentz transformation can be used precisely as in Special Relativity.

From the Lorentz transformation, Eq. (3-35), we have

$$t' = \frac{t - (V/c^2)x}{(1 - V^2/c^2)^{1/2}}.$$

We make all comparisons at the origin at x = 0. Therefore the relation for t' = γt. Where v = 0.99 c, γ = 7.09. The clock speed goes as the inverse of the time length. The clock speed of the particle is reduced by the factor 0.141.

8. CAB forms an isosceles 30°-75°-75° triangle. The base length is easier to calculate by dropping a perpendicular to AB from C. This divides the CAB triangle into a 30°-60°-90° triangle and a 15°-90°-75° triangle. If we take the length of the sides AB and AC to be 0.9 c, then the length of the base is 0.9 c sin 30° cos 15° = 0.437 c = V. C moves away from B with velocity V = 0.437 c. Einstein would predict that C's clocks were slowed by the factor $1/\gamma$ = 0.9006. The author would predict that the clocks ran at the same rate as each other [$1/\gamma = 1.000$].

9. Einstein's Special Relativity does have a clock paradox [or rather contradiction].   The author's Special Quasi-Relativity does not have a clock paradox.

Problem Set 4

1. Because the aether particles have non-zero mass, Postulate 3 says, "all non-zero mass particles accelerate at the same rate at the same potential in a gravitational field," and all other non-zero mass particles starting at rest at infinity will reach the escape velocity at every point in space, $V = (2GM/r)^{1/2}$.

2. Only red shifted to zero frequency.

3. From Eqn. (4-18) we learn the relative frequency shift of light in terms of $\Delta\varphi/c^2$. From Eqn. (4-19) we learn that

$$\frac{\Delta\varphi}{c^2} = -\frac{GM}{c^2}\left(\frac{1}{r+10^4\ ft} - \frac{1}{r}\right)$$

$$\approx -\frac{GM}{rc^2}\left(1 - \frac{10k\ ft}{r} - 1\right)$$

$$\approx + \frac{GM}{r^2 c^2} 3048 \ m \approx 3.37 \ x \ 10^{-11}.$$

A positive sign shows a red shift.

4. Same as above, only a negative sign, indicating a blue shift.

5. Length is contracted and time dilated (expanded) in the proper time frames. The opposite is true in the coordinate frame in which the Schwarzschild Line Element is calculated.

6. Because $r^2(d\theta + \sin^2\theta d\varphi^2)$ are perpendicular directions to the radius, comparable to the y and z directions in special relativity transformations, where there is no contraction or expansion.

7. With Eqns. (4-99) and (4-100), Eqn. (4-98) becomes

$$\delta\varphi = 2\pi\left(\frac{3GM}{rc^2}\right).$$

Substituting in the constants in problem 7 in units of kilograms, meters, and seconds, we obtain the perihelic shift of earth = $1.86 \ x \ 10^{-07}$ radians per orbital revolution = 3.84" per century.

8. Solving the above equation with a new radius, the perihelic shift of the satellite = $2.79 \ x \ 10^{-05}$ radians per revolution.

9. We substitute the values in problem 9 in Eq. (4-137):

$$\Delta = 2.78 \ x \ 10^{-09} \ rad = 5.74 \ x \ 10^{-04} \ \text{sec.}$$

10. R_s should be GM/c^2, not $2GM/c^2$. At GM/c^2, the light would orbit the massive object through space warp. At $2GM/c^2$, the orbit would not be closed.

11. Slower.

12. An aether is compatible with general relativity. The key velocity in converting Special Quasi-Relativity into General Quasi-Relativity is the escape velocity in the radial direction

$$v = \left(\frac{2GM}{r} \right)^{1/2}.$$

Problem Set 5

1. Relative mass increase is caused by a change in aether velocity relative to the object or observer in question. (See Chapter 3.)

2. The interior of the craft would be exposed to constant aether velocities as the craft accelerated. Thus there would be no mass increase or inertial force felt.

3. Both gravity and inertia are relativistic effects concerning aether velocity. (See Eq. (5-50) which holds true for both gravity and inertia.) Since inertia obeys the same equations as gravity, it should not be considered a fictitious force.

4. The aether inertial force formula we first derived for centrifugal force (Eq. (5-13)) gives the expected value of the gravitational force. We must conclude that the gravitational force itself is an inertial force mediated only by the aether, and has no direct action-at-a-distance force complementing it. (See pages 158-159.)

5. A gradient of aether velocity squared should accelerate aether particles as well as other more massive particles.

6. Eq. (5-50) and (5-51).

7. The time derivative of the velocity term and one half the gradient of the velocity squared term of the aether add to zero.

8. The length of the aether velocity vector remains unchanged with changes in radial position in an orbiting object in a circular orbit around a gravitational body, though the direction of the aether vector changes. Thus grad v^2 vanishes, and since the time derivative of the aether velocity vector also vanishes, the inertial force vanishes in a free falling body in a circular orbit about a gravitational body.

Problem Set 6

1. We are detecting extremely small particles of matter now. To probe smaller requires much higher energy. It is difficult to go any smaller with accelerators we can construct.
2. Deriving a system of particles from a few postulates and comparing them with known particles.

3. One must assign a radial velocity in absolute space to each elementary particle to unlock the unified particle system and unified field theory.

4. There would be no optical or mechanical way to detect a spin. Only a magnetic field could signal there is a spin of the particle. But how would mother nature know that it was spinning to assign it a magnetic field? There would

have to be some independent method of detecting relative motion. But there is none. Therefore mother nature should not assign the particle a magnetic moment. Then there would be no way of detecting spin.

5. At least two and no more than two particles must be in a ½ spin electron. At least four and no more than four particles must be in a 0 spin pion meson. Only one particle is required for the simplest 0 spin system.

6. Smaller orbits of the semions can account for heavier electrons (muons and tauons). Smaller orbits of the semions surrounding the core uniton can also account for heavier neutrons.

7. The different variety of neutrinos (electron, muon, tauon) can be accounted for by different masses of neutrinos, which in turn can be caused by different radii of the semion orbits orbiting about the pion quarton systems in the neutrinos.

8. Smaller echons are more massive; smaller electrinos are less massive.

9. $cr_e/R_0 = 3.582 \times 10^{30}$ m/sec. 3.961×10^{-05} sec. to travel 15 billion light years (about the radius of the Universe). 5.390×10^{-44} sec to go the radius of an electron.

10. No. The charge distribution shell does not have zero radius.

11. The speed of light barrier for aether particles.

12. A quarton has a greater charge to volume ratio than a semion. An octon would have 1/8 of the radius, mass, and charge of a uniton.

13. Electrinos travel at the speed of light in the aether frame, so the electrino rest frame is a forbidden frame for calculating relativistic transformations. Adding the escape velocity of gravity eliminates infinities and defines a system with interesting properties.

14.

$$R_0 = i\left(\frac{G\hbar}{c^3}\right)^{1/2} \approx i\ 1.616 \times 10^{-35}\ m;$$

$$C_0 = \lambda_0 = 2\pi R_0 \approx i\ 1.01539 \times 10^{-34}\ m.$$

$$M_0 = -i\left(\frac{\hbar c}{G}\right)^{1/2} \approx -i\ 2.17672 \times 10^{-8}\ kg.$$

The relativistically increased mass of an electron echon is about 2.39×10^{22} times the mass of an electron, and minus imaginary. The relativistically contracted circumference and radius of the electron echon are about 2.39×10^{22} times smaller than the ordinary circumference and radius of the electron, and imaginary.

15.

$$m_e = |M_0|\frac{v}{c}.$$

16. The mass M_0 is the relativistic transformation of m. In gravity only the non-relativistic rest mass is observed at a distance. Besides, M_0 is imaginary. The only place the imaginary axis intersects the real axis is at zero. The imaginary M_0 will not be detected. The real m will be detected in real units.

17. While the non-relativistic acceleration of the charge shell inward calculates to be finite, the relativistic acceleration is zero because the shell is traveling precisely at c, both inward and outward, relative to the aether particles, and cannot be accelerated any more because of the speed of light barrier. There is a speed of light barrier both inward and outward. Therefore the radius of the electrino is firmly fixed.

18. Electrons do not explode because the relativistic imaginary radii reverse the electric force, orbiting the semions. The same sign and magnitude strong gravitational force is not extra to be added to the strong electric force, but equivalent in space warp to the strong electric force in flat space. Thus the strong electric force perfectly balances with the inertial force, and the electron semions are bound in orbit with R_0. The forces being attractive in the relativistic frame constrains the forces to be attractive also in the non-relativistic frame, where they perfectly balance also. Thus the electron does not explode or implode.

19. By equating the strong gravitational force and strong electric force in fine structures, and solving for α.

20. Because their non-relativistic (inverse transformed from relativistic) electrino lengths are x transformed, whereas the y and z widths are unchanged at small imaginary widths. Thus the revolving electrinos appear as long, thin, curved lines in the non-relativistic frame of the particle. They look like curved strings. But in the relativistic frame, the electrinos are perfect, thin, spherical shells.

21. No.

22. Electrino Particle Model and Unified Field Theory.

23. The requirement that spins come in integral values times ½ℏ spin–which is a result of a single octon-anti-octon master particle origin of the universe.

24. As the mass goes up, the radius of the particle system goes down corespondingly, leaving the spin of the particles the same because the speed c is the same.

25. The quantization of intrinsic spin through electrino matter waves is similar to the Neils Bohr quantization of atomic spins and energy states due to the matter waves of electrons.

26. $2M_0 = 2(-\hbar c/G)^{1/2}$. There are a uniton and an anti-uniton—two wholes in the photon.

27. No. In both the photon rest frame and the relative rest frame of the observer, in the circular polarization case, the velocity of the photon unitons in the transverse axial direction is exactly c. (The speed of light is constant to all observers.)

28. Circular photon polarizations are when the orbit of the unitons is in the transverse plane. The elliptical polarizations are when there is an angle between the orbital plane and the transverse plane. Line polarizations occur when the orbit is parallel to the light path axis, and the longitudinal dimension is foreshortened by relativity, leaving a line across the light path axis.

29. Yes. The photons of all colors travel linearly along the light path axis at the speed of light c. But the charges in the photons travel faster than the speed of light, and faster for higher energy photons. That may be why gravity is a dispersive medium.

30. No. The strong gravitational force in warped space is the equivalent of the strong electric force in flat space. They are equivalent. You only need to count one or the other, but not both, in particle calculations.

31. Through elementary particle fusion.

32. One copy of an orbiting octon-anti-octon pair.

33. The origin of the Universe through the control of one or more copies of orbiting octon-anti-octon pairs.

34. The spin of a single semion in an electron is $\pm$ 1/4 $\hbar$. The spin of a single quarton in a pion is $\pm$ 1/8 $\hbar$. These spins are not observed because they are spins of fractional particles, and only whole particles are observable. Whole electrons have detectable spin of $\pm$ ½ $\hbar$. Whole pions have detectable spin of 0. Only whole integrals (including 0) times $\pm$ ½ $\hbar$ spin are observable in particle physics.

Problem Set 7

1. Gravity.

2. No. It would not be possible without an aether.

3. Velocity and the frame of reference you are observing from.

4. α^w.

5. Sometimes imaginary numbers, sometimes real numbers, for fine structures: imaginary numbers for the strong electric force, real numbers for the meso-electric force and fine structures; real numbers for macroscopic

structures. That is because electrinos travel faster than the speed of light, but electrons travel slower than the speed of light in their orbits in atoms, and all other macroscopic objects travel slower than the speed of light.

6. Magneton. There is no force enhancement for first two weak forces. For first two particles, the magneton aether velocity is much less than c. Then for the third particle, the magneton velocity is near c, and the third force is increased by 32. The implication is that the magneton is the mediator particle for the weak forces.

7. Because in the magnetic force and the first two particles, the radial magneton velocity is much less than c.

8. n.

9. All particles are minature mass singularities.

10. The strong force can escape a singularity through pion mediation. The weak force and gravity can escape a singularity with or without magneton force enhancement.

11. Coupling constant between adjacent levels of mass singularity without enhancement: 0.007 297 352 5376. . .   Coupling constant with enhancement between adjacent levels of black hole: 0.233 515 281 2032. . .

12. $F = m \left[\dfrac{\partial}{\partial t} P_3 + \nabla P_t \right]$

$F = m \left[\dfrac{dv}{dt} + 1/2\, \nabla (v)^2 \right]$, where $v \ll c$.

13. The strong electric formula for masses, which in turn are responsible for gravity.

14. They are simply equated in the model. They are equivalent expressions for the same force.

15. Simply mathematically exponentially fading out the strong gravitational force and ingrowing the gravitational force exponentially and taking the strength of the force reduced 1/32 due to mediation.

16. 1/32. Particle mediation by the pion.

17. 32 times.

18. An infinite number. Eight.

19. ≈ 0, ≈ 0, and ≈ 0.

20. They have the same force equation, $F = \dfrac{q_1 q_2}{4\pi\varepsilon_0 \alpha^k r^2}$, and differ only by the parameter k—0 for electric force, and 1 for the strong force.

21. Same force form. They differ only in velocity and frame of reference.

22. It is like the electric force with real radii, but it has α in the denominator like the strong force without the imaginary radii and masses. Photons, neutrinos, and neutrons.

23. Maxwell's equations. Yes, this is compatible with an aether.

24. The unified field theory should not be a series of Gauge theories because, on the lowest fundamental level, no single force is mediated by only one boson.

Problem Set 8

1. Taking the scale factors as 1 in natural units.

2. Five times for scale factors, once for strong gravity, and once for the gravity of non-zero masses. $F = ma$, $c = $ length/time, $F = \dfrac{q_1 q_2}{r^2}$, $\hbar = M_0 R_0 c$, $f = qv \times B$,

$F = \dfrac{GM_1 M_2}{R^2}$, $f = \dfrac{Gm_1 m_2}{r^2}$.

3. No, never.

4. Because of the units of measure we use.

5. 1.

6. Because of infinite relative mass increase at the speed of light c.

7. Gravity. It is the only particle force that can be defined by the 0×0 truism with standard particles from the Master Definition.

8. Covariance of physical laws.

9. Constancy of the speed of light to all observers.

10. Two.

11. Because it is a two body problem.

12. An aether, the escape velocity, and Special Quasi-Relativity in an aether.

13. No. Yes with an aether.

14. Because there are additional terms for receptor or target enhancement and mediation of the weak forces by magnetons.

15. A sum of terms—each term for a force–each term an integration of a force with respect to r or R (in other words, an energy), divided by the standard energy of that particle system in the strong relativistic frame or the non-relativistic frame, as appropriate (in other words, $E = Mc^2$ or $E = mc^2$, respectively).

16. Because there are not magnetic monopoles, and we must obtain magnetic and weak forces from the circulation of particles over an orbit of $2\pi r$ or $2\pi R$.

17. An infinite number.

18. The sum of all subscripted weak forces.

19. No. Strong gravity and strong electric force are equivalent expressions for the same force.

20. 45.

21. The terms for the strong force, the electric force, and the magnetic force.

22. Microscopically.

23. G and $\hbar$.

Problem Set 9

1. 7.

2. Yes.

3.   String Theory, Many Dimensional Theories, The Standard Model.

4. 61. 1.

5. Instead of having to start with a Big Bang, the Universe could be created a step at a time from controlled octons.

6. How to reverse the order to disorder arrow in the second law of thermodynamics, thereby giving power to heal diseases, reverse aging, and reverse decay processes. How to convert matter into antimatter, or vice versa, etc..

7.   Derivation of Constants, Creation of anti-matter, Creation of matter, Reversing the second law of thermodynamics.

8. 1 GeV.

9. Two. Entropy arrow and order arrow.

10. No. The less efficient the machine is, the wider the area it will be effective over.

Problem Set 10

1a-2a. Both of the problem decay schemes are solved in full in Appendix A, Lepton Summary Table. You may use

Appendix A to check your answers or help you if you have a problem.

1b. Beta decay, where echons are wrenched from orbit by aligning magnetic moments, followed by straight-forward recombinations.

2b. Electrino collisions and reversal of orbital spins.

Problem Set 11

1. No. Quarks have never been discovered isolated. The reason the quark hypothesis has been primarily adopted is that it is far more parsimonious than the previous uncategorized sea of hundreds of "elementary" particles. Twenty four quarks, anti-quarks, leptons, and anti-leptons are far fewer and more parsimonious than the hundreds of "elementary" particles. Thus scientists generally favor the quark hypothesis because it makes a significant reduction in the number of necessary "elementary" particles.

2. No.   Electrinos have never been detected isolated. They can never be isolated in the foreseeable future with man-made machinery, for their masses are on the order of - i 10^{-8} kg compared to 10^{-30} kg for electrons.

3. The electrino theory is preferable to the quark theory because it is far more parsimonious and unified than the quark theory. It currently takes 24 quarks, anti-quarks, leptons, and anti-leptons and 37 other elementary particles to make up known matter by means of the quark theory. It only takes one electrino pair to create all known light and matter. The whole Universe could have been created a stage at a time from only one octon-anti-octon pair. Quartons are fused octons, semions are fused quartons, and unitons are fused semions. With octons, that is all that is

needed to build the particles of the Universe. What can be more parsimonious and unified than that?

4. 8.

5. The quark model cannot balance decay schemes. The electrino chonomic model can balance every decay scheme. Chonomic equations tell not only what echons and yachons are in a particle, but in what energy states they are, what their intrinsic spins are, what their orbital spins are, and how much angular momentum is carried in or out of a particle reaction by off-centeredness in collision and lines of retreat. Chonomic equations tell if there is beta decay, simple recombinations, knocking echons to other energy states, collisions and reversing orbital spins, flipping intrinsic spins, fusion of electrinos, or annihilation of electrinos, or creation of new particles and anti-particles.

Problem Set 12

1. They can be created when they are surrounded by other sub-particles, reducing their mass.

2. About 2 GeV. Because the fusion process creates two new particles, each with about 938 MeV. Technically the fusion process could occur slightly less than 1.88 GeV. But that would be the threshold. To fuse particles in any numbers, you want about 2 GeV or more.

3. When electron semions are fused, anti-protons and anti-neutrons are formed. In our Universe they would not be stable. They would annihilate existing protons and neutrons for the release of energy in the form of photons.

4. When positron semions are fused, protons and neutrons are formed. In our Universe they would be stable.

However, unless they were produced from an energy source not indebted to pre-existing matter, it would take more energy and mass loss to produce them than their mass. So additional matter would not be created.

5. Yes. An all matter Universe could be created through electrino fusion. No, it could not be created in a Big Bang.

Problem Set 13

1. No. Time dilation and mass increase do not depend on the direction of the aether wind.

2. Mathematically the aether wind appears to come from the normal to the V-v plane of electrino orbits, even when it does not.

3. Yes. Length contraction depends on the direction of the aether wind. It does depend on θ. It also depends on φ.

4. In the limiting case (as we see in the paragraph under Eq. (4-40)), we see that the length contraction calculated for the real mass electrino orbit object is the correct magnitude for special relativity.

5. The length contraction of an electron might make little difference in the overall length contraction of a macroscopic object. What would make more difference is the length contraction of the spaces between the electrons and other particles.

Problem Set 14

1. The Universe was made to appear younger than some local stars.

2. Doppler or velocity red shifting.

3. Cepheid variable stars.

4. The answer can be absolute particle-wave duality photon, or dual photon.

5. The photon charges travel faster than the absolute velocity of light c, and different speeds for different photon energies. High energy photons can interact with low energy photons by weak force, exchanging energy, and red shifting the high energy photon.

6. Greater.

7. No.

8. Yes.

9. Spectra broadening red shift and expanding Universe red shift (pure Doppler red shift).

10. Older.

11. 1) The simplest cause of an electromagnetic wave train is a positive and negative particle orbiting each other as they travel together at the speed of light. 2) The many possible polarizations of light cannot be accounted by just one particle—the photon. But they can all be accounted by the various possible orientations of the orbital circle of the two photon sub-particles to the light path axis. 3) Polarizations show there are photon sub-particle motions perpendicular to the light path axis, and that photon charges cannot therefore travel at the absolute velocity c.

12. $V^2 = 2c^2$, irrespective of the photon energy.

13. To answer this question we need only solve Eq. (5-22) for a 10 eV photon. Converting the photon energy to MKS units we have $E_\gamma = 1.6021917 \times 10^{-18}$ J. Evaluating the complete term on the right side of the equation, we have $r_o = 1.9737253 \times 10^{-8}$ m.

Problem Set 15

1. See Chapter 10, Section III. E. This reaction demonstrates the process of natural electrino fusion. Scientists have not recognized this process because they do not have an orderly method of balancing decay schemes such as we have with the electrino fusion model. Decay schemes of quarks and leptons won't balance. Scientists have had the wrong model of particle physics.

2. Electrino fusion of the constituents of electrons and the annihilation of the resultant anti-nucleons with nucleons.

3. Electrons with 938 MeV each or more collide with like spins in the Center of Mass Frame. They fuse to anti-unitons with the help of graviton assist. They form anti-protons or anti-neutrons.

Problem Set 16

1. The Second Law of Thermodynamics. Order to disorder arrow.

2. More.

3. If the set is not in the field of an order pump, such as a positron constituent fusion reactor, the movie is playing backward. In our world each reaction is required to go from order to disorder by the second law of thermodynamics. Any movie that reverses that arrow of

time is playing backwards, or is in the field of an order pump. In a backward movie, people would walk backward. Not in a movie in the field of an order pump. The order pump does not back up the clock. It restores order.

4. Whole integral spin gravitons are depleted and more ½ and 0 spin particles are formed. The system is going to less order.

5. An old automobile would become new and rust free and stay that way. Spoiled fruit would unspoil and stay fresh until eaten. Sick people would get well, old people would become young adults again, and stay that way. Everything would stay in its fresh perfect state of order.

Problem Set 17

1. Electrinos.

2. 8.

3. Octons, quartons, semions, and unitons.

4. Octon-anti-octon orbiting pair.

5. No. The energies required to ionize the particles are far more than any current man-made machine can produce.

6. Whole particles: uniton yachons, semion echons, and quarton echons.

7. Whole particles that are made up of only one part—unitons.

8. Whole particles that are made up of more than one part—orbiting electrinos.

9. 8.

10. -, +, o, •.

Problem Set 18

1. No.

2. Electrinos constructed of charge.

3. Octon-anti-octon orbiting pair.

4. Whole particles.

5. No.

6. It is not adequate. There must be some deeper structure.

7. Yes.

8. It is better than the quark-lepton model.

9. No.

10. The quark model requires 61 elementary particles to construct all known light and matter. The electrino model requires one particle—the octon-anti-octon orbiting pair—to construct all known light, matter, and gravitons.

11. 1) It is more parsimonious. 2) Its structures are unique. 3) Its theories of special and general quasi-relativity are without contradiction. 4) It introduces and explains heretofore unexplained supernatural processes, such as how to achieve inertialess travel, how to convert matter into antimatter and vice versa, and how to reverse the order to disorder arrow in the second law of thermodynamics. 5) It explains how there can be an all

matter Universe. 6) It can better explain interstellar red shift. 7) It can better explain super nova. 8) It can explain the nature and origin of masses. 9) It can predict the masses of elementary particles (charged leptons so far). 10) It can induce the structure of all known particles. 11) It can unite all the forces. 12) It can make a model of gravity and inertia. 13) It can shed light on the origin of the Universe. 14) It can shed light on the structures within a black hole. 15) It can derive the Universe with just a few postulates. 16) It can give the researcher tools for balancing all decay schemes. 17) It can provide for the fusion of elementary particles. 18) It can provide the theory behind a new kind of power—Electrino Fusion Power (EFP). There are probably other distinct ways the electrino model is superior to the quark model.

12. 7. Particle charge, spin, parity, mass, spin feasibility, preceding particles (to avoid duplication), and decay schemes.

Problem Set 19

1. No. But they can contribute spin to a system through their orbits.

2. Extremely thin spherical shells.

3. The aether is made up of bosons (integral spin particles) which are systems of orbiting electrinos.

4. The octon is the smallest electrino. The uniton is the largest electrino. Quartons and semions and electrino anti-particles.

5. They are all fusion products of octon-anti-octon pairs and each other.

6. Through the fusion of smaller electrinos.

7. No. They are supernatural particles which must be introduced through intelligent intervention.

8. Whole particles composed of quarton, semion, and uniton electrinos.

9. Orbiting semions or quartons.

10. $-i\, 2.177 \times 10^{-08}$ kg. $i\, 1.616 \times 10^{-35}$ m.

11. $-i\, 1.088 \times 10^{-08}$ kg. $i\, 8.080 \times 10^{-36}$ m.

12. $-i\, 5.442 \times 10^{-09}$ kg. $i\, 4.040 \times 10^{-36}$ m.

13. $-i\, 2.721 \times 10^{-09}$ kg. $i\, 2.020 \times 10^{-36}$ m.

14. $>c$ domain.

15. $M_0 = -i\left(\dfrac{\hbar c}{G}\right)^{1/2}$.

16. 4 or ∞, depending on how you look at it. Coulomb force, magnetism, weak force or weak forces, and strong force.

17. No.

Problem Set 20

1. Eq. (6-52).

2. They are not all defined by the sufficient number of independent equations.

3. One.

4. In the relativistic frame, the effective mass is $2M_0$. In the non-relativistic frame, the effective mass is E/c^2. The mass of the photon is zero.

Problem Set 21

1. You might say 4—n, l, m, and s. But n is calculable in all cases from one parameter—b.

2. No.

3. 0.0128 of 1% error.

4. It does not have the advantage of a g/2-factor for the tauon.

5. e/2.

6. $\alpha^{n/b}$ factor in the denominator for the semion force.

7. Two.

8. Faster.

9. No.

10. In Bohr's Model, $C = n\lambda$. In the author's model, $b^2C = n\lambda$.

11. No. It should be impossible.

12. Imaginary radii caused by aether velocities relative to the semions faster than the speed of light c.

13. You may trace that mass back to the earlier equation for the energy of the particle (Eq. (21-17), which is intended to be the energy of a general particle, not just that of the electron.

14. The strong electric and the inertial forces.

15. 3/2.

16. $\dfrac{3b^2}{2n\alpha^{n/b}} m_e$.

17. In a series of terms with previous terms with previous n and b.

18. $n_j = n_{j-1} + j$. $b = j$.

19. Variable based on student's intuition.

20. Yes. Because the mass m is a function of α.

BIBLIOGRAPHY

Books

Adler, Ronald, Maurice Bazin, and Menahem Schiffer, *Introduction to General Relativity*. New York: McGraw-Hill Book Company, 1965.

Asimov, Isaac. *Asimov's Biographical Encyclopedia of Science and Technology*. Second Revised Edition. Garden City, New York: Doubleday & Company, Inc., 1982.

Atwater, H. A. *Introduction to General Relativity*. Oxford: Pergamon Press, 1974.

Bowler, M. G. *Gravitation and Relativity*. Oxford: Pergamon Press, 1976.

Cember, Herman. *Introduction to Health Physics*. Oxford: Pergamon Press, 1969.

Cohen, Nathan, Ph.D. *Gravity's Lens*. New York: John Wiley & Sons, Inc., 1988.

CRC Handbook of Chemistry and Physics, 73rd Edition, 1992-1993. Boca Ratan, FL: CRC Press, Inc., 1992.

CRC Handbook of Chemistry and Physics, 74th Edition, 1994-1995. Boca Ratan, FL: CRC Press, Inc., 1994.

CRC Handbook of Chemistry and Physics, 80th Edition, 1999-2000. Boca Ratan, FL: CRC Press, Inc., 1999.

Davies, P. C. W. *The Forces of Nature*, Second Edition. Cambridge: Cambridge University Press, 1986.

Encyclopaedia Britannica, 15th edition. Chicago: Encyclopaedia Britannica, Inc., 1974.

Fox, Charles. *An Introduction to the Calculus of Variations*. London: Oxford University Press, 1963.

Funk & Wagnall's New Encyclopedia. New York: Funk & Wagnalls, Inc., 1972.

Gasiorowicz, Stephen. *Quantum Physics*. New York: John Wiley & Sons, 1974.

Gell-Mann, Murray. *The Quark and the Jaguar*. New York: W.H. Freeman and Company, 1994.

Gottfried, Kurt and Victor F. Weisskopf. *Concepts of Particle Physics*, Volume 1. Oxford: Clarendon Press, 1984.

Griffiths, David. *Introduction to Elementary Particles*. New York: John Wiley & Sons, Inc., 1987.

Haken, H. and H. C. Wolf. *Atomic and Quantum Physics: An Introduction to the Fundamentals of Experiment and Theory*, Translated by W. D. Brewer. Berlin: Springer-Verlag, 1984.

Halliday, David, and Robert Resnick. *Physics* For Students of Science and Engineering, Part II, Second Edition. New York: John Wiley & Sons, Inc., 1962.

Halzen, Francis and Alan D. Martin. *Quarks & Leptons: An Introductory Course in Modern Particle Physics*. New York: John Wiley & Sons, 1984.

Hawking, Stephen. *A Brief History of Time*–From the Big Bang to Black Holes. New York: Bantum Books, 1988.

Kittel, Charles, Walter D. Knight, and Malvin A. Ruderman. *Mechanics*, Berkeley Physics Course, Volume 1. New York: McGraw-Hill Book Company, 1965.

Leighton, Robert B. *Principles of Modern Physics*. New York: McGraw-Hill Book Company, Inc., 1959.

"Light." *Encyclopaedia Britannica*, 15th edition, Macropaedia, Volume 10. Chicago: Encyclopaedia Britannica, Inc., 1974.

MØLLER, C. *The Theory of Relativity*. Oxford: Clarendon Press, date unknown to the author.

Physical Science Study Committee. *College Physics*. Printed in U.S.A.: Raytheon Education Company, 1968.

Purcell, Edward M. *Electricity and Magnetism*. Berkeley Physics Course, Volume 2. New York: McGraw-Hill Book Company, 1965.

Reitz, John R., Frederick J. Milford, and Robert W. Christy. *Foundations of Electromagnetic Theory*, Third Edition. Reading, Massachusetts: Addison-Wesley Publishing Company, 1980.

Reif, F. *Statistical Physics*, Berkeley Physics Course– Volume 5. New York: McGraw-Hill Book Company, 1967

Saxon, David S. *Elementary Quantum Mechanics*. San Francisco: Holden-Day, 1968.

Schaffner, Kenneth F. *Nineteenth-Century Aether Theories*. Oxford: Pergamon Press, 1972.

Scott, William Taussig. *The Physics of Electricity and Magnetism*. New York: John Wiley & Sons, Inc., 1966.

Swenson, Loyd S., Jr. *The Ethereal Aether*. Austin: University of Texas Press, 1972.

Trefil, James S. *From Atoms to Quarks: An Introduction to the Strange World of Particle Physics*. New York: Charles Scribner's Sons, 1980.

Van Flandern, Tom. *Open Questions in Relativistic Physics*. Edited by Franco Selleri. Montreal: Apeiron, 1998.

Computer URL addresses

Blamire, Professor John. "Atomic Structure—The mystery of. . .--. . . the quantum atom," Exploring Life@BIOdotEDU. http://www.brooklyn.cuny.edu/bc/ahp/LAD/C3/C3_elecPos_02.html.

"Bohr model." *Wikipedia*, the free encyclopedia. http://en.wikipedia.org/wiki/Bohr_model.

Watkins, Thayer. "The Relativistic Bohr Moddel of a Hydrogen-like Atom," applet-magic.com: Silicon Valley, Tornado Alley & BB Island USA. http://www.applet-magic.com/relaborh.htm.

http://physics.nist.gov/cuu/Constants/

http://www.fine-structure-constant.org/

http://www.metaresearch.org/cosmology/gps-relativity.asp.

Journals

Beardsley, T. 'Quantum Dissidents,' *Scientific American*, December 1992, pp. 39, 40.

Dudley, H. C. *Ind. Res.*, 43, 44 (Nov. 15, 1974).

Fulcher, L. P., J. Rafelski, and A. Klein, *Sci. Am. 241:6*, 150-159 (1979).

Gell-Mann, M. *Physics Letters* **8**(3) 214-5 (1 Feb. 1964).

Gilson, J. G. *Physics Essays* **9** (2) 342-53 (1996).

Hafele, J. C. and R. E. Keating. *Science 177*, 166, 167 (1972).

Hafele, J. C. and R. E. Keating. *Science 177*, 168 (1972).

Johnson, K. A. *Sci. Am. 241:1*, 115 (1979).

Kennedy and Thorndike. *Phys. Rev.*, **42**, 400 (1932).

Marmet, Paul. *Physics Essays*, Vol. 1, No. 1, pp. 24-32, 1988.

Matthews, R. 'Do Galxies Fly Through The Universe In Formation?', *Science*, Vol. 271:759, 1996.

Michelson, A. A. and E. W. Morley. *Am. J. Sci.* 34, 333 (1887).

Ogievetsky, V. And I. Polubarinov. *Ann. Phys.* (USA) **35**, 167 (1965).

Xiao, Z.J. *Comm. Th. Phys.*, 16(1):115-118 (1993).

Magazines

Lemonick, Michael D. And J. Madeleine Nash. "Unraveling Universe," *Time*, March 6, 1995, pp. 76-84.

UNITON   NEUTRON

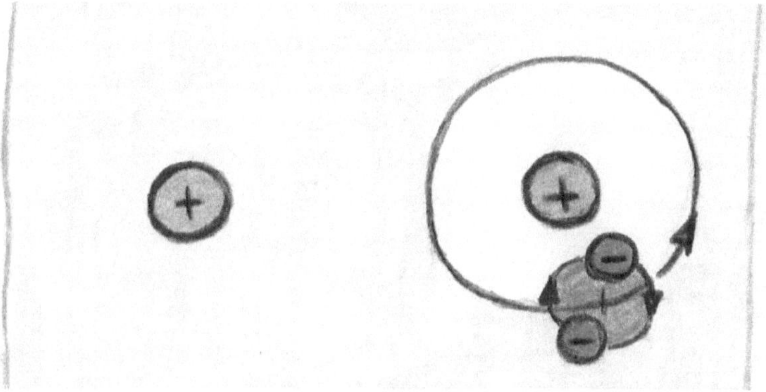

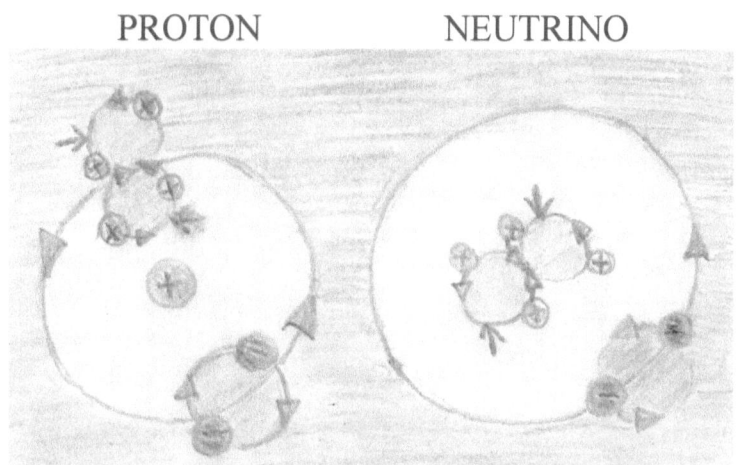

Figure 3. A uniton is a whole electrino. It is the core particle of protons and neutrons and is half of photons. They never come alone.

Figure 4. A neutron is a pair of orbiting semions orbiting about a uniton. The total charge is zero.

PROTON   NEUTRINO

Figure 5. A proton is an electron and pion orbiting a uniton.

Figure 6. A neutrino is an an electron orbiting a pion, and traveling near c.

www.ingramcontent.com/pod-product-compliance
Lightning Source LLC
Chambersburg PA
CBHW020718180526
45163CB00001B/23